Lecture Notes of the Institute for Computer Sciences, Social Informatics and Telecommunications Engineering 455

The LNICST series publishes ICST's conferences, symposia and workshops. It reports state-of-the-art results in areas related to the scope of the Institute.

LNICST reports state-of-the-art results in areas related to the scope of the Institute. The type of material published includes

- Proceedings (published in time for the respective event)
- Other edited monographs (such as project reports or invited volumes)

LNICST topics span the following areas:

- General Computer Science
- E-Economy
- E-Medicine
- Knowledge Management
- Multimedia
- Operations, Management and Policy
- Social Informatics
- Systems

Bereket H. Woldegiorgis · Kibret Mequanint ·
Mekuanint A. Bitew · Teketay B. Beza ·
Abdulkerim M. Yibre
Editors

Artificial Intelligence and Digitalization for Sustainable Development

10th EAI International Conference, ICAST 2022
Bahir Dar, Ethiopia, November 4–6, 2022
Proceedings

 Springer

Editors
Bereket H. Woldegiorgis
Bahir Dar University
Bahir Dar, Ethiopia

Kibret Mequanint
Western University
London, ON, Canada

Mekuanint A. Bitew 🆔
Bahir Dar Institute of Technology
Bahir Dar, Ethiopia

Teketay B. Beza
Bahir Dar University
Bahir Dar, Ethiopia

Abdulkerim M. Yibre
Bahir Dar University
Bahir Dar, Ethiopia

ISSN 1867-8211   ISSN 1867-822X (electronic)
Lecture Notes of the Institute for Computer Sciences, Social Informatics
and Telecommunications Engineering
ISBN 978-3-031-28724-4   ISBN 978-3-031-28725-1 (eBook)
https://doi.org/10.1007/978-3-031-28725-1

This Springer imprint is published by the registered company Springer Nature Switzerland AG
The registered company address is: Gewerbestrasse 11, 6330 Cham, Switzerland

Preface

It is our pleasure to introduce the proceedings of the tenth edition of the EAI International Conference on Advancements of Science and Technology (EAI ICAST 2022). EAI ICAST is an annual conference, the 2022 edition took place at Bahir Dar Institute of Technology, Bahir Dar University, Bahir Dar, Ethiopia. The conference covered topical science and technology issues and brought together researchers, engineers, developers, practitioners, scholars, scientists, and academicians from around the world.

The technical program of EAI ICAST 2022 consisted of seven main tracks: Track 1, Sustainable Processes for Green Technologies; Track 2, Artificial Intelligence and Digitalization for Sustainable Development; Track 3, River Basin Management and Trans-boundary Cooperation; Track 4, Agro-Mechanization and Manufacturing Systems; Track 5, Advances in Electrical and Computer Engineering; Track 6, Advances in Green Energy Technologies; and Track 7, Materials for Emerging Technologies. A total of 217 full papers were submitted, from which 90 papers were accepted in a peer reviewed process. Each paper was reviewed by on average three reviewers who are experts in the area. After a thorough evaluation process, the technical program consisted of 64 high quality full research papers in oral presentation sessions in the seven main conference tracks. In this LNICST volume, out of the 45 papers initially submitted to the tracks related to artificial intelligence and digitalization, 18 papers were accepted for publication.

The technical program of EAI ICAST 2022 also featured two general session keynote and ten track session keynote speeches along with exhibitions and poster presentations. The two keynote speakers were Sossina Haile from Materials Science and Engineering at Northwestern University, USA and Asregedew Kassa Woldesenbet from Construction Management at the Ethiopian Institute of Architecture Building Construction and City Development, Addis Ababa University. The keynote speakers shared their research and industry experience, respectively, in *electrochemistry* and *construction*. We sincerely appreciate the work of the Steering Committee chair and members; the Organizing Committee chair, Kibret Mequanint; the Organizing Committee co-chairs, Mekuanint Agegnehu and Muluken Zegeye for their constant support and guidance which ensured the success of the conference. It was also a great pleasure to work with such an excellent Organizing Committee. We are grateful to the Technical Program Committee TPC Co-chairs: Zenamarkos Bantie (Track 1), Abdulkerim Mohammed (Track 2), Birhanu Kebede (Track 3), Assefa Asmare (Track 4), Teketay Mulu (Track 5), Eshetu Getahun (Track 6), and Addisu Alemayehu (Track 7). The team performed exceptionally well to handle the peer-review process and design a high-quality technical program. We are also grateful to the conference manager, Veronika Kissova, for her support and guidance throughout the process, and all the authors who submitted their papers to the EAI ICAST 2022 conference.

We are convinced that the EAI ICAST 2022 conference provided a good forum for all scientific communities and a scientific body of knowledge we could use to discuss all science and technology aspects relevant to each track. We also expect that future EAI ICAST conferences will be as successful and stimulating, as indicated by the contributions presented in this volume.

Bereket H. Woldegiorgis

Organization

Steering Committee

Imrich Chlamtac	University of Trento, Italy
Seifu Tilahun	Bahir Dar Institute of Technology, Ethiopia
Kibret Mequanint	University of Western Ontario, Canada

Organizing Committee

General Chair

Kibret Mequanint	University of Western Ontario, Canada

General Co-chairs

Mekuanint Agegnehu	Bahir Dar Institute of Technology, Ethiopia
Muluken Zegeye	Bahir Dar Institute of Technology, Ethiopia

TPC Chair and Co-chairs

Bereket Haile	Bahir Dar Institute of Technology, Ethiopia
Muluken Zegeye	Bahir Dar Institute of Technology, Ethiopia

Sponsorship and Exhibit Chair

Bantelay Sinateyhu	Bahir Dar Institute of Technology, Ethiopia

Local Chair

Degnet Teferi	Bahir Dar Institute of Technology, Ethiopia

Workshops Chair

Ephrem Yetbarek	Bahir Dar Institute of Technology, Ethiopia

Publicity & Social Media Chair

Wubie Engdew   Bahir Dar Institute of Technology, Ethiopia

Publications Chair

Solomon Workineh   Bahir Dar Institute of Technology, Ethiopia

Web Chair

Haileyesus Demissie   Bahir Dar Institute of Technology, Ethiopia

Posters and PhD Track Chair

Nakachew Assefa   Bahir Dar Institute of Technology, Ethiopia

Panels Chair

Muluken Zegeye   Bahir Dar Institute of Technology, Ethiopia

Demos Chair

Amare Kassaw   Bahir Dar Institute of Technology, Ethiopia

Tutorials Chairs

Ephrem Yetbarek   Bahir Dar Institute of Technology, Ethiopia

Technical Program Committee

Zenamarkos Bantie   Bahir Dar Institute of Technology, Ethiopia
Abdulkerim Mohammed   Bahir Dar Institute of Technology, Ethiopia
Birhanu Kebede   Bahir Dar Institute of Technology, Ethiopia
Assefa Asmare   Bahir Dar Institute of Technology, Ethiopia
Teketay Mulu   Bahir Dar Institute of Technology, Ethiopia
Eshetu Getahun   Bahir Dar Institute of Technology, Ethiopia
Addisu Alemayehu   Bahir Dar Institute of Technology, Ethiopia

Contents

Amharic Text Complexity Classification Using Supervised Machine
Learning 1
    Gebregziabihier Nigusie and Tesfa Tegegne

Process Parameter Optimization of Single Lap-Adhesive Joint Date Palm
Fiber Reinforced Polyester Composite Using ANN-Genetic Algorism 13
    Ermias Wubete Fenta and Assefa Asmare Tsegaw

Process Parameter Optimization of Single Lap-Bolt Joint Date Palm Fiber
Reinforced Polyester Composite Using ANN-Genetic Algorism 27
    Ermias Wubete Fenta and Assefa Asmare Tsegaw

Applications of Artificial Intelligence for Fault Diagnosis of Rotating
Machines: A Review 41
    Fasikaw Kibrete and Dereje Engida Woldemichael

Finite Time Trajectory Tracking of a Mobile Robot Using Cascaded
Terminal Sliding Mode Control Under the Presence of Random Gaussian
Disturbance 63
    Adisu Safo Bosera, Ayodeji Olalekan Salau, Asrat Gedefa Yadessa,
    and Kaheli Anteneh Jembere

Artificial Intelligence-Based Breast and Cervical Cancer Diagnosis
and Management System 79
    Elbetel Taye Zewde, Mizanu Zelalem Degu,
    and Gizeaddis Lamesgin Simegn

Numerical Simulation and Optimization of a Locally Built Midibus
Structure in Quasi-static and Rollover Condition 95
    Hailemichael Solomon Addisu, Ermias Gebrekidan Koricho,
    and Adino Amare Kassie

Transfer Learning with Pre-trained CNNs for Breast Cancer Stage
Identification 127
    Tesfahunegn Minwuyelet Mengistu, Birtukan Shegaw Arega,
    and Birhanu Hailu Belay

Fuzzy Sliding Mode Controller Based Trajectory Tracking Control of Free
Flying Space Robot Manipulator System 137
    Esubalew W. Shibabw and Gerbaw Y. Tamiru

Prediction of Teff Yield Using a Machine Learning Approach 159
  Adugna Necho Mulatu and Eneyachew Tamir

Amharic Character Recognition Using Deep Convolutional Neural
Network 177
  Achamie Aynalem

Assistive Smart Cane Technology for Visually Impaired Peoples: A Review 196
  Getnet Ayele Kebede and Yosef Kassa Shiferaw

A Headphone-Based Heart Rate and Heart Rate Variability Monitoring Unit ... 209
  Gashaye Lewtie Hailu

Amharic Sentence-Level Word Sense Disambiguation Using Transfer
Learning 227
  Neima Mossa and Million Meshesha

Deep Complex-Valued Neural Networks for Massive MIMO Signal
Detection 239
  Isayiyas Nigatu Tiba and Mao youhong

Design and Performance Analysis of a Multi-level Fuzzy-Based Stabilizer
to Dampen Low-Frequency Oscillation in Single-Machine Infinite Bus
Systems 252
  Tenaw Ayew Mezigebu and Belachew Bantyirga Gessesse

Super Twisting Sliding Mode Controller for Trajectory Tracking Control
of Autonomous Ground Vehicle System 272
  Tamiru Takele, Tefera Terefe, and Sam Sun Ma

Trajectory Tracking of a Two-Wheeled Mobile Robot Using Backstepping
and Nonlinear PID Controller 290
  Lencho Duguma Fufa and Endalew Ayenew

Author Index 305

Amharic Text Complexity Classification Using Supervised Machine Learning

Gebregziabihier Nigusie[1]([✉]) and Tesfa Tegegne[2]

[1] ICT4D Research Center, Faculty of Computing, Bahir Dar Institute of Technology, Bahir Dar University, Bahir Dar, Ethiopia
gerenigusie138@gmail.com

[2] ICT4D Research Center, Bahir Dar Institute of Technology, Bahir Dar University, Bahir Dar, Ethiopia

Abstract. Amharic documents tremendously increase after the proliferation of the internet. It uses a variety of lexicons to organize the document. Some of them may not be familiar to second language learners and low literacy readers which can cause difficulty to comprehend the idea. Text complexity is focused on how difficult or easy a text is to read and understand based on the reader's level of knowledge. The appropriateness of text for a certain learner group needs to be in line with their proficiency level. A document that contains complex lexicons can also reduce the performance of NLP tasks such as machine translation. Studying the complexity classification model for the Amharic text helps in solving text complexity for a target population and NLP applications. In this paper, we have developed a complexity classification model for Amharic texts using supervised machine learning. For the experiment, 5126 sentences are used. TFIDF and BOW with bigram language modeling are applied for vectorizing the text document and, Support Vector Machine (SVM), Random forest (RF), and Naïve Bayes (NB) algorithms are used for the experiment. SVM has better classification accuracy with a result of 87.1% using bag-of-words (BOW) feature extraction and 10-fold cross-validation. The RF and NB algorithms score an accuracy of 83% and 80.3% respectively. For error analysis, we have used Mean Square Error (MSE) and Root Mean Square Error (RMSE) metrics. In this study, we have addressed the classification of Amharic text complexity. The simplification process of such identified complex texts is our recommendation for future research works.

Keywords: Text complexity · Supervised classification · Lexical complexity

1 Introduction

Natural language processing (NLP) is one of the emerging areas in the machine learning research community (Santucci et al. 2020). It is applicable in many application areas such as text classification for automatically understanding, processing, and categorizing text data, Information extraction, and sentiment analysis. To present language learners and low literacy readers with texts suitable to their level the morphological, lexical,

B. H. Woldegiorgis et al. (Eds.): ICAST 2022, LNICST 455, pp. 1–12, 2023.
https://doi.org/10.1007/978-3-031-28725-1_1

syntactic, and discursive complexity of a text is to be considered. NLP became interested in automatically classifying the complexity of a text, typically using lexicon features is a key solution for presenting documents appropriate to concerned bodies (Kurdi 2019). Text documents utilize a wide variety of vocabularies, some of those words seem to be unfamiliar to low literacy readers and increase document complexity. This complexity is the degree of difficulty in reading and comprehending a text, which can be determined based on a variety of characteristics such as familiarity of words, knowledge demands, and the educational background of readers. The appropriateness of a text for a certain learner group needs to be in line with the proficiency level of the learners (Knapp and Antos 2016).

Lexical complexity is one of the text complexity issues that can happen due to the existence of unfamiliar words in the document and it plays a critical role in readability, particularly for children second language learners and poor readers because of their laborious interpreting and word recognition skills (Gala and Ziegler 2016). Classifying documents that contain complex lexicons which are considered hard to understand for a target population is a vital step for text simplification (Shardlow et al. 2020). Text complexity classification help to provide documents that can appropriate for certain readers and to make information more accessible to a large variety of people with low literacy levels including children, non-native speakers, and people with cognitive disabilities (Rello et al. 2013) furthermore text complexity classification is a valuable preprocessing stage for different NLP tasks, such as machine translation (Sulem et al. 2018), relation extraction.

The Amharic language is a Semitic family and morphologically rich language which is largely spoken in Ethiopia. Like other languages, the Amharic language is also one research area for many NLP applications. Amharic documents can be organized using a variety of words (lexicons). Some of the lexicons in the document may not be familiar to certain readers which can cause difficulty to comprehend the idea of the content for second language learners and low literacy readers (Belete et al. 2015). Ethiopia Early Grade Reading Assessment study which was conducted in 2010 and 2018 for 6 languages in Ethiopia, such as Amharic, Somali, Tigrinya, Afan Oromo, Sidamigna, and Harrigan. The study targeted grade 2 and grade 3 students. The assessments included letter sound fluency, naming fluency of unfamiliar words, reading comprehension, and the ability of listening comprehension. Fidel naming fluency in grade 3 scores are significantly higher than those of grade 2 however children in all languages have limited skills in reading and understanding new and unfamiliar words (Yigzaw 2018). To overcome the issue of text complexity many researches are conducted for different languages such as Text Complexity Classification Based on Linguistic Information for Italian text (Santucci et al. 2020), Efficient Measuring of Readability to Improve Documents Accessibility for Arabic Language Learners (Bessou and Chenni 2021). However, the complexity of text depends on the language script, structure, and morphology. So, studying the complexity classification model for the Amharic language helps in solving text complexity for a target population. It can also help to improve the performance of NLP applications, such as parsing, information extraction, and Machine translation (Sulem et al. 2018). Furthermore, classifying Amharic text complexity is the base for future research work on text simplification. Due to unavailability of such a complexity classification model for Amharic documents, we have developed an Amharic text complexity classification using a supervised machine learning method. In this paper, we have compared different supervised machine learning algorithms for classifying Amharic documents that contain complex lexicon using Term Frequency-Inverse Document Frequency (TFIDF) and Bag-of-Word (BOW) feature extraction techniques.

2   Related Work

Recently due to the increase in the availability of text documents, machine learning-based text classification becomes one of the key techniques for organizing text data (Gasparetto et al. 2022), by exploiting a supervised learning method to assign predefined labels to documents based on the prospect suggested through a trained set of labels and documents (Zhang et al. 2008). One application area of those supervised machine learning algorithms is text complexity classification. Measures appropriateness of text to particular readers widely in the education field to select texts that match a learner's understanding level and to support educationalists in conscripting textbooks and curricula that much with students (Review 2021). Text complexity measurement and classification are key in a variety of NLP applications such as sentiment analysis, text simplification, and automatic translation for non-native readers of the language farther more the suitability of a text for a certain learner group should be in line with the proficiency level of the learners (Knapp and Antos 2016).

Measuring the complexity of a text using a supervised classification model by using a dataset of texts produced by linguistics and language experts for evaluating the language abilities of non-native speakers of Italian (Santucci et al. 2020). They have used 692 sentences which have 336,022 tokens and 29,983 unique tokens collected from certification materials. Based on their evaluation Random Forest and Support Vector Machine resulted better than other models with an accuracy of 72.5% and 71.7% respectively. Automatic text complexity classification model they proposed helps to introduce objectivity in teaching tasks. The use of semantic features and automatic augmentation for syntactic is uncovered tasks in the study.

The study conducted in reading for Ethiopia's achievement developed monitoring and evaluation (Read 2019), which concerns the subtasks of familiar words reading, new words reading, and reading comprehension of early-grade students. For the study, the data is collected from 459 schools and evaluated by 17,879 students in grades 2 and 3. Their result helps to determine students' understanding of the text and their ability to answer factual questions and make inferences based on what they read. So, using familiar words concerning student's grade level is help students easily read and recognize a text which is their part of the conclusion. Supervised machine learning methods for identifying Arabic text complexity using both count and TF-IDF feature representation techniques and applied NB, LR, SVM, and RF (Bessou and Chenni 2021). Based on the author's experimental result SVM achieves the optimal result using TF-IDF Vectors trained by a combination of word-based unigrams and bigrams. The accuracy of the selected model was 87.14%. Covering additional domains for exploring deeper features like syntactic and semantic features by increasing the corpus is their future research direction.

3   Methodology

In this research work, we have followed an experimental research design for Amharic text complexity classification for manipulating the effect of different variables such as dataset size, text preprocessing, and feature representation technique on the result of the accuracy of such Amharic text complexity classification task. The following phases are the main components of our work dataset collection, dataset annotation, preprocessing, word representation, train machine learning model, and evaluate the performance of the model.

3.1   Amharic Text Dataset

The dataset for the Amharic text complexity classification task using a machine learning model is collected from academic textbooks (from grade 6th to grade 12th), fiction, and social media sources. These sources contain complex text identified by linguistics and book authors (Endalemaw et al. 2012). We have used such resources because, for appropriate data collection it is better to determine what types of data and sources are needed (Muhammad and Kabir 2018). Such dataset collection process is the main component of our research work that needs to be collected and analyzed appropriately. In addition to this, we have conducted a sample survey evaluated by three Amharic lingusts. The survey contains six pages of Amharic text which was randomly taken from written material such as student textbooks, news, and fiction for the reader to identify the sentence that contains unfamiliar words from the paragraph.

As we evaluated we found that 123 sentencess are indentified that contains complex terms by all three annotators. They have identified sentences that contain words like ወናፍ, ቀማታኖ, and ለኖሳስ. Then based on the respondents we have collected a total of 5126 sentences, half of such sentences contain a complex word that is labeled as complex and the rest are labeled as non-complex texts. Then before we vectorize the text data and train the machine learning model we have applied different preprocessing stages for it.

Text Preprocessing. This stage is a very common task in NLP applications even the way of preprocessing is depending on the type of dataset and the language. We have used different preprocessing stages for our dataset because we have collected the dataset from different sources which contains noise in various forms like special characters, punctuation, different writing formats, and common words (the words which are not content bearing). To develop an optimized model, appropriate data are required, and preprocessing is a vital part of acquiring such data (Woo et al. 2020).

Tokenization. At this stage segment the annotated Amharic dataset into a list of tokens. The process is performed for both the training and testing dataset and removing special characters like ', ፡, ፣, ፤, '!'because the existence of these characters does not have a significant contribution to increase the performance of the classification model rather it increases irrelevant features at the time of model training.

Stop-word Removal. Eliminating those stop-words from our dataset not only saves time but also reduces the size and vector space of the text. In many natural language processing applications, an appropriate stop-word extraction technique is required

(Kaur 2018). Our aim in this stage is to remove low-level information, allowing us to focus on the most important information. The existence of these stopwords results in reduce the performance of the model (Qiang et al. 2016). Words like sle(ስለ), wede(ወደ), new(ነው), neber(ነበር)are removed from our tokenized dataset.

Normalization. Some Amharic words can be written in a different format for the same representation and function (homophones). To reduce such word variation, we have transformed those words into a single representation (homophone normalization). Due to historical sound changes, some of the graphemes that represent different sounds in Ge'ez are now pronounced the same in Amharic. For example, the phoneme /h/ can be represented by the <ሀ>, <ሐ>, <ኀ>, and <ኅ> series of graphemes (Zupon 2019) to reduce such Fidel variation in Amharic words we have applied this normalization stage.

Morphological Analysis. Morphological analysis of highly inflected languages is a non-trivial task and Amharic is one of the most morphologically complex languages (Goebel 2014). At this stage, we have reduced morphological variants of Amharic tokens to their representative morpheme by removing affixes. To do this morpheme extraction process we have used the hybrid technique of our root analyzer algorithm with HornMorpho (Gasser 2011). The reason for a hybrid of such methods is to handle words that are not analyzed by HornMorpho and to enable the analyzer to work with document-level analysis.

Feature Extraction. To build a machine learning model for Amharic text complexity classification, it is necessary to apply feature extraction operations on text data, in order to transform it into computer understandable format (Gasparetto et al. 2022). This stage is a very common part of the machine learning model building process because to train the models the feature should be represented in numeric format. We have converted the preprocessed text to numeric format using Term Frequency Inverse Document Frequency (TFIDF) which is the most commonly used feature weighting technique for many research (Das et al. 2021) and Bag of Words (BOW) with bi-gram language modeling to handle the context and order of the tokens. Those techniques are common and popular feature representation techniques for supervised machine learning models.

N-gram Language Modeling: The N-gram language model helps to predict the probability of a given N-word within any sequence of words in the language. We have used this language modeling for handling the sequence of words in the document and experimented with selected machine learning models based on the feature extracted using TFIDF and BOW (Katona et al. 2021). Specifically for our study, we have used the bigram feature by considering the sequence of two adjacent elements of tokens (Nurmaini et al. 2021), for the arrangement represents of n-gram for n = 2 (Gbenga et al. 2021). The frequency distribution of each word in the training dataset is considered using such bigram language modeling and is used for context handling. The bigram model that we have used approximates the probability of a word given all the previous words, by using only the conditional probability of the last preceding word (using two words to predict the probability)

$$p = (W_n | W_{n-1}) = \frac{p(wn - 1, wn)}{p(wn - 1)} \tag{1}$$

where P is the conditional probability of the chosen feature (word) w.

3.2 Classification Model

Train machine learning models for classifying the document as complex or noncomplex were the next task after the dataset was represented in the form of a vector. Since the text dataset is converted to a multi-dimensional numeric vector representation by computing the linguistic features of text, it is now possible to train the required model. We have applied the classification model to analyze text and then assign a set of predefined tags or categories (complex or noncomplex) based on the context of the document. For this binary classification task, supervised machine learning algorithms are used. We have selected such supervised machine learning models because, under limited computational complexity of data, classical machine learning models outperform than deep learning models (Li et al. 2021). To compare and select the appropriate classification algorithm, we have trained Support Vector Machine, Random Forest, and Naïve Bayes, using selected BOW and TFID feature extraction techniques with bigram language modeling.

4   Experiment

We have conducted an experiment on three supervised machine learning algorithms to compare and select appropriate classification models using 5126 Amharic sentences. The 85/15 dataset split rule is applied to experiment the models. We have selected this 85/15 split instead of 90/10, 80/20, 75/25 because when we experimented using these different split ratios, during the 85/15 ratio we have a balanced feature distribution (37126 features for training and 6442 features for testing) and we have got better classification accuracy. So this split ratio is selected rather than others. For cross-validation, we have used 10-fold cross-validation technique to take single validation data in every 10 intervals of training data. It is a preferred validation technique for our model because when we increase the k-fold size the validation size is too small and it is not preferred to measure the progressive improvement of the model. The dataset distribution for the experiment is summarized in Table 1. To handle the context by assigning a weight to every term that appears in each document bi-gram language modeling was applied. The experiment was conducted using Support Vector Machine, Random Forest, and Naïve Bayes by setting some hyperparameters. These selected models are trained using TFIDF and BOW feature extraction with bi-gram language modeling. The performance of those models is evaluated using precision, recall, f1-score, and accuracy.

Table 1. Dataset distribution for the experiment

Dataset split	Dataset size (sentences)	Total tokens	Unique features
Training data	4,357	37,126	7,011
Testing data	769	6,442	2,527

4.1 Experiment-1 Support Vector Machine

We have used degree of optimization (C = 0.9), flexibility control of the decision boundary (degree = 1) to train the SVM model, small value for boundary decision is applied, because we have binary classification operation which does not need higher degree of flexible decision boundary, and linear kernel which is common kernel type for classifying the two categories belonging classification sub-task (Gasparetto et al. 2022). The training performance of the model is validated using 10-fold cross-validation. The overall training accuracy of Support Vector Machine is summarized in Table 2.

Table 2. SVM experimental result

Feature extraction	N-gram	K-fold cross validation	Test accuracy	Validation accuracy
BOW	2	10-fold	87.1%	85%
TFIDF	2	10-fold	83.4%	81%

4.2 Experiment-2 Random Forest

Random forest is an ensemble learning algorithm that are robust to noise (Coşkun et al. 2011). We have experimented random forest algorithm using similar dataset with SVM. The classifier has many advantages i.e. It handles more input variables and which is lighter than other ensemble algorithms (Rodriguez-Galiano et al. 2012). In the training phase, the required hyperparameters such as the number of trees the model build before averaging the predictions (n_estimators = 10), random state of 3, to control the random sample for each sub node are considered see detail experimental result of RF in Table 3.

Table 3. Random Forest experimental result

Feature extraction	N-gram	K-fold cross validation	Test accuracy	Validation accuracy
BOW	2	10-fold	83%	82.5%
TFIDF	2	10-fold	68%	78%

4.3 Experiment-3 Naïve Bayes

In addition to SVM and RF, Navie Bayes is also trained for such text complexity classification tasks. The algorithm is working based on conditional probability and independence assumptions (Coşkun et al. 2011). Its assumption is based on the occurrence or non-occurrence of a particular attribute. The result presented in Table 4 is the classification accuracy of NB model using two feature extraction methods.

Table 4. NB experimental result

Feature extraction	N-gram	K-fold cross validation	Test accuracy	Validation accuracy
BOW	2	10-fold	80.4%	78.2%
TFIDF	2	10-fold	78.8%	76.7%

5   Result and Discussion

Amharic written materials such as academic textbooks utilize a wide variety of vocabularies to organize the document, some of those words seem to be unfamiliar to low literacy readers. The presence of unfamiliar words in sentences decreases the reading performance of low literacy readers by 18% (Sauvan et al. 2020), and 73% of the review indicated that increasing text complexity, decreases the reading rate and reading comprehension (Spencer et al. 2019). The appropriateness of a text for a certain learner needs to be in line with their proficiency level (Knapp and Antos 2016). To detect this text complexity issue for one of morphologically reach language Amharic we have developed Amharic text complexity classification model using supervised machine learning. We have conducted experiments on SVM, RF, and NB using 5126 Amharic sentences collected from different sources and annotated by experts. TFIDF and BOW feature extraction techniques are applied for vectorizing the dataset. To consider the context during feature extraction bi-grams language modeling is used because TFIDF and BOW do not capture semantics co-occurrences of tokens. Based on the experimental result the models have better classification accuracy using BOW feature extraction that 87.1% (SVM), 83 (RF), and 80.4% (NB).

To see the improvement progress of the models we have trained them using 40 iterations of sampling (we have increased the training data size by 60 in each iteration) with 10-fold cross-validation. As shown in Fig. 1 the training performance of the models is converge to similar progressive update results after the dataset size reaches 3750, because the models handle maximum feature to be train (Mukhamediev et al. 2021). Similarly as shown in Fig. 2 the training loss is also improved (decreased) untile it reaches the maximum feature.

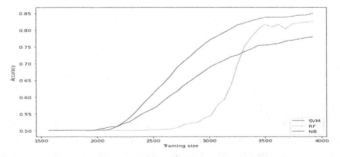

Fig. 1. Training accuracy of SVM, RF and NB

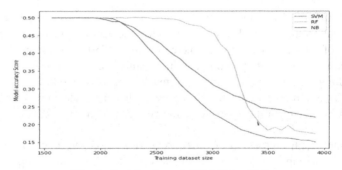

Fig. 2. Training loss of SVM, RF and NB

When we see the overall classification performance of the models, the SVM has better classification accuracy than the other two models (see in Table 5). The reason for SVM model has better accuracy for Amharic text complexity classification is that it has convex optimization to find global minimum and our dataset has high dimensional feature that SVM is aproperate for such feature (Pappu and Pardalos 2013). It has also the ability for automatically minimize over-fit problems (Jakkula 2011). The overall classification performance of these three selected supervised machine learning algorithms is summarized in Table 5.

Table 5. Classification performance of three selected models using BOW feature extraction

Model	Precision	Recall	f1-score	Testing accuracy	Validation accuracy
SVM	**88%**	**87%**	**87%**	**87.1%**	**85%**
RF	87%	83%	83%	83%	82.5%
NB	81%	80%	80%	80.4%	78.2%

As we have seen recently conducted works for classifying text complexity such as automatic classification of text complexity for Italian language (Santucci et al. 2020), their experimental result on RF and SVM scores an accuracy of 74.1% and 72.7%. Reasonably we can say that the model that we have developed for Amharic text complexity classification task achieves state-of-the-art result. In our experiment, the accuracy of SVM is better than RF because our desired problem is binary class classification task that SVM with the linear kernel is better to easily separate two classes using a hyperplane.

5.1 Error Analysis

In some cases, these machine learning models are potential of susceptible to bias and some error prediction. To identify such causes for further model improvement evaluating uncertainty of the experimental result of the model is important (Chhetri 2012). In this section, we have discussed the error analysis of Amharic text complexity classification models. When we evaluate the model using confusion matrix and error metrics such as MSE and RMSE it has some falsely predicted results, which is caused by the algorithm's internal nature and the dataset nature. When the dataset is increased the training performance of selected algorithms are increased up to 3750 dataset size. Beyond this, the models are not significantly improving their training accuracy due to the maximum feature handling capacity of such supervised machine learning models.

The other point that cause the model falsely predicted the sentence is the existence of major tokens vector in the opposite of the actual label training dataset, for example,   the   sentence በተለይ አዲስ አበባ የአፍሪካ መዲና ሆና እያለች የዕዳትና የአሳት አደጋ ችግሮች የሚፈታተናትናት ።its actual label is complex however the models predict it as noncomplex because of the tokens አፍሪካ, ጽዳት, እሳት, አደጋ and ፈተነexisted more frequently in the noncomplex training dataset, which means from the total of 7 tokens in a sentence (አዲስ አበባ አፍሪካ መዲና ጽዳት እሳት አደጋ ፈተነ)after preprocessed 5 of the tokens vector is high in noncomplex data than complex data that cause the model to miss classified it. To evaluate the overall error rate of these models we have used confusion matrix, MSE, and RMSE evaluation metrics. Such metrics help us to measure the amount of error the machine learning models do by assessing the average squared difference between the actual labeled value and predicted values. The smaller the value of the MSE, the machine learning is the best fit (Khan and Noor 2019). By using the mathematical representation.

$$MSE = \frac{y = \hat{y}}{n} \tag{2}$$

$$RMSE = \sqrt{MSE} \tag{3}$$

The models are tested using 769 unseen sentences and the result evaluation of these models using confusion matrix, mean square error and root mean square error are visualized in Table 6 below.

Table 6. Error analysis result of selected algorithms.

Model	False positive	False negative	MSE	RMSE
SVM	18	81	13%	36%
RF	2	129	17%	41%
NB	93	58	20%	44%

As presented in the table above (Table 4), from the total of 769 test datasets 99 sentences are misclassified by the SVM model, 131 sentences falsely predicted by the RF, and 151 sentences by the NB model. Some of the reasons for such misclassifications are like token duplication in both complex and non-complex datasets.

6   Conclusion and Future Work

In this study, we have developed Amharic text complex classification model by employing three supervised machine learning algorithms (Support Vector Machine, Random

Forest, and Naïve Bayes) with different hyperparameter setups. For the experiment of this study, we have used 5126 sentences. The feature of the dataset tokens represented using TFIDF and BOW with the n-gram language modeling feature extraction technique. The experiment results indicate that SVM (87.1%) has better classification accuracy compared to RF (83%) and NB (80.4%) algorithms. Both SVM and RF achieves state of art classification accuracy when we compare the work conducted on Automatic Classification of Text Complexity with the accuracy of SVM (72%) and RF (73%). To compute the error rate of the selected models we have used mean square error and root mean square error analysis and we have got the misclassification rate of 13% SVM, 17% RF, and 20% NB of mean square error. In this study, we have focused on the classification of text complexity, the simplification process of such detected complex texts through substituting the complex lexicon with its simpler equivalent, and address the syntactic (considering spelling and grammar) and morphological complexity of Amharic text are our recommendations for future research works.

References

Bessou, S., Chenni, G.: Efficient measuring of readability to improve documents accessibility for arabic language learners. J. Digit. Inf. Manage. **19**(3), 75–82 (2021). https://doi.org/10.6025/jdim/2021/19/3/75-82

Chhetri, K.B.: Computation_of_Errors_and_Their_Analysis_on_Physic. Himalayan Phys. **3**, 76–86 (2012)

Coşkun, C., Doç, Y., Baykal, A.: Comparison of classification algorithms in data mining on an example. Sci. Program. **116**(22), 51–58 (2011)

Das, M., Kamalanathan, S., Alphonse, P.J.A.: A comparative study on TF-IDF feature weighting method and its analysis using unstructured dataset. In: COLINS-2021: 5th International Conference on Computational Linguistics and Intelligent Systems, April 22–23, Kharkiv, Ukraine, 5571, p. 2 (2021)

Gala, N., Ziegler, J.: Reducing lexical complexity as a tool to increase text accessibility for children with dyslexia. In: Proceedings of the Workshop on Computational Linguistics for Linguistic Complexity, 4, pp. 59–66 (2016). https://www.aclweb.org/anthology/W16-4107

Gasparetto, A., Marcuzzo, M., Zangari, A.: A survey on text classification algorithms: from text to predictions. Information **13**(83), 1–39 (2022)

Gasser, M.: HornMorpho: a system for morphological processing of Amharic, Oromo, and Tigrinya. In: Conference on Human Language Technology for Development, pp. 94–99 (2011)

Gbenga, J.R., Aderemi, T.E., Balogun, A.: Investigative Study of Unigram and Bigram Features for Short Message Spam Detection, vol. 254, pp. 70–81. Springer, New York (2021). https://doi.org/10.1007/978-3-030-80216-5

Goebel, R.: Advances in natural language. In: 9th International Conference on NLP, PolTAL 2014 Warsaw, Poland, September 17–19, 2014 Proceedings (2014)

Jakkula, V.: Tutorial on Support Vector Machine (SVM). School of EECS, Washington State University, pp. 1–13 (2011). http://www.ccs.neu.edu/course/cs5100f11/resources/jakkula.pdf

Katona, E., Buda, J., Bolonyai, F.: Using N-grams and statistical features to identify hate speech spreaders on Twitter. In: CLEF 2021 – Conference and Labs of the Evaluation Forum, September 21–24, Bucharest, Romania, pp. 1–9 (2021)

Kaur, J.: Stopwords removal and its algorithms based on different methods. Int. J. Adv. Res. Comput. Sci. **9**(5), 81–88 (2018). https://doi.org/10.26483/ijarcs.v9i5.6301

Khan, M., Noor, S.: Performance analysis of regression-machine learning algorithms for predication of runoff time. Agrotechnology **8**(1), 1–12 (2019). https://doi.org/10.35248/2168-9881.19.8.187

Knapp, K., Antos, G.: Handbook of Second Language Assessment, pp. 1–437 (2016). https://doi.org/10.1515/9781614513827

Kurdi, M.Z.: Text complexity classification based on linguistic information: application to intelligent tutoring of ESL. J. Data Mining Digit. Human. 1–40 (2019)

Li, Q., et al.: A survey on text classification: from traditional to deep learning. ACM Trans. Intell. Syst. Technol. **37**(4), 111–139 (2021)

Muhammad, S., Kabir, S.: Methods of data collection. ResearchGate, pp. 201–276 (2018)

Mukhamediev, R.I., Symagulov, A., Kuchin, Y., Yakunin, K.: From classical machine learning to deep neural networks: a simplified scientometric review. Appl. Sci. **11**, 1–26 (2021)

Nurmaini, S., et al.: Bigram feature extraction and conditional random fields model to improve text classification clinical trial document. TELKOMNIKA Telecommun. Comput. Electron. Control **19**(3), 886–892 (2021). https://doi.org/10.12928/telkomnika.v19i3.18357

Pappu, V., Pardalos, P.: High Dimensional Data Classification. ResearchGate, 1–34 (2013). https://doi.org/10.1007/978-1-4939-0742-7

Qiang, J., Li, Y., Zhu, Y., Yuan, Y., Wu, X.: Lexical simplification with pretrained encoders. In: AAAI 2020 – 34th AAAI Conference on Artificial Intelligence, pp. 8649–8656 (2016)

Read, M.: Reading for Ethiopia. S Achievement Developed Monitoring Usaid Reading for Ethiopia. S Achievement Developed Monitoring. Usaid, pp. 1–20

Rello, L., Baeza-Yates, R., Bott, S., Saggion, H.: Simplify or help? Text simplification strategies for people with Dyslexia. In: Co-Located with the 22nd International World Wide Web Conference. W4A2013 – Technical May 13–15, 2013, Rio de Janeiro, Brazil

Review, S.: Levels of reading comprehension in higher education: systematic review and meta-analysis **12** (2021). https://doi.org/10.3389/fpsyg.2021.712901

Rodriguez-Galiano, V.F., Ghimire, B., Rogan, J., Chica-Olmo, M., Rigol-Sanchez, J.P.: An assessment of the effectiveness of a random forest classifier for land-cover classification. ISPRS J. Photogramm. Remote. Sens. **67**(1), 93–104 (2012). https://doi.org/10.1016/j.isprsjprs.2011.11.002

Santucci, V., Santarelli, F., Forti, L., Spina, S.: Automatic classification of text complexity. Appl. Sci. **10**, 1–19 (2020). https://doi.org/10.3390/app10207285

Shardlow, M., Cooper, M., Zampieri, M.: CompLex : a new corpus for lexical complexity prediction from likert scale data **11**, 1–6 (2016)

Belete, Z., Mlkt, Z., Bezabh, E., Chekol, T.: Amharic Teacher Guide Grade-7. FDRE Minister of Education and ABKME Education Bureau, pp. 1–247 (2015)

Sulem, E., Abend, O., Rappoport, A.: Semantic structural evaluation for text simplification. In: NAACL HLT 2018 – 2018 Conference of the North American Chapter of the Association for Computational Linguistics: Human Language Technologies – Proceedings of the Conference, 1, pp. 685–696 (2018). https://doi.org/10.18653/v1/n18-1063

Woo, H., Kim, J., Lee, W.: Validation of text data preprocessing using a neural network model. Hindawi **2020**, 1–9 (2020). https://doi.org/10.1155/2020/1958149

Zhang, W., Yoshida, T., Tang, X.: Knowledge-based systems text classification based on multi-word with support vector machine. Knowl. Based Syst. **21**(8), 879–886 (2008). https://doi.org/10.1016/j.knosys.2008.03.044

Zupon, A.: Text normalization for low-resource languages of Africa. ResearchGate, 1–10 (2021)

Process Parameter Optimization of Single Lap-Adhesive Joint Date Palm Fiber Reinforced Polyester Composite Using ANN-Genetic Algorism

Ermias Wubete Fenta[(✉)] and Assefa Asmare Tsegaw

Bahir Dar Institute of Technology, Bahir Dar, Ethiopia
ermiw2010@gmail.com

Abstract. Adhesive joining of composite materials is rapidly increasing in different engineering application areas such as aerospace, maritime and automotive, due to its potential for lightweight products. However, the use of adhesive joining for this purpose might lead to failure when a tensile load is acting on the composite. This work focus on the process parameters optimization of single lap adhesive joint Date palm fiber reinforced polyester composite (DPFRPC) to improve its joint strength. The study was conducted experimentally by making single-lap adhesive joining of DPFRPC under tensile testing. The key parameters influencing the adhesively joint's performance such as overlapping length (24, 40, and 56 mm), width (20, 28, and 36 mm), and adhesive thickness (0.5, 0.75, 1 mm) were studied using L_9 orthogonal array experimental design. Artificial neural network (ANN) modeling tool was utilized to relate input and output parameters. The best parameter combinations were found using a genetic algorithm (GA) optimization technique. Using this technique, the optimum parameters of single lap adhesive joint DPFRPC were, 56 mm overlapping length, 36 mm width, and 0.95 mm adhesive thickness, with a load carrying capacity of 9.48 kN.

Keywords: DPFRPC · Single lap · Adhesive joint · ANN · GA · Tensile strength

1 Introduction

Natural fiber reinforced composites (NFRPCs) materials are increasingly used in different engineering application areas such as aerospace, maritime and automotive industries, due to their lightweight and good specific mechanical properties [1–3]. From natural fibers, date palm fiber (DPF) specifically Phoenix dactyliferal, is the cheapest fiber with good physical and mechanical properties. All portions of the tree can be used to extract fibers [4]. This plant is founded abundantly in Bahir Dar, Ethiopia. But these date palms are fired and thrown as waste after cultivation, leads to environmental pollution and leads to illness [5]. Using this DPF as a composite material can solve these problems. When DPF and polyester are combined, gives a date palm fiber-reinforced composite (DPFRPC) with properties that are distinct from the ingredients [6–8].

© ICST Institute for Computer Sciences, Social Informatics and Telecommunications Engineering 2023
Published by Springer Nature Switzerland AG 2023. All Rights Reserved
B. H. Woldegiorgis et al. (Eds.): ICAST 2022, LNICST 455, pp. 13–26, 2023.
https://doi.org/10.1007/978-3-031-28725-1_2

In many NFRPC application domains, NFRPCs are used in joined form for assembling purposes due to the increase in size and geometric complexity of structures made it impractical and expensive for the composite structure to be manufactured in a single shot continuous molding process [9]. This can be solved through adhesive joining, mechanical fastening, and a combination of the two [10]. The selection of the type of composite joining depends on the application area, the load needed to be transferred, and the weight of the joint [10, 11].

Adhesive joining of NFRPC involves the use of a variety of adhesives to attach the composite. This approach has a lengthy history, having been used in the aerospace sector in the 1970s and early 1980s [12]. Adhesive joining is widely accepted as a potential substitute for mechanical joints in modern industries (marine, automotive, aeronautical, construction, and so on) for different applications [13, 14]. With adhesive joining, there is no need to drill the composite; instead, alternative adhesives are used [15]. The type of adhesives utilized in the composite joining has an impact on its tensile strength. Epoxies, acrylics, polyester, urethanes, and other adhesives are used to assemble composites [16].

There are various types of adhesive joints, including single lap, double lap, scarf type, shim insert, strap type, stepped lap, and others. Single and double lap joints are the most popular and appropriate adhesive joining methods. Due to its simple geometry and great structural efficiency, single lap is employed more than double lap [12, 16]. This research focuses on a single lap joint for a uniform cross-section in the NFRPC application area as shown in Fig. 1.

Fig. 1. Single lap adhesive joining of composite material

The joint configuration has an effect on adhesive joint quality during composite adhesive joining [12, 17]. However, geometric parameter optimization of the adhesive joining of composite had failed miserably, which leads to excessive or insufficient use of joining geometry parameters. This made the joined to be either overweighted or weak to resist the applied load. Nevertheless, optimizing overlap width, length, and adhesive thickness of joint was not studied adequately yet. Hence, this research work aims to optimize the joining geometry process parameters of adhesive joining of DPFRPC.

The joining geometry parameters in the adhesive joining process have an impact on the strength of a joined composite. Overlap length, width, adhesive thickness, and type of adhesive are the most important factors that help to achieve a high joint quality [18]. Overlap width is a more significant design feature than overlap length, even though joints with bigger surfaces have better strength. Due to stress concentration at the joint's ends, joint strength improves slightly with overlap length up to a limit, then remains constant. A joint with a long and narrow bond area is inferior to one with a short and wide bond area [18].

2   Materials and Methods

2.1   Manufacturing of Composites

Date palm fiber reinforced polyester composite (DPFRPC) was manufactured using date palm fiber and polyester. The fiber was extracted from the rachis of date palm tree, from Bahir Dar, Ethiopia, due to its availability and better physical and mechanical properties. The fiber was extracted using a biological approach and then chemically treated with 1% NaOH alkali for 5 h to improve the fiber quality. Hand lay-up technique was used to make the composite, due to its simplicity, with low-cost, easy processing. The composite mold was made of wood with dimensions of $200 \times 360 \times 3.5$ mm^3, as shown in Fig. 2. In this study, the composite was made under 30% fiber loading with unidirectional orientation. Figure 2 represents the steps involved in the fabrication of DPFRPC.

(a)   (b

(c)   (d)

Fig. 2. DPF preparation process; (a) rachis of date palm tree, (b) extracted DPF, (c) over its mold for composite making, (d) fabricated composite

The composite was made by coating the mold with gel to provide a weak bond between the mold and the DPFRPC during demolding. Then polyester was poured over the mold followed by putting the DPF over it, followed by rolling to remove air gaps. Finally, the layup was then covered with a concrete block capable of exerting 12.5 kPa pressure. After 5 h, the DPFRPC was demolded.

2.2   Adhesive Joining of DPFRPC

Following the fabrication of DPFRPC, single lap adhesive joining of DPFRPC was made. In this study polyester (parent matrix) that was used to make DPFRPC was also employed to make adhesive joining, due to its non-reactivity with the parent composite and helps to make comparable strength with the adherend. The polyester has been applied to the joining area of the specimens at the specified geometric area and adhesive thickness. Three joining parameters (overlap length, overlap width, and adhesive thickness) were considered with their corresponding levels as shown in Table 1. These variables were chosen based on how frequently they occurred and how they affected the tensile strength of adhesive joining. According to a review of the literature, the strength of an adhesively bonded composite is significantly influenced by the overlap length, overlap breadth, and adhesive thickness of the prior experiments [12, 19–22]. The specimen thickness and free length were 3.5 mm and 130 mm respectively. Figure 3 shows single-lap adhesive joint DPFRPC specimens.

Table 1. Parameters and levels of DPFRPC bolted joining

No.	Factors	Levels		
		Level 1	Level 2	Level 3
1	Overlap length, OL (mm)	24	40	56
2	Overlap width, OW (mm)	20	28	36
3	Adhesive thickness, t (mm)	0.5	0.75	1

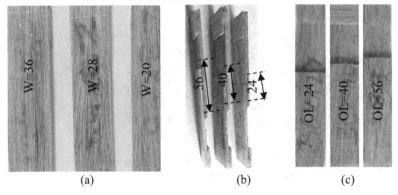

(a)   (b)   (c)

Fig. 3. Adhesive joining; (a) width of specimen, (b) overlap length side view of specimen, and (c) overlap length side view of the specimen

2.3 Design of Experiment

The tensile properties of DPFRP were characterized using fiber loadings of 30% under unidirectional fiber orientations. After the tensile properties of DPFRPC were determined, an orthogonal array was used to identify the significant variables with the least number of trials possible, saving both time and money. Three factors and levels were used as shown in Table 1, which are overlap length, width, and adhesive thickness. As a result, the L_9 orthogonal array was used, as shown in Table 2.

Table 2. Design of experiment using L$_9$ orthogonal array

No.	Overlap length	Overlap width	Adhesive thickness
1	24	20	0.5
2	24	28	0.75
3	24	36	1
4	40	20	0.75
5	40	28	1
6	40	36	0.5
7	56	20	1
8	56	28	0.5
9	56	36	0.75

2.4 Tensile Test

The tensile property of DPFRPC and single lap adhesive joint DPFRPC were determined using tensile testing machines. Tensile testing was carried out with a UTM: WAW-600D, as illustrated in Fig. 4.

Fig. 4. Tensile testing machine, UTM: WAW-600D

2.5 Model Development and Optimization Using ANN-GA Approach

The optimization technique helps to get a combination of levels of parameters of single lap adhesive joining of DPFRPC that will result in the best tensile load carrying capacity. For this study, artificial neural network (ANN) modeling and the genetic algorism (GA)

optimization approach were utilized to identify the appropriate joining geometry parameters. Because it's highly accurate modeling, predicting, and optimizing tool and it can
be used by other researchers for different targets. ANN takes input data, trains itself to
detect patterns, and then predicts the output for a new collection of similar data by using
the network shown in Fig. 5. ANN model was made using three layers; input layers, hidden layer, and output layers. The input layer contains the input process parameters (OL,
OW, t), whereas the output layer contains the failure load (Fl). The hidden layer consists
of many interconnected neurons that have been determined through training, testing, and
validation. The sum of inputs is transferred as output by each neuron's transfer function
(activation function). The weight value is assigned to each connection [23].

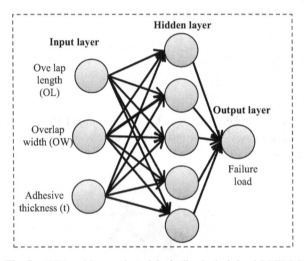

Fig. 5. ANN architectural model of adhesively joined DPFRPC

In this work, GA was employed to optimize the ANN model's input space to get a set
of optimum levels of process parameters. The fitness function of previously developed
ANN model was used for GA to determine the optimal solutions using the flow chart
shown in Fig. 6. The constraints used for GA to optimize the process parameters were
overlap length from 24 mm to 56 mm, overlap width from 20 mm to 36 mm, and adhesive
thickness from 0.5 mm to 1 mm.

3   Results and Discussions

3.1 Tensile Properties of DPFRPC

The tensile properties of DPFRPC, which was made under 30% fiber loading with unidirectional fiber orientation, were determined through tensile testing. From the experiment,
the ultimate tensile strength of DPFRPC was found to be 145 MPa as shown in Fig. 7.

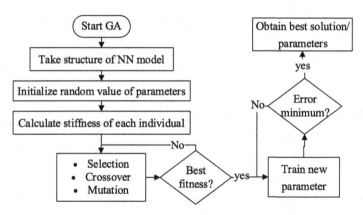

Fig. 6. GA model flow chart

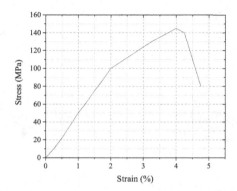

Fig. 7. Stress-strain curves DPFRPC

3.2 Tensile Properties of Joined DPFRPC Subjected to Tensile Load

Based on the design of experiments, experimental testing was conducted to examine the influence of geometric joining variables of single lap adhesive joint DPFRPC. From the result, different joint geometries show different load-carrying capacities. From the experiment, a maximum of 9.3 kN load caring capacity exists in experiment number 9 as shown in Table 3.

The effect of input parameters with its load carrying capacity was related through surface plots as shown in Fig. 8. Each plot has been made by taking the lowest levels of the remaining parameters.

The higher load carrying capacity was found at maximum overlap length and width because the bonding area increases with an increase in overlap width and length help to sustain higher loads and vice versa. On the other hand, maximum load carrying capability was found at medium adhesive layer thickness, because excessive adhesive layer thickness resulted in a thick bond line, which caused tiny cracks in the manufacturing due to vacancies.

Table 3. Experimental results of single lap adhesive joining of DPFRPC

No.	Overlap length (mm)	Overlap width (mm)	Adhesive thickness (mm)	Tensile failure load (kN)	Failure behavior
1	24	20	0.5	3.2	Thin layer cohesive
2	24	28	0.75	5.6	Fiber tear
3	24	36	1	6.5	Adhesive
4	40	20	0.75	5.2	Cohesive
5	40	28	1	5.7	Cohesive
6	40	36	0.5	7.2	Adhesive
7	56	20	1	4.9	Adhesive
8	56	28	0.5	7.8	Stock break
9	56	36	0.75	9.3	Stock break

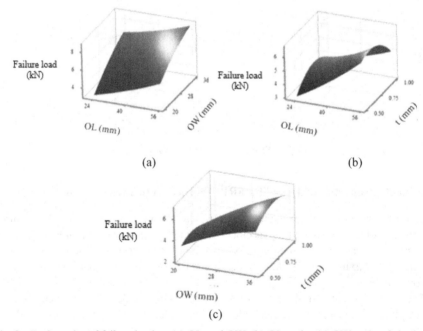

(a)   (b)

(c)

Fig. 8. Surface plot of failure load vs (a) OL and OW, (b) OL and t, (c) OW and t of single lap adhesive joint DPFRPC

3.3 Failure Behavior of Joined DPFRPC

The adhesive joint fails as the load applied over it exceeds its load-carrying capacity. Structural failure, cohesive failure, fiber-tear failure, and adhesive failure occurred, as shown in Fig. 9. Adhesive failure, Fig. 9 (3,6, and 7), as one side bonded more than the other side. Cohesive failure, Fig. 9 (4 and 5), where the adhesive thickness was at a higher level, the adhesive was too weak to resist the applied load. The stock failure, Fig. 9 (8 and 9), where the adherent fails rather than the joint and adhesive joint was strong. Fiber tear failure, Fig. 9 (1 and 2), due to the weak bond during DPFRPC manufacturing.

Fig. 9. Failure mechanism of adhesive joined DPFRPC

3.4 Modeling of ANN

The experimental findings were used to create an artificial neural network model. The developed ANN was trained with 9 sets of input (OL, OW, t), and output (Fl) parameters that emerged from the experiments conducted. Two-thirds of the samples were used for training, while the remaining one-third were used for testing and validation of the model using trainlm. After several trials, the best ANN model for single lap adhesive joint DPFRPC was determined to be 3-5-1-1 (three for input layer neurons, five for first hidden layer neurons, one for second hidden layer neuron, and one for output layer neuron) as shown in Fig. 10. The activation functions for convergence in the first and second hidden layers, respectively, were the hyperbolic tangent and the linear transfer function.

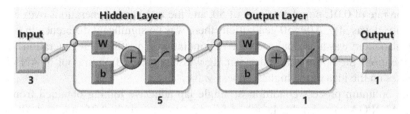

Fig. 10. Architecture of ANN model of single lap adhesive joint DPFRPC

The training performance curve for neural networks is depicted in Fig. 11 below, where adhesive joining ANN model convergence to mean square error (MSE) was determined to be 0.15049 made within 1000 epochs or iterations. However, MSE reaches saturation in the third iteration. Beyond this threshold, when an iteration increases, MSE begins to deviate from its optimal value.

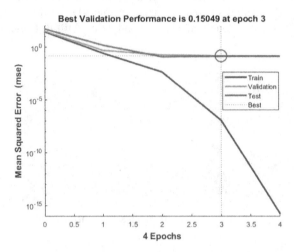

Fig. 11. Convergence of MSE during training of the ANN model

The training, validation, and testing patterns were investigated in the regression plot as shown in Fig. 12. The regression study revealed that the correlation coefficient (R) for single lap adhesive joint DPFRPC was 0.9896, implying that the experimental and expected responses are highly correlated.

There was relatively little variation between the experimental and predicted ANN model, as shown in Fig. 13, with a maximum error of 0.53427%.

3.5 Optimization of Process Parameters by GA

GA always seeks to minimize the objective function, so the negative of the proposed fitness function was minimized. Convergence was used to decide on the GA parameter setting. The GA-specific parameters were as follows: probability of crossover of 0.8, mutation rate of 0.01, population size of 50, and the number of generations over which GA evolved was 300. After 50 generations there was no significant difference in fitness value, also there was a gradual decrease in population size after each generation. After 80 generations the optimal adhesive joining was designated from a pool of ANN/GA responses on the idea of the highest fitness value.

The optimum process condition of single lap adhesive joining obtained from the hybrid ANN/GA were overlap length of 56 mm, overlap width of 36 mm, and adhesive thickness of 0.982 mm. These optimum parameters were used by ANN to predict the optimum output (failure load), which was found to be 9.48 kN.

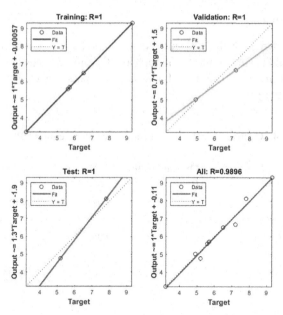

Fig. 12. ANN model simulation results in comparison with experimental results during training, validation, and testing.

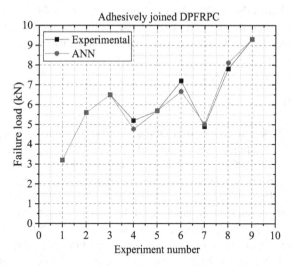

Fig. 13. Comparison of experimental and ANN predicted failure load values of adhesively joined DPFRPC

3.6 Confirmation Test

Based on the optimized ANN-GA result, a confirmation test was undertaken as shown in Table 4 to validate the model. Although 0.982 mm of adhesive thickness provided the greatest strength, 0.95 mm of adhesive thickness was chosen for the precise manufacturing of the joint.

Table 4. Conformation test result single lap adhesive joint DPFRPC

Experiment	Overlap length (mm)	Overlap width (mm)	Adhesive thickness (mm)	Failure load (kN)
GA optimal solution	56	36	0.982	10.114
Experimental solution	56	36	0.95	9.48
Error	–	–	–	0.634

The optimum process parameter levels found through GA were predicted with the developed ANN model. From the confirmation experiment, a maximum of 9.48kN load carrying capacity obtained with an error of 0.634, indicates the confirmation test output values agreed well with the ANN-GA model predicted values. From the experimental study, the maximum tensile strength of unjointed DPFRPC was 145 MPa. Therefore, the adhesively joined DPFRPC showed less strength when it was compared to unjointed DPFRPC parent material.

4 Conclusion

In this study, DPFRPC was made from DPF and polyester matrix through hand layup approach at the fiber loading and orientation of 30% and 0/0° respectively. A single lap adhesive joining of DPFRPC was made using L9 orthogonal array design of experiments with three parameters and levels of overlapping length (24, 40, and 56 mm), width (20, 28, and 36 mm), and adhesive thickness (0.5, 0.75, 1 mm). ANN model was developed to relate these input parameters and failure load through training, testing, and validation of the model. The best ANN architecture of adhesively joined DPFRPC was determined to be 3-5-1-1 with an activation function of hyperbolic tangent and the linear transfer function in the first and second hidden layers, respectively. GA optimization technique was used by taking ANN model result as an input to get the optimum parameters. Using this technique, the optimum process parameters of single lap adhesive joint DPFRPC were, 56 mm overlapping length, 36 mm width, and 0.95 mm adhesive thickness, with a load carrying capacity of 9.48 kN. The test result indicates that, the increase in overlap length and overlap width the load carrying capacity of single lap adhesive joining of DPFRPC. However, it increases up to the optimum level of adhesive thickness and starts to fall beyond that point of thickness.

References

1. Mann, G.S., Singh, L.P., Kumar, P., Singh, S.: Green composites: a review of processing technologies and recent applications. J. Thermoplast. Compos. Mater. **33**(8), 1145–1171 (2020)
2. Rajak, D.K., Pagar, D.D., Menezes, P.L., Linul, E.: Fiber-reinforced polymer composites: manufacturing, properties, and applications. Polymers **11**(10), 1667 (2019)
3. Srinivas, K., Naidu, A.L., Bahubalendruni, M.R.: A review on chemical and mechanical properties of natural fiber reinforced polymer composites. Int. J. Perform. Eng. **13**(2), 189 (2017)
4. Johnson, R.D.J., Arumugaprabu, V., Ko, T.J.: Mechanical property, wear characteristics, machining and moisture absorption studies on vinyl ester composites – a review. Silicon **11**(5), 2455–2470 (2019)
5. Indracanti, M., Mekonnen, D.T., Tsegaw, M.: Molecular characterization of some landraces and varieties of date palm (*Phoenix dactylifera* L.) from Afar region of Ethiopia using ISSR markers (2019)
6. Asim, M., Jawaid, M., Khan, A., Asiri, A.M., Malik, M.A.: Effects of date palm fibres loading on mechanical, and thermal properties of date palm reinforced phenolic composites. J. Market. Res. **9**(3), 3614–3621 (2020)
7. Bendada, A., Boutchicha, D., Khatir, S., Magagnini, E., Capozucca, R., Abdel Wahab, M.: Mechanical characterization of an epoxy panel reinforced by date palm petiole particle. Steel Compos. Struct. **35**(5), 627–634 (2020)
8. Chihaoui, B., Serra-Parareda, F., Tarrés, Q., Espinach, F.X., Boufi, S., Delgado-Aguilar, M.: Effect of the fiber treatment on the stiffness of date palm fiber reinforced PP composites: macro and micromechanical evaluation of the Young's modulus. Polymers **12**(8), 1693 (2020)
9. Galińska, A.: Mechanical joining of fibre reinforced polymer composites to metals – a review. Part I: bolted joining. Polymers **12**(10), 2252 (2020). https://doi.org/10.3390/polym12102252
10. El Zaroug, M., Kadioglu, F., Demiral, M., Saad, D.: Experimental and numerical investigation into strength of bolted, bonded and hybrid single lap joints: effects of adherend material type and thickness. Int. J. Adhes. Adhes. **87**, 130–141 (2018)
11. Zhang, J., Xie, Q., Xie, Y., Zhou, L., Wang, Z.: Investigation of mechanical performances of composite bolted joints with local reinforcements. Sci. Eng. Compos. Mater. **25**(1), 75–83 (2018)
12. Jeevi, G., Nayak, S.K., Abdul Kader, M.: Review on adhesive joints and their application in hybrid composite structures. J. Adhes. Sci. Technol. **33**(14), 1497–1520 (2019)
13. Galińska, A., Galiński, C.: Mechanical joining of fibre reinforced polymer composites to metals – a review. Part II: riveting, clinching, non-adhesive form-locked joints, pin and loop joining. Polymers **12**(8), 1681 (2020). https://doi.org/10.3390/polym12081681
14. Naik, R., Panda, S., Racherla, V.: A new method for joining metal and polymer sheets in sandwich panels for highly improved interface strength. Compos. Struct. **251**, 112661 (2020)
15. Romanov, V.S., Heidari-Rarani, M., Lessard, L.: A parametric study on static behavior and load sharing of multi-bolt hybrid bonded/bolted composite joints. Compos. B: Eng. **217**, 108897 (2021)
16. Fazel, D., Kadivar, M.H., Zohoor, H., Farid, M., Hematiyan, M.R.: Failure procedure in epoxy adhesive joining composite plates. Iran. J. Sci. Technol. Trans. Mech. Eng. **45**(2), 337–350 (2020). https://doi.org/10.1007/s40997-020-00379-0
17. Shaikh, S., Anekar, N., Kanase, P., Patil, A., Tarate, S.: Single lap adhesive joint (SLAJ): a study. Int. J. Eng. Technol. **7**, 64–70 (2017)
18. Antunes, D.P., et al.: Development of a drop weight machine for adhesive joint testing. J. Test. Eval. **49**(3), 1651–1673 (2019)

19. Choudhury, M.R., Debnath, K.: Experimental analysis of tensile and compressive failure load in single-lap adhesive joint of green composites. Int. J. Adhes. Adhes. **99**, 102557 (2020)
20. Delzendehrooy, F., Ayatollahi, M., Akhavan-Safar, A., da Silva, L.: Strength improvement of adhesively bonded single lap joints with date palm fibers: effect of type, size, treatment method and density of fibers. Compos. B: Eng. **188**, 107874 (2020)
21. Khalili, S., Mokhtari, M.: Numerical study of adhesive single-lap joints with composite adherends subjected to combined tension–torsion Loads. J. Adhes. **91**(3), 214–234 (2015)
22. Shang, X., Marques, E., Machado, J., Carbas, R., Jiang, D., da Silva, L.: Review on techniques to improve the strength of adhesive joints with composite adherends. Compos. B: Eng. **177**, 107363 (2019)
23. Hramov, A.E., et al.: Artificial neural network detects human uncertainty. Chaos Interdiscipl. J. Nonlin. Sci. **28**(3), 033607 (2018)

Process Parameter Optimization of Single Lap-Bolt Joint Date Palm Fiber Reinforced Polyester Composite Using ANN-Genetic Algorism

Ermias Wubete Fenta[✉] and Assefa Asmare Tsegaw

Bahir Dar Institute of Technology, Bahir Dar, Ethiopia
ermiw2010@gmail.com

Abstract. Natural fiber reinforced polymer composites are widely employed in automotive, aerospace, and civil applications due to their high strength-to-weight ratios and these applications require joining composite. Bolt joining of composite materials is the most prevalent way of joining, due to its efficiency of transferring load and ease of disassembly. However, bolt joining of composite is largely influenced by geometrical parameters such as edge to diameter ratio (E/D), width to diameter ratio (W/D), and fiber orientation. This work emphases on the process parameters optimization of single lap bolt joint date palm fiber reinforced polyester composite (DPFRPC) to improve the joint strength. The study was conducted experimentally by making single lap bolt joining of DPFRPC under tensile testing. The important factors affecting the performance of the adhesively joint such as E/D (1.5, 2.5, and 3.5), W/D (2.5, 3.5, and 4.5), and fiber orientation (0/0°, 45/−45°, and 0/90°) was studied using L_9 orthogonal array experimental design. Artificial neural network (ANN) was used to model the experimental results. Genetic algorithm (GA) optimization technique was used to determine the optimum process parameters. Using this technique, the optimum parameters of single lap bolt joint DPFRPC were, 3.5 E/D, 4.5 W/D, and 56.5° fiber orientation, with a load carrying capacity of 9.52 kN.

Keywords: DPFRPC · Single lap · Bolt joining · ANN · GA · Tensile strength

1 Introduction

In today's world, there are many more composite materials available. Researchers are more interested in natural fiber reinforced polymer matrix composite material because of its high strength-to-weight ratio [1, 2]. Amongst natural fibers, date palm fiber (DPF) specifically Phoenix dactyliferal, is the most cost-effective fiber with good physical and mechanical characteristics. The fiber can be extracted from all parts of date palm tree [3]. It is founded abundantly in Bahir Dar, Ethiopia. When date palm fiber and polyester are combined, gives date palm fiber reinforced composite (DPFRPC) with properties that are distinct from the ingredients [4, 5].

© ICST Institute for Computer Sciences, Social Informatics and Telecommunications Engineering 2023
Published by Springer Nature Switzerland AG 2023. All Rights Reserved
B. H. Woldegiorgis et al. (Eds.): ICAST 2022, LNICST 455, pp. 27–40, 2023.
https://doi.org/10.1007/978-3-031-28725-1_3

In a variety of application domains, natural fiber reinforced composites (NFRPCs) are used in joined form for assembling purposes, since the making of long and complex geometry structure composite using a single mold is impractical and expensive [6]. NFRPCs cannot be welded due to their electrical nonconductivity. Adhesive joining, mechanical fastening, and hybrid joining can all be used to tackle this problem [7]. The type of composite joining is determined by the application area, the load to be transferred, and the weight of the joint [7, 8].

Bolt joining is the most prevalent way of joining several materials such as metal, nonmetals, and composites to each other and other materials [9, 10]. Furthermore, unlike adhesive joining, bolted joints have no environmental impact and transfer higher stresses between the joined structures. Several previous works evidenced that net tension, shear out, and cleavage failures exist under this joining method. Bolt joining can achieve 40–80% joint efficiency during joining of NFRPCs. Bolt joining of NFRPCs can be designed over the adhesive joint, if weight is not an issue [11]. In adhesive joining, there is no delamination due to the absence of holes and weight decrease. But it can't be used for structures that require disassembly [12, 13].

The most prevalent and appropriate bolt joining methods are single and double lap joints. Single lap is preferred over double lap because of its simple geometry and high structural efficiency [6]. This research focuses on a single lap bolt joint of DPFRPC as indicated in Fig. 1 for a uniform cross-section in the NFRPC application area.

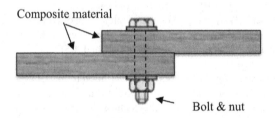

Fig. 1. Single lap bolt joining of composite material.

Drilling a hole is required, and damage happens at the beginning and end of the drilling operation [14]. The fiber layers peel up as the drill enters the NFRPCs. However, the bottom fiber layers are pushed out as the drill enters the bottom of the NFRPCs [6]. In general, inappropriate drilling causes delamination, resin erosion, and fiber breakout in composites. Low feed and high drilling speed are the optimum conditions for drilling to overcome these damages [15]. To avoid fretting in the clearance hole, the bolt must fit snugly in this joint. Interference fittings can cause composite delamination, thus using washers helps to protect the clearance hole [16].

During composite material bolt joining, the joint configuration has an impact on bolt joint strength [17, 18]. However, geometric parameter optimization of composite bolt joining failed horribly, resulting in overuse or underuse of joining geometry parameters. As a result, the joint became overweight or weak (unable to withstand the applied load). Hence, this study focuses to optimize joining geometry process parameters of bolt joint DPFRPC. The strength of bolt joint composite is highly influenced by joint geometry parameters such as composite thickness, type of lap, number of lap, number of bolt, hole diameter, W/D ratio, E/D ratio, and composite thickness [19]. Most research found that geometrical variables like W/D, E/D, and fiber orientation had an impact on the failure process [4, 20]. The optimum process parameters for strong single lap bolt joint DPFRPC are discovered in this study.

2 Materials and Methods

2.1 Manufacturing of the Composite

In this study, date palm fiber reinforced composite was manufactured using date palm fiber and polyester. The fiber was extracted from the rachis of date palm tree, from Bahir Dar, Ethiopia, due to its availability and better physical and mechanical properties. The fiber was extracted using a biological approach and then chemically treated with 1% NaOH alkali for 5 h to improve the fiber quality. Due to its simplicity, low cost, and ease of processing, the composite was manufactured utilizing the hand lay-up approach. The composite mold was made of wood with dimensions of $200 \times 360 \times 3.5$ mm^3, as shown in Fig. 2(c). In this study the composite was made under 20%, 30% and 40% fiber loading with unidirectional orientation. The steps involved in fabricating DPFRPC are depicted on Fig. 2.

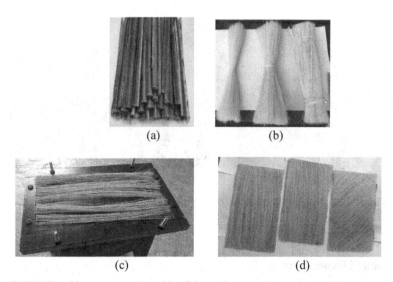

(a)   (b)

(c)   (d)

Fig. 2. DPFRPC making process; (a) rachis of date palm tree, (b) extracted DPF, (c) over its mold for composite making, (d) fabricated composite.

The composite was made by coating the mold with gel to provide a weak bond between the mold and the DPFRPC during demolding. Then a polyester was poured over the mold followed by putting the DPF over it, followed by rolling to remove air gaps. Finally, the layup was then covered with a concrete block capable of exerting 12.5 kPa pressure. After 5 h, the DPFRPC was demolded.

2.2 Bolt Joining of DPFRPC

Bolt joining was made after drilling DPFRPC specimen using drilling machine. To achieve a nice drilled hole, a fast drill speed, low feed, and drill diameter are needed [6]. In this study, the hole was made using 8 mm HSS drill bit under a rotational speed of 2000 RPM and feed of 45 mm/min. Finally, the bolt joint was made using M8 bolt. The bolt joint strength is highly affected by fiber orientation, E/D, and W/D [21]. Three joining parameters such as W/D (2.5, 3.5, and 4.5) and E/D (1.5, 2.5, and 3.5) were selected for this study as shown in Table 1 and Fig. 3. The specimen thickness and free length were kept constant at 3.5 and 130 mm, respectively.

Table 1. Parameters and levels of single lap bolt joint DPFRPC

No.	Factors	Levels		
		Level 1	Level 2	Level 3
1	Fiber orientation (°)	0/0	45/−45	0/90
2	Edge to diameter ratio (E/D)	1.5	2.5	3.5
3	Width to diameter ratio (W/D)	2.5	3.5	4.5

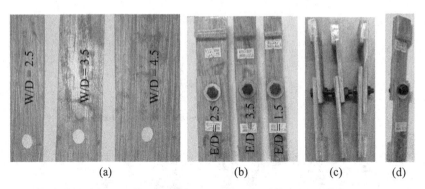

Fig. 3. Bolt joint specimens (a) W/D, (b) E/D, (c) side, and (d) isometric view

2.3 Design of Experiment

The tensile properties of DPFRPC were characterized using fiber loadings of 20%, 30%, and 40% under unidirectional fiber orientations. After the tensile properties of DPFRPC were determined, an orthogonal array was used to identify the significant variables with the least number of trials possible, saving both time and money. Three factors with three levels were selected for single lap bolt joining of DPFRPC (E/D, W/D, and fiber orientation) as shown in. As a result, L_9 orthogonal array design of the experiment was adopted in Table 2. In order to ensure that all levels of all components are taken into account equally, L_9 orthogonal arrays were chosen.

Table 2. Experimental design of single lap bolt joint DPFRPC using L_9 orthogonal array

No.	Edge to diameter ratio (E/D)	Width to diameter ratio (W/D)	Fiber orientation (°)
1	1.5	2.5	0/0
2	1.5	3.5	45/−45
3	1.5	4.5	0/90
4	2.5	2.5	45/−45
5	2.5	3.5	0/90
6	2.5	4.5	0/0
7	3.5	2.5	0/90
8	3.5	3.5	0/0
9	3.5	4.5	45/−45

2.4 Tensile Test

The tensile properties of DPFRPC and single lap bolt joint DPFRPC were determined using UTM: WAW-600D tensile testing machine, as illustrated in Fig. 4. The WAW-600D is a computer-controlled hydraulic universal testing machine that uses hydraulic loading and computer displays for ease of use.

Fig. 4. Tensile test of single lap bolt joint DPFRPC using UTM: WAW-600D

2.5 Model Development and Optimization Using ANN-GA Approach

To determine the appropriate joining geometry parameters, artificial neural network (ANN) modeling and genetic algorism (GA) optimization method was used for this study, because this method is highly accurate modeling, predicting and optimizing tool, and the developed model can be used by other researchers for different targets [6, 22]. ANN takes input data, trains itself to detect patterns, and then predicts the output for a new collection of similar data by using the network shown in Fig. 5. ANN model was made using three layers; input layers, hidden layer, and output layers. The input layer contains the input process parameters (FO, E/D, W/D), whereas the output layer contains the failure load (Fl). The hidden layer is made up of many linked neurons determined through training, testing, and validation. The sum of inputs is transferred as output by each neuron's transfer function (activation function) and the weight value is assigned to each connection [22].

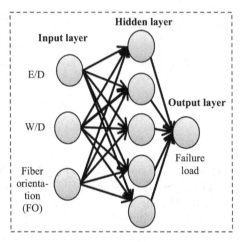

Fig. 5. ANN architectural model of bolt joint DPFRPC

In this study, GA was used to optimize the input space of the ANN model to obtain a set of optimum process parameter levels. Using the flow chart illustrated in Fig. 6, the fitness function of a previously created ANN model was used for GA to discover the optimal solutions. The constraints used for GA were E/D from 1.5 to 3.5, W/D from 2.5 to 4.5, and fiber orientation from 0 to 90 (degree). The GA parameters settings were

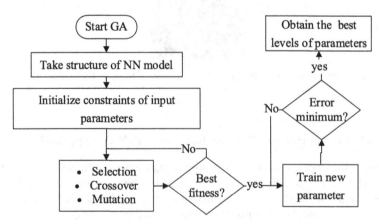

Fig. 6. GA model flow chart

determined based on past research and convergence. The GA-specific parameters are as follows: probability of crossover of 0.8, mutation rate of 0.01, population size of 50, and the number of generations over which GA evolved was 300.

3   Results and Discussions

3.1   Tensile Properties of DPFRPC

The tensile strength of DPF and polyester was determined experimentally before DPFRPC was made, and they were found to be 230 MPa and 45 MPa respectively. Ultimate tensile strength and stiffness of polyester resin were 79.13% and 62.2% lower than DPF respectively. The tensile properties of DPFRPC, which was made under 20%, 30%, and 40% fiber loading with unidirectional fiber orientation, was determined through tensile testing. From the experiment, the strongest DPFRPC was found at 30% fiber loading, 145 MPa, which was 58.6% lower than DPF and 66.8% higher than polyester matrix as shown in Fig. 7. There was 3.6 and 18.6% increase in tensile strength for 30% fiber loading compared to 40% and 20% fiber loading respectively. The tensile strength increases up to optimum fiber loading and starts to fall beyond that level of loading. This indicates that the DPFRPC joint has to be made using 30% fiber loading.

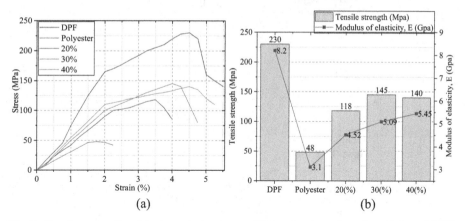

Fig. 7. Tensile properties (a) stress-strain curves (b) tensile strength and modulus of DPF, polyester, and DPFRPC with various DPF loadings.

3.2   Tensile Properties of Joined DPFRPC Subjected to Tensile Load

The experimental testing was carried out to analyze the effect of geometric joining variables of single lap bolted joint DPFRPC based on the design of experiments. From the result, different joint geometries show different load-carrying capacities. A maximum of 9.5 kN load caring capacity was found in experiment number 9 as shown in Table 3.

Table 3. Experimental results of single lap bolt joint DPFRPC

No.	E/D	W/D	Fiber orientation (°)	Failure load (kN)	Mode of failure
1	1.5	2.5	0/0	4.7	Net tension
2	1.5	3.5	45/−45	5.8	Net tension
3	1.5	4.5	0/90	7	Shear out
4	2.5	2.5	45/−45	4.8	Net tension
5	2.5	3.5	0/90	6	Composite failure
6	2.5	4.5	0/0	9.5	Cleavage tension
7	3.5	2.5	0/90	4	Net tension
8	3.5	3.5	0/0	9	Net tension
9	3.5	4.5	45/−45	9.45	Composite failure

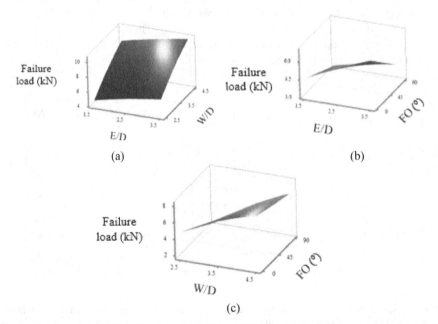

Fig. 8. Surface plot of failure load vs (a) W/D and E/D, (b) E/D and fiber orientation, (c) W/D and fiber orientation of single lap bolt joint DPFRPC

The effect of input parameters with its load carrying capacity were related through surface plots as indicated in Fig. 8, shows the surface plots of responses of failure load with E/D, W/D and fiber orientation.

The highest load-carrying capacity was found at maximum E/D and W/D and minimum fiber orientation. An increase in E/D ratio and W/D ratio tends to increase load-carrying capacity (failure load), whereas the decreasing trend of failure load was observed with increased values of fiber orientation. This might be due to the fact that as then E/D and W/D increase, the joint area increases which help to sustain higher loads, whereas increased fiber orientation reduces the composite's strength, causing it to fail under a lesser load.

3.3   Failure Behavior of Joined DPFRPC

Post failure analysis of the fracture surfaces was examined after testing in terms of failure type (shear out, net tension, bearing). Figure 9 shows the failure single lap bolt joint of DPFRPC.

Fig. 9. The failure mechanisms of single lap bolt joint DPFRPC

According to the results of the experiment, net-tension failure occurs when the specimen's W/D ratio was too low (Fig. 9 experiment number 1, 2, 4,7, and 8). As the W/D ratio decreases, the side length decreases, reducing the force applied area. Shear-out failure was discovered along the shear-out plane on the hole boundary (Fig. 9 experiment number 3). It occurs when the E/D of the DPFRPC was small. On the other hand, composite failure (Fig. 9 experiment number 5 and 9) exists when the load acted on the specimen is beyond the load carrying capacity of DPFRPC. The combination of shear out and net tension (cleavage tension failure) exists, Fig. 9 experiment number 6, when E/D and W/D are nearly equal. The frequently existed failure was net tension, so the W/D has a significant effect than other parameters on the tensile strength of single lap bolt joining DPFRPC. Net tension and shear out can be avoided by increasing edge and width of specimen, whereas bearing exists in all types of failure before it fractures, and any change in the geometry will not be enough to avoid it.

3.4   ANN Modeling

The results of the experiments were used to develop an artificial neural network model. The developed ANN was trained with 9 sets of input (E/D, W/D, FO), and output (Fl) parameters emerged from the experiments conducted. The model was trained with two-thirds of the samples, and the remaining one-third was utilized for testing and validation with trainlm. The best ANN architecture for a single lap bolt joint DPFRPC was discovered after multiple attempts to be 3-6-1-1 (three neurons for the input layer, five neurons for the first hidden layer, one neuron for the second hidden layer, and one neuron for the output layer) as shown in Fig. 10. The activation functions for convergence in the first and second hidden layers, respectively, were the hyperbolic tangent and the linear transfer function.

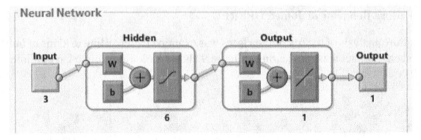

Fig. 10. Architecture of ANN model of single lap bolt joint DPFRPC

The performance of neural network training was conducted as shown in Fig. 11 curve, with the convergence to mean square error (MSE) of 0.045012 achieved after 1000 iterations or epochs. However, MSE reaches saturation in the second epoch itself. If an iteration increases beyond this, MSE begins to deviate from its optimal value.

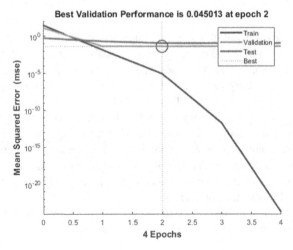

Fig. 11. Convergence of MSE during training of the ANN model

The network was validated by plotting the relationship between the network's outputs and the targets in a regression plot. The training, validation, and testing patterns were investigated in this regression plot as shown in Fig. 12. The correlation coefficient (R) for DPFRPC bolt joining was 0.996 in regression analysis, indicating a reasonable correlation between experimental and projected ANN response. The correlation coefficient (R) for single lap bolt joint DPFRPC was 0.996 in regression analysis, indicating a reasonable correlation between experimental and projected ANN response.

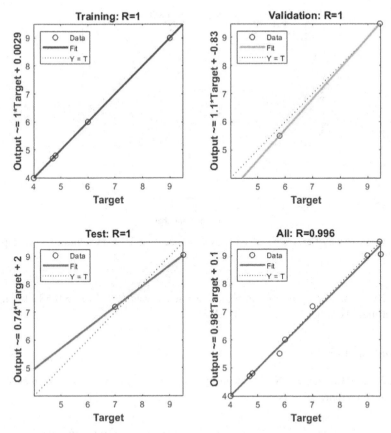

Fig. 12. ANN model simulation results in comparison with experimental results during training, validation, and testing.

There was relatively little variation between the experimental and predicted values of the ANN model, as shown in Fig. 13. From the analysis, the maximum error exists at experiment number six which was 0.4578%, and a minimum of zero error at experiment number eight. The error was within the acceptable range [22], so the model is accurate to relate the input and output parameters. The fitness function of this model can be used for genetic algorism to get the optimum input process parameters.

3.5   Optimization of Process Parameters by GA

The negative of the intended fitness function, obtained from ANN, was minimized because GA always minimizes the objective function. The GA parameters settings were determined based on convergence. The GA-specific parameters were 0.8 probability of crossover, 0.01% mutation rate, 50 population size, and 300 generations of evolution. With 56 generations, the mean fitness value of GA approaches the dominant fitness. According to GA, the best process parameters are 3.5 E/D, 4.5 W/D, and 56.4534° of

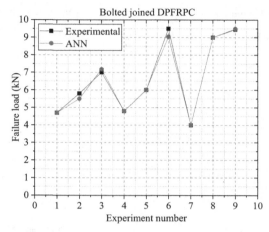

Fig. 13. Comparison of experimental and ANN predicted failure load values of single lap bolt joined DPFRPC

fiber orientation. The optimum output (failure load) was found to be 9.584 kN using these optimum settings, as predicted by ANN.

3.6 Confirmation Test

Based on the optimized ANN-GA result, a confirmation test was undertaken as shown in Table 4 to validate the model.

Table 4. Confirmation test result for bolt joined DPFRPC

Experiment	E/D	W/D	Fiber orientation (°)	Failure load (kN)
GA optimal solution	3.5	4.5	56.4534	9.584
Experimental solution	3.5	4.5	56.5	9.52
Error	–	–	–	0.064

The optimum process parameters levels found through GA were predicted with the developed ANN model. From the confirmation experiment, a maximum of 9.52 kN load-carrying capacity existed with an error of 0.064. So, the confirmation test output values agreed well with the ANN-GA model predicted values.

4 Conclusion

In this study, DPFRPC was made from DPF and polyester matrix through hand layup approach at the fiber loading of 20%, 30%, and 40% under unidirectional fiber orientation. A single lap bolt joining of DPFRPC was made using L9 orthogonal array design of experiments with three parameters and levels of E/D (1.5, 2.5, and 3.5), W/D (2.5, 3.5, and 4.5), and fiber orientation (0/0°, 45/−45°, and 0/90°). ANN model was developed to relate these input parameters and failure load through training, testing, and validation of the model. The best ANN architecture of single lap bolt joined DPFRPC was determined to be 3-6-1-1with an activation function of hyperbolic tangent and the linear transfer function in the first and second hidden layers, respectively. GA optimization technique was used by taking ANN model result as an input to get the optimum process parameters. Using this technique, the optimum process parameters of single lap bolt joint DPFRPC were, E/D of 3.5, W/D of 4.5, and fiber orientation of 56.5°, with a load-carrying capacity of 9.52 kN. An increase in E/D ratio and W/D ratio tends to increase load-carrying capacity (failure load), whereas the decreasing trend of failure load was observed with increased values of fiber orientation. This might be due to the fact that as then E/D and W/D increase, the joint area increases that helping to sustain higher loads, whereas increased fiber orientation reduces the composite's strength, causing it to fail under a lesser load.

References

1. Nurazzi, N., et al.: A review on natural fiber reinforced polymer composite for bullet proof and ballistic applications. Polymers **13**(4), 646 (2021)
2. Srinivas, K., Naidu, A.L., Bahubalendruni, M.R.: A review on chemical and mechanical properties of natural fiber reinforced polymer composites. Int. J. Perform. Eng. **13**(2), 189 (2017)
3. Johnson, R.D.J., Arumugaprabu, V., Ko, T.J.: Mechanical property, wear characteristics, machining and moisture absorption studies on vinyl ester composites – a review. Silicon **11**(5), 2455–2470 (2019)
4. Alarifi, I.M.: Investigation into the morphological and mechanical properties of date palm fiber-reinforced epoxy structural composites. J. Vinyl Add. Tech. **27**(1), 77–88 (2021)
5. Chihaoui, B., Serra-Parareda, F., Tarrés, Q., Espinach, F.X., Boufi, S., Delgado-Aguilar, M.: Effect of the fiber treatment on the stiffness of date palm fiber reinforced PP composites: macro and micromechanical evaluation of the young's modulus. Polymers **12**(8), 1693 (2020)
6. Galińska, A.: "Mechanical joining of fibre reinforced polymer composites to metals – a review. Part I: bolted joining. Polymers **12**(10), 2252 (2020)
7. El Zaroug, M., Kadioglu, F., Demiral, M., Saad, D.: Experimental and numerical investigation into strength of bolted, bonded and hybrid single lap joints: effects of adhered material type and thickness. Int. J. Adhes. Adhes. **87**, 130–141 (2018)
8. Zhang, J., Xie, Q., Xie, Y., Zhou, L., Wang, Z.: Investigation of mechanical performances of composite bolted joints with local reinforcements. Sci. Eng. Compos. Mater. **25**(1), 75–83 (2018)
9. Hu, X., Haris, A., Ridha, M., Tan, V., Tay, T.: Progressive failure of bolted single-lap joints of woven fibre-reinforced composites. Compos. Struct. **189**, 443–454 (2018)
10. Feng, N.L., Malingam, S.D., Irulappasamy, S.: Bolted joint behavior of hybrid composites. In: Failure Analysis in Biocomposites, Fibre-Reinforced Composites and Hybrid Composites, pp. 79–95. Elsevier (2019). https://doi.org/10.1016/B978-0-08-102293-1.00004-8
11. Fiore, V., Calabrese, L., Scalici, T., Bruzzaniti, P., Valenza, A.: Experimental design of the bearing performances of flax fiber reinforced epoxy composites by a failure map. Compos. B: Eng. **148**, 40–48 (2018)

12. Galińska, A., Galiński, C.: Mechanical joining of fibre reinforced polymer composites to metals – a review. Part II: riveting, clinching, non-adhesive form-locked joints, pin and loop joining. Polymers **12**(8), 1681 (2020)

13. Naik, R., Panda, S., Racherla, V.: A new method for joining metal and polymer sheets in sandwich panels for highly improved interface strength. Compos. Struct. **251**, 112661 (2020)

14. Kudus, M.H.A., Ratnam, M.M., Akil, H.M.: Factors affecting hole quality during drilling of natural fiber-reinforced composites: a comprehensive review. J. Reinf. Plast. Compos. **40**(9–10), 391–405 (2021). https://doi.org/10.1177/0731684420970650

15. Shipsha, A., Burman, M.: Failure mechanisms in NCF composite bolted joints: experiments and FE model. Compos. B: Eng. **192**, 107950 (2020)

16. Kerur, S., Shivakumar, S.: Optimization of adhesive joining process parameters on single lap shear strength properties of hybrid metal polymer composite joints, p. 020004

17. Jeevi, G., Nayak, S.K., Abdul Kader, M.: Review on adhesive joints and their application in hybrid composite structures. J. Adhes. Sci. Technol. **33**(14), 1497–1520 (2019)

18. Shaikh, S., Anekar, N., Kanase, P., Patil, A., Tarate, S.: Single lap adhesive joint (SLAJ): a study. Int. J. Eng. Technol. **7**, 64–70 (2017)

19. Kim, J., Lopez-Cruz, P., Heidari-Raran, M., Lessard, L., Laliberté, J.: An experimental study on the mechanical behaviour of bonded and hybrid bonded-bolted composite joints using digital image correlation (DIC) technique. Compos. Struct. 114544 (2021)

20. Hachaichi, A., Kouini, B., Kian, L.K., Asim, M., Jawaid, M.: Extraction and characterization of microcrystalline cellulose from date palm fibers using successive chemical treatments. J. Polym. Environ. **29**(6), 1990–1999 (2021). https://doi.org/10.1007/s10924-020-02012-2

21. Choudhury, M.R., Debnath, K.: Experimental analysis of tensile and compressive failure load in single-lap bolted joint of green composites. Compos. Struct. **225**, 111180 (2019)

22. Hramov, A.E., et al.: Artificial neural network detects human uncertainty. Chaos Interdiscipl. J. Nonlin. Sci. **28**(3), 033607 (2018)

Applications of Artificial Intelligence for Fault Diagnosis of Rotating Machines: A Review

Fasikaw Kibrete[1,2](✉) ⓘ and Dereje Engida Woldemichael[2] ⓘ

[1] Department of Mechanical Engineering, University of Gondar, P.O. Box 196, Gondar, Ethiopia
fasikaw.kibrete@uog.edu.et
[2] Department of Mechanical Engineering, College of Electrical and Mechanical Engineering,
Addis Ababa Science and Technology University, P.O. Box 16417, Addis Ababa, Ethiopia
dereje.engida@aastu.edu.et

Abstract. Rotating machines are commonly used mechanical equipment in various industrial applications. These machines are subjected to dynamic and harsh operating conditions over a long time leading to various types of mechanical failures, thereby resulting in undesirable downtime. Consequently, research on fault diagnosis is practically significant to enhance the safety of machinery. Over the years, several fault diagnosis methods have been developed for rotating machines. Of these, artificial intelligence-based diagnosis methods have gained increasing attention due to their reliability, robustness in performance, and capability for adaptation. However, the selection of suitable artificial intelligence methods for specific types of faults or machines is still dependent on the experience of users. The recent research achievements in intelligent fault diagnosis are not reviewed, and future research directions are not clearly stated. To fill these gaps, this paper provides a review of artificial intelligence techniques applied for fault diagnosis of rotating machines, with a special emphasis given to deep learning methods published in the last five years (2017–2022). The research challenges and some possible prospects in this field are discussed to provide valuable guidelines for future research development. The present work can be extended to review the applications of transfer learning for fault diagnosis of rotating machines.

Keywords: Artificial intelligence · Deep learning · Fault diagnosis · Machine learning · Rotating machine

1 Introduction

In modern industrial systems, there is an increasing trend toward the need for more reliable machines. Rotating machines are commonly used mechanical equipment in various industrial applications. They accounts for more than 90% of industrial machines [1]. As these machines usually operate under dynamic and harsh conditions for a long time, they often suffer from various types of mechanical failures. Any type of failure in rotating machines, even minor failure, cannot be accepted as it can significantly affect the entire system, and can even lead to undesirable downtime, huge economic losses and serious safety problems [2, 3]. Consequently, research on fault diagnosis is practically

© ICST Institute for Computer Sciences, Social Informatics and Telecommunications Engineering 2023
Published by Springer Nature Switzerland AG 2023. All Rights Reserved
B. H. Woldegiorgis et al. (Eds.): ICAST 2022, LNICST 455, pp. 41–62, 2023.
https://doi.org/10.1007/978-3-031-28725-1_4

significant to enhance the reliability of machines, reduce economic losses, and avoid safety problems [4, 5].

Over the last decades, numerous methods have been presented to diagnose the faults of rotating machines. These methods fall into three broad classes: model-based methods, statistical methods, and artificial intelligence-based methods [6]. Model-based methods are formed based on the physical characteristics of a monitored machine with the necessary assumptions to establish an explicit mathematical model [7]. However, it is challenging to establish an exact mathematical model for complex systems [8]. The high complexity of industrial faults and the cost of model-based methods limit their applicability in fault diagnosis of machines. Statistical methods assume that historical data can be used to establish the fault modes and the future mechanism of machine failure [9]. However, this assumption might not hold in practical scenarios because the failure mechanisms of machines are complex, nonlinear, and involve the coupling of different physical processes. Nowadays, artificial intelligence-based fault diagnosis methods are the focus of academic and industrial research for overcoming the problems in the fault diagnosis of complex industrial machines [10]. The primary reason is that artificial intelligence methods are instrumental if they can be improved as compared with other methods. Artificial intelligence can be easily extended and modified. These methods can also be made adaptive by integrating new data [11].

Motivated by the advantages of artificial intelligence methods, intelligent fault diagnosis methods have gained great attention in recent decades. Therefore, this paper provides a review of artificial intelligence methods applied for fault diagnosis of rotating machines, with a special emphasis given to deep learning methods published from 2017 to 2022. This paper analyzes the strengths and weaknesses of each method, so as to give valuable guidance for researchers in selecting an appropriate intelligent method for specific applications instead of choosing randomly. The research challenges in this field are also discussed to provide possible research directions for further exploration.

The remainder of this paper is organized as follows. Section 2 provides a general overview of intelligent fault diagnosis of rotating machines. Section 3 presents a detailed review of the applications of deep learning methods in the fault diagnosis of rotating machines. Section 4 discusses the observations of the review, research challenges, and future direction in this area. Finally, conclusions are drawn in Sect. 5.

2 Overview of Intelligent Fault Diagnosis of Rotating Machines

In the last decades, traditional machine learning methods have been widely applied in the intelligent fault diagnosis of rotating machines. These methods mainly comprise three consecutive steps: data acquisition, feature extraction, and fault classification [12, 13]. In the data acquisition stage, a variety of signals such as vibration, acoustic emission, noise, temperature, etc., are acquired from target machines by sensor systems [14]. In the feature extraction stage, fault-sensitive information from sensor signals is manually extracted using different types of signal processing methods [15]. Such processes rely too much on the step of feature extraction, which requires prior signal processing knowledge and diagnosis experience [16, 17]. Finally, the extracted features are fed into the traditional machine learning methods for classification [18, 19]. However, the traditional machine learning methods are designed for specific types of faults or machines and therefore are case dependent and not used for general applications [20]. Moreover, these methods are not efficient for processing high-dimension data [21]. In general, traditional intelligent diagnosis methods have low diagnosis performance for machines that operate under adverse and complex conditions [22]. These reasons can seriously restrict the applicability of traditional machine learning methods for rotating machine fault diagnosis.

Recently, deep learning methods have received great interest and achieved significant successes in machine fault diagnosis, which overcomes the limitations of traditional machine learning methods [23, 24]. Deep learning-based fault diagnosis methods can extract the learnable features from large amounts of sensor data directly by constructing deep network architecture with multiple layers of linear and non-linear transformations and performs an end-to-end fault diagnosis [25, 26]. In the following section, the most common deep learning methods have been discussed.

3 Deep Learning Methods in Fault Diagnosis of Rotating Machines

This section reviews the applications of the most common deep learning methods and their corresponding variants in fault diagnosis of rotating machines.

3.1 Convolutional Neural Network (CNN)

Convolutional neural network (CNN) is a biologically inspired feed-forward neural network used to extract local features from the raw sensor data to perform classification [27]. The typical convolutional neural network model consists of multiple hidden layers, namely, the convolution layer, pooling layer, and fully connected layer [28]. The convolution layer is composed of a series of learnable filters (also known as kernels) that can extract different features of input data to generate new feature maps as the input to the next layer. The pooling layer is the down-sampling layer which decreases the size of the input and the number of parameters, and thus it can decrease the number of computations and prevent overfitting. The fully connected layer is used to compute the class scores [29].

Convolutional neural network (CNN) was originally designed for processing two-dimensional (2D) or three-dimensional (3D) input data such as images and video frames [30]. The traditional CNN is not suitable for fault diagnosis of mechanical equipment since most measured signals are one-dimensional (1D) signals. Thus, the input 1D data needs to be converted into 2D data through some methods to complete the feature extraction and classification [31]. Studies by [32, 33] proposed CNN-based fault diagnosis methods by converting the original 1D signals into 2D images for different machine diagnosis tasks. However, the process of converting the original signal is time-consuming, and may certainly cause the loss of faulty data. The emergence of a one-dimensional convolutional neural network (1D-CNN) provides a feasible solution to avoid the above problems. Compared with 2D-CNN, 1D-CNN has a simpler and more compact network structure, and it can effectively diagnose the faults of machines with limited training data. Using one-dimensional vibration signals as input data, researchers often utilize a 1D-CNN to diagnose the faults of different rotating machine components, such as bearings

[12, 20, 34], automobile engines [35], and gearboxes [36]. Abdeljaber et al. [37] used a one-dimensional convolutional neural network (1D-CNN) for structural damage detection based on vibration signals. Yin et al. [38] combined 1D-CNN with self-normalizing neural networks (SNN) to improve the diagnosis accuracy and generalization capability of rotating machine fault diagnosis.

Many researchers have made significant efforts on developing novel CNN-based models and have achieved considerable progress. Jia et al. [39] developed a deep normalized CNN for imbalanced fault classification of machines from mechanical vibration signals. In [40, 41], an adaptive deep CNN model was to diagnose the faults of rolling bearings. Kolar et al. [42] proposed a multi-channels deep CNN model for rotary machine fault diagnosis from the raw vibration data. Sun et al. [43] presented a convolutional discriminative feature learning (CDFL) approach to diagnose the faults of the motor. Dilated CNN methods have been used for bearing fault diagnosis from raw vibration signals [44, 45]. Liu [46] developed a dislocated time series CNN to diagnose the faults of an induction motors. Zhang et al. [47] utilized a CNN model with wide first-layer kernels for rolling bearing fault diagnosis using one-dimensional vibration data. Chen et al. [48] developed a novel deep capsule network with stochastic delta rule (DCN-SDR) for rolling bearing fault diagnosis. Ye and Yu [49] proposed a deep morphological CNN for fault diagnosis of the gearbox. Wang et al. [50] developed a novel multiple-input, multiple-task CNN method for roller bearing fault diagnosis. Studies [51–53] proposed a hierarchical convolutional neural network (HCCN) for fault diagnosis of different rotating machine components. Zhang et al. [54] developed an intelligent method based on multi-level information fusion and hierarchical adaptive CNN to diagnose the faults of centrifugal blowers. Jiang et al. [55] developed a multiscale convolutional neural network (MSCNN) for fault diagnosis of wind turbine gearboxes. Wang et al. [56] proposed a cascade CNN with progressive optimization for motor fault diagnosis under dynamic working conditions.

3.2 Recurrent Neural Network (RNN)

Recurrent neural network (RNN) is the deepest neural network with both feedforward connections and internal feedback connections between network layers. Varying from feedforward neural networks like CNN, RNN can exploit temporal information from multiple sequential data because of its internal memory. Neurons of RNN can not only receive information from other neurons but also receive their information to form a network structure with loops. RNN has more advantages in exploiting temporal information. Thus, it has been widely utilized in machine fault diagnosis. Hu et al. [57] utilized an improved deep RNN for rotating machine fault diagnosis. Huang et al. [58] proposed the RNN-based variational auto-encoder (VAE) for motor fault detection. However, RNN has gradient vanishing and exploding problems, thus it has inherent limitations in capturing long-term information [59, 60]. To overcome the limitations of the recurrent neural network, researchers have proposed long short-term memory (LSTM) [61], gated recurrent unit (GRU), and other improved RNN models.

As an improved recurrent neural network, LSTM resolves the problems of gradient vanishing and exploding, and captures long-term dependencies and nonlinear dynamics of time series data [62]. As a result, the LSTM model with memory function has gained increasing attention in machine fault diagnosis. For instance, Yin et al. [63] developed an optimized fault diagnosis method based on the cosine loss LSTM neural network for the wind turbine gearbox. Yang et al. [64] developed an improved long short-term memory model to diagnose the faults of electromechanical actuators. However, LSTM cannot make full use of data since it can only process data in one direction [65]. Furthermore, unidirectional LSTM has a relatively high network complexity, thus the training process takes a long time [66]. Bi-directional LSTM is an improvement of LSTM that can address the limitations of unidirectional LSTM. Bi-directional LSTM can extract features from both forward and backward directions. Cao et al. [67] developed a novel intelligent method based on deep bi-directional LSTM diagnose the faults of wind turbine gearboxes. Han et al. [68] combined Bi-LSTM and a Capsule Network with a CNN for rotating machine fault diagnosis. The bi-directional LSTM was used to complete the feature denoising and fusion, which was extracted by a convolutional neural network and used a capsule network to achieve the fault diagnosis for insufficient training data. Li et al. [69] utilized Bi-LSTM to detect the faults of rolling bearings. Thus, Bi-LSTM has higher diagnosis accuracy and efficiency than unidirectional LSTM.

Compared with the LSTM model, a gated recurrent unit (GRU) can better handle large training data with a simple network structure and fewer parameters, thus it greatly reduces the calculation efficiency [70]. Liu et al. [71] utilized a GRU to diagnose the faults of rolling bearings. Tao et al. [72] utilized a multilayer GRU method for fault diagnosis of spur gear from vibration signals. To verify its superiority, the proposed method was compared with LSTM, multilayer LSTM, and support vector machine (SVM). Besides the most basic GRU structure, Bidirectional GRU (Bi-GRU) has also been employed for fault diagnosis. Bi-GRU can learn information from both forward and backward directions of the input data at the same time. For this reason, Lv et al. [73] proposed a new heterogeneous Bi-GRU method based on fusion health indicators. Zhao et al. [74] utilized a local feature-based GRU network for bearing fault diagnosis. This method used an enhanced bi-directional gated recurrent unit to extract high-level features from vibration data.

Recurrent neural network and its improvement have also been combined with other machine learning methods. Fu et al. [75] combined a CNN with a LSTM to monitor and warn of the fault of wind turbine gearbox bearing using temperature data. Zhao et al. [76] also combined bi-directional LSTM with CNN to address tool wear prediction tasks. In this study, convolutional neural network was used to extract local features from the sequential input, and bi-directional long short-term memory was used to encode the temporal information. Qiao et al. [77] combined a deep CNN with LSTM to complete an end-to-end bearing fault diagnosis under variable loads and different noise interferences. Liao et al. [78] developed a fault diagnosis method for hydroelectric generating units based on one-dimensional convolutional neural network and GRU from the raw vibration signal collected under different operational conditions.

3.3 Generative Adversarial Network (GAN)

In practical engineering scenarios, the faulty data collected from the target machine is usually more limited than the normal data, i.e., the model training data is highly imbalanced. The deep learning method trained with imbalanced data is prone to poor generalization performance. The generative adversarial network is a well-known data generative model inspired by the game theory that can address the data imbalance problem [79]. The generative adversarial network (GAN) model is mainly composed of a Generator and a Discriminator [80]. The generator receives the original data to generate new data that have a similar distribution to the real data, thus expanding the training dataset. The generated new data are imported to the discriminator with the original data to predict whether the input data are real or false data [81]. As a result, GAN has been successfully employed to fault diagnosis. For instance, Liu et al. [82] developed a fault diagnosis method based on global optimization generative adversarial network to solve the unbalanced data problem of rolling bearings. Ding et al. [83] proposed a novel fault diagnosis method for rotating machines based on GAN, and validated the effectiveness through small sample rolling bearing and gearbox datasets.

Recently, researchers have made many improvements and developed a wide variety of generative adversarial network variants. For instance, Yan et al. [84] developed a fault detection and diagnosis method that utilizes the conditional Wasserstein GAN to overcome the imbalanced data problem for air handling units. Zheng et al. [85] proposed a conditional GAN model with a dual discriminator for imbalanced rolling bearing fault diagnosis. Studies by [86–89] used a deep convolutional GAN for fault diagnosis of rotating machines with imbalanced data. Luo et al. [90] utilized a conditional deep convolutional GAN to address the data imbalance problem in machine fault diagnosis. Shao et al. [91] utilized an auxiliary classifier GAN to generate fake sensor signals to solve unbalanced fault data problems and diagnose the faults of the induction motor. Xiong et al. [92] utilized a Wasserstein gradient-penalty GAN with a deep auto-encoder (DAE) to diagnose the faults of rolling bearing. However, Wasserstein gradient-penalty GAN has the shortcomings of vanishing gradient and mode collapse. As a result, Li et al. [93] proposed a rotating machine fault diagnosis model based on a deep Wasserstein GAN with gradient penalty for the imbalanced data problem. Zareapoor et al. [94] developed a new model named Minority oversampling GAN for class-imbalanced fault diagnosis. Zi et al. [95] proposed a novel multitask redundant lifting adversarial network (MRLAN), and the results confirm its satisfactory performance under sharp speed fluctuation and little data. Liu et al. [96] developed a variational auto-encoding GAN model with deep regret analysis for bearing fault diagnosis. The study by [97] combined an auxiliary classifier GAN with a stacked denoising auto-encoder for fault diagnosis of rolling bearing. Liu et al. [98] proposed a categorical adversarial auto-encoder (CatAAE) for fault diagnosis of rolling bearings under different working conditions and achieved satisfactory performance and high clustering indicators even in different working conditions.

While GAN and its extensions have yielded certain success for the imbalanced training dataset problem, there are still some practical problems that need further exploration. For example, sometimes GAN generates no reasonable data due to the lack of auxiliary information in the deep features of input data. Besides, to create sufficient fault data, a generative adversarial network consumes a huge computing resources and takes a long training time. Consequently, it is practically significant to develop novel fault diagnosis methods based on GAN to overcome these real problems.

3.4 Auto-Encoder (AE)

Auto-encoder (AE) is an unsupervised feed-forward neural network that uses a back-propagation algorithm to learn discriminative features in an unsupervised manner by minimizing reconstruction error between the input data and the output data [99, 100]. The typical AE consists of the input layer, hidden layer, and output layer. The input and hidden layers form the encoder network, whereas the hidden and output layers form the decoder network [101]. The encoder transforms the high-dimensional input data into low-dimensional hidden features and the decoder reconstructs the input data from the learned hidden features [102].

Compared with supervised deep learning methods like CNN and RNN, the auto-encoder possesses the properties of unsupervised learning, high-efficiency training, simple structure, and easy implementation. As a result, auto-encoders had been applied for fault diagnosis of bearings [103], electric motors [104, 105], turbines [106], and other components. However, the standard auto-encoder (AE) has limited feature extraction capability for fault diagnosis tasks due to the lack of label data [107]. Moreover, in most practical situations, the measured signals are always polluted by heavy background noises, which decreases the performance of the standard auto-encoder. To overcome the aforementioned challenges, several variants of the auto-encoder have been introduced into rotating machine fault diagnosis. The common variants are denoising auto-encoder (DAE), sparse auto-encoder (SAE), contractive auto-encoder (CAE), and variational auto-encoder (VAE) [108]. The AE, DAE, SAE, and CAE can be stacked to extract deep features with better representative ability, which are named stacked auto-encoder (SAE), stacked denoising auto-encoder (SDAE), stacked sparse auto-encoder (SSAE), and stacked contractive auto-encoder (SCAE), respectively. The following subsection reviews their applications in machine fault diagnosis.

Stacked Auto-Encoder (SAE). The structure of the stacked auto-encoder (SAE) is composed of multiple auto-encoders stacked that can extract more implicit features from high-dimensional complex data and reduce the dimensionality of input data than a single auto-encoder [109]. In SAE, the output of the formerly hidden layer is used as the input to the next hidden layer [110]. Since SAE is an unsupervised learning method, it cannot be directly employed for machine fault diagnosis. Consequently, a classification layer is usually added at the end of the network structure of the model. In this context, Liu et al. [111] proposed a stacked auto-encoder (SAE) based deep learning method for gearbox fault diagnosis. Studies by [112, 113] utilized SAE to develop new methods for fault diagnosis of roller bearings. Karamti et al. [114] developed a fault diagnosis method based on stacked auto-encoders for diagnosing rotating system faults with imbalanced samples. An et al. [115] developed a batch-normalized stacked auto-encoder method for intelligent fault diagnosis of rotating machines. The effectiveness of this method was validated through motor bearing and gearbox datasets. Shao et al. [116] also proposed an improved SAE to diagnose the faults of rotating machines, and its effectiveness was validated through sun gear and roller bearing datasets.

Denoising Auto-Encoder (DAE). The denoising auto-encoder (DAE) is an AE obtained by adding noise to the input data with some statistical characteristics to increase the anti-noise capability [117]. DAE can automatically extract robust features from corrupted and partially destroyed data, so it is more suitable for fault diagnosis of different rotating machines. For instance, studies by [118, 119] employed a DAE to diagnose the faults of rolling bearings. Lu et al. [120] applied the stacked denoising auto-encoder (SDAE) for rolling bearing fault diagnosis. Zhao et al. [121] developed a deep learning method using a SDAE for motor fault diagnosis. Chen and Li [122] applied a deep neural network based on a SDAE to diagnose the faults of the rotor system. J. Yu [123] proposed a manifold regularized SDAE (MRSDAE) for planetary gearbox vibration signals. Zhan et al. [124] also utilized a SDAE combined with a SVM classifier for a permanent magnet synchronous motor used in an electric vehicle. Xu et al. [125] proposed an intelligent fault diagnosis method for metro traction motor bearings based on an improved SDAE. Xiao et al. [126] proposed a noisy domain adaptive marginal SDA for fault diagnosis of gear and motor using acoustic signals. Godói et al. [127] proposed a new denoising convolutional AE method configuration employed to the condition monitoring of rotating machines. Zhao et al. [128] combined a one-dimensional denoising convolutional auto-encoder (DCAE) with a 1D-CNN for rotating machine fault diagnosis under noisy environments. Although DAE can extract robust features and achieve remarkable results in fault diagnosis, it takes more time to select the most suitable corruption level and corrupt the input data into corrupted inputs of the DAE. Moreover, the extracted features by DAE may consist of some useless features for fault diagnosis.

Sparse Auto-Encoder (SAE). As an extension of AE, the sparse auto-encoder (SAE) is a widely used auto-encoder that introduces the sparse penalty term, adding constraints to the hidden layer for a concise expression of the input data [129]. Compared with other deep learning methods, SAE is superior in extracting sparser features, highly discriminative, and useful for classification. As a result, many researchers have widely used SAE for fault diagnosis of rotating machines. For instance, Xin et al. [130] combined a sparse auto-encoder with softmax regression to diagnose the fault of the attachment on the blades of the marine current turbine. Zhao et al. [131] proposed a semi-supervised deep SAE with local and non-local information for fault diagnosis of rotating machines. Kim et al. [132] utilized a sparse SAE to develop a new fault diagnosis method for the gearbox. Qi et al. [133] developed an intelligent fault diagnosis method based on a SSAE, and its effectiveness was validated through rolling bearing and gearbox vibration datasets. Sun et al. [101] developed a novel intelligent diagnosis method of automatic feature learning and classification of rotating machines based on SSAE. Studies by [134–137] proposed new fault diagnosis methods based on sparse stacked denoising AE for bearing fault diagnosis. Zhang et al. [138] also developed a stacked pruning sparse denoising AE method for rolling bearing fault diagnosis. Wen et al. [139] proposed a new fault diagnosis method based on stack pruning sparse denoising auto-encoder and CNN to detect and categorized the actuator damage fault of the unmanned aerial vehicle, and showed good fault diagnosis accuracy in an actual high noise environment. Jia et al. [140] developed a local connection network (LCN) constructed by normalized sparse autoencoder (NSAE) for fault diagnosis of rotating machines. The superiority of the proposed NSAE-LCN was verified using gearbox and bearing datasets. However, the accuracy and generalization ability of sparse stacked auto-encoder is affected by its hyperparameter settings and there is no clear rule for determining the optimal hyperparameter values, which heavily depends on experimental experience. Moreover, the standard learning method employed in sparse stacked auto-encoder is time-consuming.

Contractive Auto-Encoder (CAE). The contractive auto-encoder (CAE) is a well-known AE variant that can automatically learn more robust features and is thus suitable for dealing with noise-overwhelmed signals. The robustness of the data description is obtained by adding a contractive penalty to the error function of the reconstruction. This penalty is used for penalizing the attribute sensitivity in the input variations. CAE can handle noisy data without knowing noise intensity and has been applied successfully for robust feature extraction and fault classification. Qi et al. [141] proposed a new deep fusion network that combined the SSAE and CAE for fault diagnosis of bearing and gearbox. Fu et al. [142] also proposed a deep contractive auto-encoding network (DCAEN) for fault diagnosis of bearing. Shen et al. [143] applied a stacked contractive auto-encoder (SCAE) for feature extraction and fault diagnosis of rotating machines. Gao et al. [144] proposed a new ensemble deep CAE method for machine fault diagnosis under noisy environments. The effectiveness proposed method was verified using bearing, gearbox, and self-priming centrifugal pump datasets. However, CAE still has higher reconstruction errors during the encoding and decoding process of input features to the network that cause difficulty to capture the useful information within the feature space.

Variational Auto-Encoder (VAE). As a generation model, variational auto-encoder (VAE) can augment the dataset by generating meaningful synthetic data similar to the original real data and has been successfully employed in fault diagnosis of different rotating machines [145]. In [146–148], a variational auto-encoder has been employed to create fault data of bearings. Sun et al. [149] developed a novel fault diagnosis method called conditional variational auto-encoder generative adversarial network for planetary gearboxes to solve small sample problems. However, the data augmented by the variational auto-encoder is not always real data. Thus, how to make the data samples generated by variational auto-encoder more real is still a challenge that requires further exploration.

3.5 Deep Belief Network (DBN)

Deep belief network (DBN) is a probability generation model composed of several layers of restricted Boltzmann machines (RBMs), where the output of the previously hidden layer is utilized as the input of the next layer [150] and the last layer is the backpropagation neural network. The training process of DBN comprises two stages: forward unsupervised greedy layer-by-layer pre-training and backward supervised fine-tuning process. The forward pre-training phase is an unsupervised training process that aims to extract features from bottom to top layer-by-layer. After the pre-training of multiple RBMs, the fine-tuning phase is then utilized with a backpropagation algorithm to optimize the parameters and structure of the pre-trained network to further enhance the classification accuracy. In fine-tuning, the weights and biases of every layer are updated continuously until the iteration reaches the limit [151].

Since deep belief network (DBN) is suitable for processing one-dimensional data, its applications in fault diagnosis are reported frequently. For instance, Shang et al. [152] proposed a diagnosis method based on DBN for rolling bearings, which reduces the complicated network structure to some extent. Qin et al. [153] proposed a new fault diagnosis method using a DBN for planetary gearboxes of wind turbines. Yan et al. [154] also proposed a rotor unbalance fault diagnosis method using a multi-deep belief network model with multi-sensor information. Han et al. [155] combined the DBN model with wavelet packet energy entropy and multi-scale permutation entropy to diagnose gear faults. The authors of [156] proposed a new condition monitoring method for rolling bearings by using the DBN model optimized by the multi-order fractional Fourier transform filtering algorithm and the sparrow search algorithm. Zhang et al. [157] applied a DBN algorithm to diagnose the fault of the power system, and enhanced the ability of feature extraction and fault classification by enhancing the network model. Yu et al. [158] proposed a novel fault diagnosis method by hybridizing DBN with Dempster-Shafer theory for diagnosing the wind turbine system.

The performance of deep belief networks in machine fault diagnosis depends heavily on their structure. To obtain an optimal network structure with high performance and training speed, researchers utilized various optimization techniques. In the literature [159, 160], the network structure and learning rate of DBN were optimized by using the PSO algorithm, and the diagnosis accuracy was improved. Wen et al. [161] combined the deep belief network with a fuzzy mean clustering algorithm for rolling bearing fault diagnosis without using data labels. Gao et al. [162] optimized the network architecture of a deep belief network using a salp swarm algorithm and utilized it for rolling bearing fault diagnosis. Similarly, Kamada et al. [163] used the neuron generation annihilation and layer generation algorithm to propose the adaptive structure learning method of restricted Boltzmann machine and deep belief network, and achieved remarkable success. Shen et al. [164] developed an improved hierarchical adaptive DBN optimized by Nesterov momentum (NM) for bearing fault diagnosis.

4   Discussion, Existing Challenges and Future Directions

As seen from the review provided, traditional machine learning methods and deep learning methods are widely applied in intelligent machine fault diagnosis. Intelligent fault diagnosis methods based on traditional machine learning have been widely investigated in the field of fault diagnosis of rotating machines, but they have limitations in processing massive amounts of data as useful features are extracted manually with prior expert experience. Different from traditional machine learning methods, deep learning methods can extract abstract features from massive and heterogeneous mechanical signals with the help of their multilayer nonlinear mapping ability to perform an end-to-end fault diagnosis. Table 1 reveals the strengths and weaknesses of deep learning methods applied in fault diagnosis of industrial machines.

Although deep learning methods have achieved tremendous success in fault diagnosis, there are still some practical problems that need further exploration.

1. Most existing deep learning methods often need a sufficient amount of labeled data for model training, which achieves great results in laboratory experiments since

Table 1. Strengths and weaknesses of deep learning methods.

Deep learning method	Strengths	Weaknesses
CNN	1. Good for multi-dimensional data 2. Good in extracting local features 3. Suitable to process image data	1. Complex architecture, and hence hard to train 2. Require sufficient training data 3. Overfitting problem
RNN	1. Suitable for sequential data 2. Detect changes over time	1. Difficult to train and implement since the architecture is complex 2. Gradient vanishing and exploding problem
GAN	1. Efficient for small dataset	1. Difficult to train 2. Limited data generation ability in case of very small training data
AE	1. Easy to implement 2. Takes small training time 3.Does not require labeled data	1. Requires lots of training data 2. Requires more information compared to relevant information 3. Ineffective when errors are present in the first few layers
DBN	1. Most suitable for one-dimensional data 2. Extract large-scale features from the input data 3. High performance in handling complex data without data preparation steps 4. Difficulties in avoiding poor local optima and optimization are mitigated	1. Training is very slow because of complex initialization and computationally expensive 2. Multiple stochastic hidden layers lead to inferences and intractable learning

there is sufficient labeled data. However, it is hard to acquire massive data or even impossible in practical industrial scenarios as most machines operate in healthy conditions.

2. Existing deep learning methods can recognize faults accurately with the assumption that the training dataset and the testing dataset are drawn from the same machine under the same working conditions. This assumption may not hold in many real cases due to variations in machine working conditions, interference of environmental noise, etc., which leads to significant diagnosis performance deterioration.

3. In practical industrial applications, the collected sensor signals from rotating machines are usually polluted by various forms of noise, thereby reducing the performance of the existing fault diagnosis methods.

4. For long-term monitoring, it is essential to achieve early fault detection of rotating machines. However, in real practice, it is quite difficult to realize the faults at the earliest stage due to the weakness of impulse signals and the interference of environmental noise.

5. Furthermore, rotating machines operate for a long time under changeable conditions, and compound faults may occur simultaneously. Not only this, multiple rotating machine components may fail at a time. Most existing studies have ignored the existence of simultaneous fault problems.

Therefore, it is of great significance to resolve these practical problems and advance intelligent diagnosis methods for promising employment in modern industrial applications. The following are some possible research directions given to researchers, readers, and engineers who aim to contribute to the advancement of artificial intelligence in the fault diagnosis of rotating machines.

Active research area toward promising results.

1. The emergence of transfer learning provides a feasible solution to overcome the abovementioned gaps. Different from deep learning, transfer learning targets to extract knowledge obtained in the source domain and transfer it to resolve a different but similar problem in another domain task. Therefore, transfer learning is becoming an active research area in the field of intelligent machine fault diagnosis.

2. Based on the review provided, some deep learning methods have strong feature extraction capabilities and others have limitations in fault classification. To break the limitation of a single method, researchers still have great possibilities to propose hybrid deep learning-based fault diagnosis methods for rotating machines.

5   Conclusions

This paper reviewed the applications of artificial intelligence methods for the diagnosis of the faults of rotating machines. The observations, research gaps, and some new research prospects in this research area are discussed. From the review, it is concluded that deep learning methods have better feature learning ability, better adaptability, and a more flexible network structure as compared with conventional machine learning methods. However, their applicability in fault diagnosis is highly restricted by the amount and quality of the training data, the variation of operating conditions, the disturbance of background noise, the weakness of early failure detection, and the occurrence of hidden simultaneous faults. To address these limitations, transfer learning is becoming a hot research topic in machine fault diagnosis. In addition, new intelligent diagnosis methods are needed to be able to combine the advantages of both methods in the future. In the future, the authors will continue to review the applications of transfer learning for the diagnosis of faults of rotating machines.

References

1. Han, T., Zhang, L., Yin, Z., Tan, A.C.C.: Rolling bearing fault diagnosis with combined convolutional neural networks and support vector machine. Measurement **177**, 109022 (2021). https://doi.org/10.1016/j.measurement.2021.109022

2. Liang, P., Deng, C., Wub, J., Yang, Z.: Intelligent fault diagnosis of rotating machinery via wavelet transform, generative adversarial nets and convolutional neural network. Meas. J. Int. Meas. Confed. **159**, 107768 (2020). https://doi.org/10.1016/j.measurement.2020.107768

3. Youcef Khodja, A., Guersi, N., Saadi, M.N., Boutasseta, N.: Rolling element bearing fault diagnosis for rotating machinery using vibration spectrum imaging and convolutional neural networks. Int. J. Adv. Manuf. Technol. **106**(5–6), 1737–1751 (2019). https://doi.org/10.1007/s00170-019-04726-7

4. AlShorman, O., et al.: Sounds and acoustic emission-based early fault diagnosis of induction motor: a review study. Adv. Mech. Eng. **13**(2), 1–19 (2021). https://doi.org/10.1177/1687814021996915

5. Eren, L., Ince, T., Kiranyaz, S.: A generic intelligent bearing fault diagnosis system using compact adaptive 1D CNN classifier. J. Signal Process. Syst. **91**(2), 179–189 (2018). https://doi.org/10.1007/s11265-018-1378-3

6. You, D., et al.: Intelligent fault diagnosis of bearing based on convolutional neural network and bidirectional long short-term memory. Shock Vib. **2021**, 1–12 (2021). https://doi.org/10.1155/2021/7346352

7. Zhang, Y., et al.: A simultaneous fault diagnosis method based on cohesion evaluation and improved BP-MLL for rotating machinery. Shock Vib. **2021**, 1–12 (2021). https://doi.org/10.1155/2021/7469691

8. Lee, J.-H., Pack, J.-H., Lee, I.-S.: Fault Diagnosis of induction motor using convolutional neural network. Appl. Sci. **9**(15), 2950 (2019). https://doi.org/10.3390/app9152950

9. Wen, L., Li, X., Gao, L., Zhang, Y.: A new convolutional neural network-based data-driven fault diagnosis method. IEEE Trans. Ind. Electron. **65**(7), 5990–5998 (2018). https://doi.org/10.1109/TIE.2017.2774777

10. Cheng, Y., Lin, M., Wu, J., Zhu, H., Shao, X.: Intelligent fault diagnosis of rotating machinery based on continuous wavelet transform-local binary convolutional neural network. Knowledge-Based Syst. **216**, 106796 (2021). https://doi.org/10.1016/j.knosys.2021.106796

11. Siddique, A., Yadava, G.S., Singh, B.: Applications of artificial intelligence techniques for induction machine stator fault diagnostics: review. In: IEEE International Symposium on Diagnostics for Electric Machines, Power Electronics and Drives, SDEMPED 2003 - Proceedings, pp. 29–34 (2003). https://doi.org/10.1109/DEMPED.2003.1234543

12. Peng, D., Liu, Z., Wang, H., Qin, Y., Jia, L.: A novel deeper one-dimensional CNN with residual learning for fault diagnosis of wheelset bearings in high-speed trains. IEEE Access **7**, 10278–12093 (2019). https://doi.org/10.1109/ACCESS.2018.2888842

13. Jiang, H., Li, X., Shao, H., Zhao, K.: Intelligent fault diagnosis of rolling bearings using an improved deep recurrent neural network. Meas. Sci. Technol. **29**(6), 065107 (2018). https://doi.org/10.1088/1361-6501/aab945

14. Tiboni, M., Remino, C., Bussola, R., Amici, C.: A review on vibration-based condition monitoring of rotating machinery. Appl. Sci. **12**(3), 972 (2022). https://doi.org/10.3390/app12030972

15. Zhong, J., Kin, P., Yang, Z.: Fault diagnosis of rotating machinery based on multiple probabilistic classifiers. Mech. Syst. Signal Process. **108**, 99–114 (2018). https://doi.org/10.1016/j.ymssp.2018.02.009

16. Long, Y., Zhou, W., Luo, Y.: A fault diagnosis method based on one-dimensional data enhancement and convolutional neural network. Meas. J. Int. Meas. Confed. **180**, 109532 (2021). https://doi.org/10.1016/j.measurement.2021.109532

17. Chen, Z., Mauricio, A., Li, W., Gryllias, K.: A deep learning method for bearing fault diagnosis based on cyclic spectral coherence and convolutional neural networks. Mech. Syst. Signal Process. **140**, 106683 (2020). https://doi.org/10.1016/j.ymssp.2020.106683

18. Gundewar, S.K., Kane, P.V.: Condition monitoring and fault diagnosis of induction motor. J. Vib. Eng. Technol. **9**(4), 643–674 (2020). https://doi.org/10.1007/s42417-020-00253-y

19. Liang, C., Chen, C., Liu, Y., Jia, X.: A novel intelligent fault diagnosis method for rolling bearings based on compressed sensing and stacked multi-granularity convolution denoising auto-encoder. IEEE Access **9**, 154777–154787 (2021). https://doi.org/10.1109/ACCESS. 2021.3129061

20. Chen, C.C., Liu, Z., Yang, G., Wu, C.C., Ye, Q.: An improved fault diagnosis using 1d-convolutional neural network model. Electron. **10**(1), 1–19 (2021). https://doi.org/10.3390/electronics10010059

21. Patil, S.S., Pardeshi, S.S., Patange, A.D., Jegadeeshwaran, R.: Deep learning algorithms for tool condition monitoring in milling: a review. J. Phys: Conf. Ser. **1969**(1), 012039 (2021). https://doi.org/10.1088/1742-6596/1969/1/012039

22. Qiao, H., Wang, T., Wang, P., Zhang, L., Xu, M.: An adaptive weighted multiscale convolutional neural network for rotating machinery fault diagnosis under variable operating conditions. IEEE Access **7**, 118954–118964 (2019). https://doi.org/10.1109/ACCESS.2019. 2936625

23. Zhou, F., Sun, T., Hu, X., Wang, T., Wen, C.: A sparse denoising deep neural network for improving fault diagnosis performance. SIViP **15**(8), 1889–1898 (2021). https://doi.org/10. 1007/s11760-021-01939-w

24. Hoang, D.-T., Kang, H.-J.: Rolling element bearing fault diagnosis using convolutional neural network and vibration image. Cogn. Syst. Res. **53**, 42–50 (2019). https://doi.org/10. 1016/j.cogsys.2018.03.002

25. Inyang, U., Petrunin, I., Jennions, I.: Health condition estimation of bearings with multiple faults by a composite learning-based approach. Sensors **21**(13), 4424 (2021). https://doi.org/ 10.3390/s21134424

26. Zhang, Y., Zhou, T., Huang, X., Cao, L., Zhou, Q.: Fault diagnosis of rotating machinery based on recurrent neural networks. Meas. J. Int. Meas. Confed. **171**, 108774 (2021). https:// doi.org/10.1016/j.measurement.2020.108774

27. Chen, Z., Gryllias, K., Li, W.: Mechanical fault diagnosis using convolutional neural networks and extreme learning machine. Mech. Syst. Signal Process. **133**, 106272 (2019). https://doi.org/10.1016/j.ymssp.2019.106272

28. Kiranyaz, S., Avci, O., Abdeljaber, O., Ince, T., Gabbouj, M., Inman, D.J.: 1D convolutional neural networks and applications: a survey. Mech. Syst. Signal Process. **151**, 107398 (2021). https://doi.org/10.1016/j.ymssp.2020.107398

29. Wang, Q., Zhao, B., Ma, H., Chang, J., Mao, G.: A method for rapidly evaluating reliability and predicting remaining useful life using two-dimensional convolutional neural network with signal conversion. J. Mech. Sci. Technol. **33**(6), 2561–2571 (2019). https://doi.org/10. 1007/s12206-019-0504-x

30. Wu, C., Jiang, P., Ding, C., Feng, F., Chen, T.: Intelligent fault diagnosis of rotating machinery based on one-dimensional convolutional neural network. Comput. Ind. **108**, 53–61 (2019). https://doi.org/10.1016/j.compind.2018.12.001

31. Oh, J.W., Jeong, J.: Convolutional Neural Network and 2-D Image Based Fault Diagnosis of Bearing without Retraining. PervasiveHealth: Pervasive Computing Technologies for Healthcare, 134–138 (2019). https://doi.org/10.1145/3314545.3314563

32. Ding, X., He, Q.: Energy-fluctuated multiscale feature learning with deep ConvNet for intelligent spindle bearing fault diagnosis. IEEE Trans. Instrum. Meas. **66**(8), 1926–1935 (2017). https://doi.org/10.1109/TIM.2017.2674738

33. Yuan, L., Lian, D., Kang, X., Chen, Y., Zhai, K.: Rolling bearing fault diagnosis based on convolutional neural network and support vector machine. IEEE Access **8**, 137395–137406 (2020). https://doi.org/10.1109/ACCESS.2020.3012053

34. Wang, H., Liu, C., Du, W., Wang, S.: Intelligent diagnosis of rotating machinery based on optimized adaptive learning dictionary and 1DCNN. Appl. Sci. **11**(23), 11325 (2021). https://doi.org/10.3390/app112311325

35. Du, C., et al.: Research on fault diagnosis of automobile engines based on the deep learning 1D-CNN method. Eng. Res. Express **4**(1), 015003 (2022). https://doi.org/10.1088/2631-8695/ac4834

36. Qian, W., Li, S., Wang, J., An, Z., Jiang, X.: An intelligent fault diagnosis framework for raw vibration signals: adaptive overlapping convolutional neural network. Meas. Sci. Technol. **29**(9), 095009 (2018). https://doi.org/10.1088/1361-6501/aad101

37. Abdeljaber, O., Avci, O., Kiranyaz, S., Gabbouj, M., Inman, D.J.: Real-time vibration-based structural damage detection using one-dimensional convolutional neural networks. J. Sound Vib. **383**, 154–170 (2017). https://doi.org/10.1016/j.jsv.2016.10.043

38. Yang, J., Yin, S., Chang, Y., Gao, T.: A fault diagnosis method of rotating machinery based on one-dimensional, self-normalizing convolutional neural networks. Sensors **20**(14), 3837 (2020). https://doi.org/10.3390/s20143837

39. Jia, F., Lei, Y., Lu, N., Xing, S.: Deep normalized convolutional neural network for imbalanced fault classification of machinery and its understanding via visualization. Mech. Syst. Signal Process. **110**, 349–367 (2018). https://doi.org/10.1016/j.ymssp.2018.03.025

40. Fuan, W., Hongkai, J., Haidong, S., Wenjing, D., Shuaipeng, W.: An adaptive deep convolutional neural network for rolling bearing fault diagnosis. Meas. Sci. Technol. **28**(9), 095005 (2017). https://doi.org/10.1088/1361-6501/aa6e22

41. Islam, M.M.M., Kim, J.M.: Automated bearing fault diagnosis scheme using 2D representation of wavelet packet transform and deep convolutional neural network. Comput. Ind. **106**, 142–153 (2019). https://doi.org/10.1016/j.compind.2019.01.008

42. Kolar, D., Lisjak, D., Pająk, M., Pavković, D.: Fault diagnosis of rotary machines using deep convolutional neural network with wide three axis vibration signal input. Sensors **20**(14), 4017 (2020). https://doi.org/10.3390/s20144017

43. Sun, W., Zhao, R., Yan, R., Shao, S., Chen, X.: Convolutional discriminative feature learning for induction motor fault diagnosis. IEEE Trans. Ind. Inform. **13**(3), 1350–1359 (2017). https://doi.org/10.1109/TII.2017.2672988

44. Khan, M.A., Kim, Y.-H., Choo, J.: Intelligent fault detection using raw vibration signals via dilated convolutional neural networks. J. Supercomput. **76**(10), 8086–8100 (2018). https://doi.org/10.1007/s11227-018-2711-0

45. Li, H., Zhang, Q., Qin, X., Yuantao, S.: Raw vibration signal pattern recognition with automatic hyper-parameter-optimized convolutional neural network for bearing fault diagnosis. Proc. Inst. Mech. Eng. Part C J. Mech. Eng. Sci. **234**(1), 343–360 (2020). https://doi.org/10.1177/0954406219875756

46. Liu, R., Meng, G., Yang, B., Sun, C., Chen, X.: Dislocated time series convolutional neural architecture: an intelligent fault diagnosis approach for electric machine. IEEE Trans. Ind. Inform. **13**(3), 1310–1320 (2017). https://doi.org/10.1109/TII.2016.2645238

47. Zhang, W., Peng, G., Li, C., Chen, Y., Zhang, Z.: A new deep learning model for fault diagnosis with good anti-noise and domain adaptation ability on raw vibration signals. Sensors **17**(2), 425 (2017). https://doi.org/10.3390/s17020425

48. Chen, T., Wang, Z., Yang, X., Jiang, K.: A deep capsule neural network with stochastic delta rule for bearing fault diagnosis on raw vibration signals. Measurement **148**, 106857 (2019). https://doi.org/10.1016/j.measurement.2019.106857

49. Ye, Z., Yu, J.: Deep morphological convolutional network for feature learning of vibration signals and its applications to gearbox fault diagnosis. Mech. Syst. Signal Process. **161**, 107984 (2021). https://doi.org/10.1016/j.ymssp.2021.107984

50. Wang, Y., Yang, M., Li, Y., Xu, Z., Wang, J., Fang, X.: A multi-input and multi-task convolutional neural network for fault diagnosis based on bearing vibration signal. IEEE Sens. J. **21**(9), 10946–10956 (2021). https://doi.org/10.1109/JSEN.2021.3061595

51. Lu, C., Wang, Z., Zhou, B.: Intelligent fault diagnosis of rolling bearing using hierarchical convolutional network based health state classification. Adv. Eng. Inform. **32**, 139–151 (2017). https://doi.org/10.1016/j.aei.2017.02.005

52. Liu, Y.-Z., Zou, Y.-S., Jiang, Y.-L., Hui, Y., Ding, G.-F.: A novel method for diagnosis of bearing fault using hierarchical multitasks convolutional neural networks. Shock Vib. **2020**, 1–14 (2020). https://doi.org/10.1155/2020/8846822

53. Yang, J., et al.: A hierarchical deep convolutional neural network and gated recurrent unit framework for structural damage detection. Inf. Sci. (Ny) **540**, 117–130 (2020). https://doi.org/10.1016/j.ins.2020.05.090

54. Zhang, Y., Li, C., Wang, R., Qian, J.: A novel fault diagnosis method based on multi-level information fusion and hierarchical adaptive convolutional neural networks for centrifugal blowers. Meas. J. Int. Meas. Confed. **185**, 109970 (2021). https://doi.org/10.1016/j.measurement.2021.109970

55. Jiang, G., He, H., Yan, J., Xie, P.: Multiscale convolutional neural networks for fault diagnosis of wind turbine gearbox. IEEE Trans. Ind. Electron. **66**(4), 3196–3207 (2019). https://doi.org/10.1109/TIE.2018.2844805

56. Wang, F., Liu, R., Hu, Q., Chen, X.: Cascade convolutional neural network with progressive optimization for motor fault diagnosis under nonstationary conditions. IEEE Trans. Ind. Informatics **17**(4), 2511–2521 (2021). https://doi.org/10.1109/TII.2020.3003353

57. Li, X., Jiang, H., Hu, Y., Xiong, X.: Intelligent fault diagnosis of rotating machinery based on deep recurrent neural network. In: Proceedings - 2018 International Conference on Sensing, Diagnostics, Prognostics, and Control, SDPC 2018, no. 51475368, pp. 67–72 (2019). https://doi.org/10.1109/SDPC.2018.8664931

58. Huang, Y., Chen, C.H., Huang, C.J.: Motor fault detection and feature extraction using RNN-based variational autoencoder. IEEE Access **7**, 139086–139096 (2019). https://doi.org/10.1109/ACCESS.2019.2940769

59. Sun, H., Zhao, S.: Fault Diagnosis for Bearing Based on 1DCNN and LSTM. Shock Vib. **2021**, 1–17 (2021). https://doi.org/10.1155/2021/1221462

60. Eltotongy, A., Awad, M.I., Maged, S.A., Onsy, A.: Fault detection and classification of machinery bearing under variable operating conditions based on wavelet transform and CNN. In: 2021 International Mobile, Intelligent, and Ubiquitous Computing Conference, MIUCC 2021, pp. 117–123 (2021). https://doi.org/10.1109/MIUCC52538.2021.9447673

61. Yang, R., Huang, M., Lu, Q., Zhong, M.: Rotating machinery fault diagnosis using long-short-term memory recurrent neural network. IFAC-PapersOnLine **51**(24), 228–232 (2018). https://doi.org/10.1016/j.ifacol.2018.09.582

62. Wang, P., Zhou, Y., Zhang, L.: Fault diagnosis method of rotating machinery based on CEEMDAN-LSTM model. J. Phys. Conf. Ser. **2173**(1), 012057 (2022). https://doi.org/10.1088/1742-6596/2173/1/012057

63. Yin, A., Yan, Y., Zhang, Z., Li, C., Sánchez, R.-V.: Fault diagnosis of wind turbine gearbox based on the optimized LSTM neural network with cosine loss. Sensors **20**(8), 2339 (2020). https://doi.org/10.3390/s20082339

64. Yang, J., Guo, Y., Zhao, W.: Long short-term memory neural network based fault detection and isolation for electro-mechanical actuators. Neurocomputing **360**, 85–96 (2019). https://doi.org/10.1016/j.neucom.2019.06.029

65. Liu, J., Li, Q., Yang, H., Han, Y., Jiang, S., Chen, W.: Sequence fault diagnosis for PEMFC water management subsystem using deep learning with t-SNE. IEEE Access **7**, 92009–92019 (2019). https://doi.org/10.1109/ACCESS.2019.2927092

66. Ma, S., Cai, W., Liu, W., Shang, Z., Liu, G.: A lighted deep convolutional neural network based fault diagnosis of rotating machinery. Sensors **19**(10), 2381 (2019). https://doi.org/10.3390/s19102381

67. Cao, L., Qian, Z., Zareipour, H., Huang, Z., Zhang, F.: Fault diagnosis of wind turbine gearbox based on deep bi-directional long short-term memory under time-varying non-stationary operating conditions. IEEE Access 7, 155219–155228 (2019). https://doi.org/10.1109/ACCESS.2019.2947501

68. Han, T., Ma, R., Zheng, J.: Combination bidirectional long short-term memory and capsule network for rotating machinery fault diagnosis. Meas. J. Int. Meas. Confed. 176, 109208 (2021). https://doi.org/10.1016/j.measurement.2021.109208

69. Li, X., Zhang, W., Ding, Q.: Understanding and improving deep learning-based rolling bearing fault diagnosis with attention mechanism. Signal Process. 161, 136–154 (2019). https://doi.org/10.1016/j.sigpro.2019.03.019

70. Li, X., Jiang, H., Xiong, X., Shao, H.: Rolling bearing health prognosis using a modified health index based hierarchical gated recurrent unit network. Mech. Mach. Theory 133, 229–249 (2019). https://doi.org/10.1016/j.mechmachtheory.2018.11.005

71. Liu, H., Zhou, J., Zheng, Y., Jiang, W., Zhang, Y.: Fault diagnosis of rolling bearings with recurrent neural network-based autoencoders. ISA Trans. 77, 167–178 (2018). https://doi.org/10.1016/j.isatra.2018.04.005

72. Tao, Y., Wang, X., Sánchez, R., Yang, S., Bai, Y.U.N.: Spur gear fault diagnosis using a multilayer gated recurrent unit approach with vibration signal. IEEE Access 7, 56880–56889 (2019). https://doi.org/10.1109/ACCESS.2019.2914181

73. Chang, Y., Chen, J., Lv, H., Liu, S.: Heterogeneous bi-directional recurrent neural network combining fusion health indicator for predictive analytics of rotating machinery. ISA Trans. (2021). https://doi.org/10.1016/j.isatra.2021.04.024

74. Zhao, R., Wang, D., Yan, R., Mao, K., Shen, F., Wang, I.: Machine health monitoring using local feature-based gated recurrent unit networks. IEEE Trans. Ind. Electron. 65(2), 1539–1548 (2018). https://doi.org/10.1109/TIE.2017.2733438

75. Fu, J., Chu, J., Guo, P., Chen, Z.: Condition monitoring of wind turbine gearbox bearing based on deep learning model. IEEE Access 7, 57078–57087 (2019). https://doi.org/10.1109/ACCESS.2019.2912621

76. Zhao, R., Yan, R., Wang, J., Mao, K.: Learning to monitor machine health with convolutional Bi-directional LSTM networks. Sensors 17(2), 273 (2017). https://doi.org/10.3390/s17020273

77. Qiao, M., Yan, S., Tang, X., Xu, C.: Deep convolutional and LSTM recurrent neural networks for rolling bearing fault diagnosis under strong noises and variable loads. IEEE Access 8, 66257–66269 (2020). https://doi.org/10.1109/ACCESS.2020.2985617

78. Liao, G.P., Gao, W., Yang, G.J., Guo, M.F.: Hydroelectric generating unit fault diagnosis using 1-D convolutional neural network and gated recurrent unit in small hydro. IEEE Sens. J. 19(20), 9352–9363 (2019). https://doi.org/10.1109/JSEN.2019.2926095

79. Li, X., Jiang, H., Liu, S., Zhang, J., Xu, J.: A unified framework incorporating predictive generative denoising autoencoder and deep Coral network for rolling bearing fault diagnosis with unbalanced data. Meas. J. Int. Meas. Confed. 178, 109345 (2021). https://doi.org/10.1016/j.measurement.2021.109345

80. Fu, Q., Wang, H.: A novel deep learning system with data augmentation for machine fault diagnosis from vibration signals. Appl. Sci. 10(17), 5765 (2020). https://doi.org/10.3390/app10175765

81. Goodfellow, B.I., et al.: Generative adversarial networks. Commun. ACM 63(11), 139–144 (2014). https://doi.org/10.1145/3422622

82. Zhou, F., Yang, S., Fujita, H., Chen, D., Wen, C.: Deep learning fault diagnosis method based on global optimization GAN for unbalanced data. Knowledge-Based Syst. 187, 104837 (2020). https://doi.org/10.1016/j.knosys.2019.07.008

83. Ding, Y., Ma, L., Ma, J., Wang, C., Lu, C.: A generative adversarial network-based intelligent fault diagnosis method for rotating machinery under small sample size conditions. IEEE Access **7**, 149736–149749 (2019). https://doi.org/10.1109/ACCESS.2019.2947194

84. Yan, K., Huang, J., Shen, W., Ji, Z.: Unsupervised learning for fault detection and diagnosis of air handling units. Energy Build. **210**, 109689 (2020). https://doi.org/10.1016/j.enbuild.2019.109689

85. Zheng, T., Song, L., Wang, J., Teng, W., Xu, X., Ma, C.: Data synthesis using dual discriminator conditional generative adversarial networks for imbalanced fault diagnosis of rolling bearings. Meas. J. Int. Meas. Confed. **158**, 107741 (2020). https://doi.org/10.1016/j.measurement.2020.107741

86. Viola, J., Chen, Y.Q., Wang, J.: FaultFace: deep convolutional generative adversarial network (DCGAN) based Ball-Bearing failure detection method. Inf. Sci. (Ny) **542**, 195–211 (2021). https://doi.org/10.1016/j.ins.2020.06.060

87. Sabir, R., Rosato, D., Hartmann, S., Gühmann, C.: Signal Generation using 1d Deep convolutional generative adversarial networks for fault diagnosis of electrical machines. In: 2020 25th International Conference on Pattern Recognition (ICPR), pp. 3907–3914 (2021). https://doi.org/10.1109/ICPR48806.2021.9413119

88. Zhang, D., Ning, Z.Q., Yang, B., Wang, T.Y., Ma, Y.J.: Fault diagnosis of permanent magnet motor based on DCGAN-RCCNN. Energy Rep. **8**, 616–626 (2022). https://doi.org/10.1016/j.egyr.2022.01.226

89. Xie, Y., Zhang, T.: Imbalanced learning for fault diagnosis problem of rotating machinery based on generative adversarial networks. In: Proceedings of the 37th Chinese Control Conference, pp. 6017–6022 (2018). https://doi.org/10.23919/ChiCC.2018.8483334

90. Luo, J., Huang, J., Li, H.: A case study of conditional deep convolutional generative adversarial networks in machine fault diagnosis. J. Intell. Manuf. **32**(2), 407–425 (2020). https://doi.org/10.1007/s10845-020-01579-w

91. Shao, S., Wang, P., Yan, R.: Generative adversarial networks for data augmentation in machine fault diagnosis. Comput. Ind. **106**, 85–93 (2019). https://doi.org/10.1016/j.compind.2019.01.001

92. Xiong, X., Hongkai, J., Li, X., Niu, M.: A Wasserstein gradient-penalty generative adversarial network with deep auto-encoder for bearing intelligent fault diagnosis. Meas. Sci. Technol. **31**(4), 045006 (2020). https://doi.org/10.1088/1361-6501/ab47df

93. Luo, J., Zhu, L., Li, Q., Liu, D., Chen, M.: Imbalanced fault diagnosis of rotating machinery based on deep generative adversarial networks with gradient penalty. Processes **9**(10), 1751 (2021). https://doi.org/10.3390/pr9101751

94. Zareapoor, M., Shamsolmoali, P., Yang, J.: Oversampling adversarial network for class-imbalanced fault diagnosis. Mech. Syst. Signal Process. **149**, 107175 (2021). https://doi.org/10.1016/j.ymssp.2020.107175

95. Shi, Z., Chen, J., Zi, Y., Zhou, Z.: A novel multitask adversarial network via redundant lifting for multicomponent intelligent fault detection under sharp speed variation. IEEE Trans. Instrum. Meas. **70**, 1 (2021). https://doi.org/10.1109/TIM.2021.3055821

96. Liu, S., Jiang, H., Wu, Z., Li, X.: Rolling bearing fault diagnosis using variational autoencoding generative adversarial networks with deep regret analysis. Meas. J. Int. Meas. Confed. **168**, 108371 (2021). https://doi.org/10.1016/j.measurement.2020.108371

97. Wu, C., Zeng, Z.: A fault diagnosis method based on auxiliary classifier generative adversarial network for rolling bearing. PLoS ONE **16**, 1–21 (2021). https://doi.org/10.1371/journal.pone.0246905

98. Liu, H., Zhou, J., Xu, Y., Zheng, Y., Peng, X., Jiang, W.: Unsupervised fault diagnosis of rolling bearings using a deep neural network based on generative adversarial networks. Neurocomputing **315**, 412–424 (2018). https://doi.org/10.1016/j.neucom.2018.07.034

99. Shao, H., Jiang, H., Li, X., Wu, S.: Intelligent fault diagnosis of rolling bearing using deep wavelet auto-encoder with extreme learning machine. Knowledge-Based Syst. **140**, 1–14 (2018). https://doi.org/10.1016/j.knosys.2017.10.024

100. Luo, X., Li, X., Wang, Z., Liang, J.: Discriminant autoencoder for feature extraction in fault diagnosis. Chemom. Intell. Lab. Syst. **192**, 103814 (2019). https://doi.org/10.1016/j. chemolab.2019.103814

101. Sun, J., Yan, C., Wen, J.: Intelligent bearing fault diagnosis method combining compressed data acquisition and deep learning. IEEE Trans. Instrum. Meas. **67**(1), 185–195 (2018). https://doi.org/10.1109/TIM.2017.2759418

102. Liu, R., Yang, B., Zio, E., Chen, X.: Artificial intelligence for fault diagnosis of rotating machinery: a review. Mech. Syst. Signal Process. **108**, 33–47 (2018). https://doi.org/10. 1016/j.ymssp.2018.02.016

103. Mao, W., Feng, W., Liu, Y., Zhang, D., Liang, X.: A new deep auto-encoder method with fusing discriminant information for bearing fault diagnosis. Mech. Syst. Signal Process. **150**, 107233 (2021). https://doi.org/10.1016/j.ymssp.2020.107233

104. Principi, E., Rossetti, D., Squartini, S., Piazza, F.: Unsupervised electric motor fault detection by using deep autoencoders. IEEE/CAA J. Autom. Sin. **6**(2), 441–451 (2019). https://doi. org/10.1109/JAS.2019.1911393

105. Wang, B., Shen, C., Xu, K., Zheng, T.: Turn-to-turn short circuit of motor stator fault diagnosis in continuous state based on deep. IET Electr. Power Appl. **13**(10), 1598–1606 (2019). https://doi.org/10.1049/iet-epa.2019.0101

106. Zhang, Y., Li, X., Gao, L., Chen, W., Li, P.: Intelligent fault diagnosis of rotating machinery using a new ensemble deep auto-encoder method. Measurement **151**, 107232 (2020). https:// doi.org/10.1016/j.measurement.2019.107232

107. Cui, M., Wang, Y., Lin, X., Zhong, M.: Fault diagnosis of rolling bearings based on an improved stack autoencoder and support vector machine. IEEE Sens. J. **21**(4), 4927–4937 (2021). https://doi.org/10.1109/JSEN.2020.3030910

108. Aamir, M., Mohd Nawi, N., Wahid, F., Mahdin, H.: A deep contractive autoencoder for solving multiclass classification problems. Evol. Intel. **14**(4), 1619–1633 (2020). https:// doi.org/10.1007/s12065-020-00424-6

109. Shi, C., Panoutsos, G., Luo, B., Liu, H., Li, B.: Using multiple-feature-spaces-based deep learning for tool condition monitoring in ultraprecision manufacturing. IEEE Trans. Ind. Electron. **66**(5), 3794–3803 (2019). https://doi.org/10.1109/TIE.2018.2856193

110. Pan, Y., He, F., Yu, H.: A novel enhanced collaborative autoencoder with knowledge distillation for top-N recommender systems. Neurocomputing **332**, 137–148 (2019). https://doi. org/10.1016/j.neucom.2018.12.025

111. Liu, G., Bao, H., Han, B.: A stacked autoencoder-based deep neural network for achieving gearbox fault diagnosis. Math. Probl. Eng. **2018**, 1–10 (2018). https://doi.org/10.1155/2018/ 5105709

112. Sun, M., Wang, H., Liu, P., Huang, S., Wang, P., Meng, J.: Stack autoencoder transfer learning algorithm for bearing fault diagnosis based on class separation and domain fusion. IEEE Trans. Ind. Electron. **69**(3), 3047–3058 (2022). https://doi.org/10.1109/TIE.2021.3066933

113. Xiang, Z., Zhang, X., Zhang, W., Xia, X.: Fault diagnosis of rolling bearing under fluctuating speed and variable load. Measurement **138**, 162–174 (2019). https://doi.org/10.1016/j.mea surement.2019.01.063

114. Karamti, H., Lashin, M.M.A., Alrowais, F.M., Mahmoud, A.M.: A new deep stacked architecture for multi-fault machinery identification with imbalanced samples. IEEE Access **9**, 58838–58851 (2021). https://doi.org/10.1109/ACCESS.2021.3071796

115. Wang, J., Li, S., An, Z., Jiang, X., Qian, W., Ji, S.: Batch-normalized deep neural networks for achieving fast intelligent fault diagnosis of machines. Neurocomputing **329**, 53–65 (2019). https://doi.org/10.1016/j.neucom.2018.10.049

116. Shao, H., Xia, M., Wan, J., de Silva, C.W.: Modified stacked autoencoder using adaptive morlet wavelet for intelligent fault diagnosis of rotating machinery. IEEE/ASME Trans. Mech. **27**(1), 24–33 (2022). https://doi.org/10.1109/TMECH.2021.3058061

117. Che, C., Wang, H., Fu, Q., Ni, X.: "Intelligent fault prediction of rolling bearing based on gate recurrent unit and hybrid autoencoder. Proc. IMechE Part C J. Mech. Eng. Sci. **235**(6), 1106–1114 (2021). https://doi.org/10.1177/0954406220941037

118. Meng, Z., Zhan, X., Li, J., Pan, Z.: An enhancement denoising autoencoder for rolling bearing fault diagnosis. Meas. J. Int. Meas. Confed. **130**, 448–454 (2018). https://doi.org/10.1016/j.measurement.2018.08.010

119. Yi, G., Cao, J., Song, X., Yao, J.: A denoising autoencoder-based bearing fault diagnosis system for time-domain vibration signals. Wirel. Commun. Mob. Comput. **2021**, 1–7 (2021). https://doi.org/10.1155/2021/9790053

120. Lu, C., Wang, Z.Y., Qin, W.L., Ma, J.: Fault diagnosis of rotary machinery components using a stacked denoising autoencoder-based health state identification. Signal Process. **130**, 377–388 (2017). https://doi.org/10.1016/j.sigpro.2016.07.028

121. Zhao, X., Wu, J., Zhang, Y., Shi, Y., Wang, L.: Fault diagnosis of motor in frequency domain signal by stacked de - noising auto - encoder. Comput. Mater. Contin. **57**(2), 223–242 (2018). https://doi.org/10.32604/cmc.2018.02490

122. Chen, Z., Li, Z.: Fault diagnosis method of rotating machinery based on stacked denoising autoencoder. J. Intell. Fuzzy Syst. **34**(6), 3443–3449 (2018). https://doi.org/10.3233/JIFS-169524

123. Yu, J.: Evolutionary manifold regularized stacked denoising autoencoders for gearbox fault diagnosis. Knowledge-Based Syst. **178**, 111–122 (2019). https://doi.org/10.1016/j.knosys.2019.04.022

124. Xu, X., Feng, J., Zhan, L., Li, Z., Qian, F., Yan, Y.: Fault diagnosis of permanent magnet synchronous motor based on stacked denoising autoencoder. Entropy **23**(3), 339 (2021). https://doi.org/10.3390/e23030339

125. Xu, Y., Li, C., Xie, T.: Intelligent diagnosis of subway traction motor bearing fault based on improved stacked denoising autoencoder. Shock Vib. **2021**, 1–9 (2021). https://doi.org/10.1155/2021/6656635

126. Xiao, D., Qin, C., Yu, H., Huang, Y., Liu, C.: Unsupervised machine fault diagnosis for noisy domain adaptation using marginal denoising autoencoder based on acoustic signals. Measurement **176**, 109186 (2021). https://doi.org/10.1016/j.measurement.2021.109186

127. Godói, L., Nóbrega, E.: Denoising convolutional autoencoder configuration for condition monitoring of rotating machines. J. Braz. Soc. Mech. Sci. Eng. **43**(1), 1–13 (2021). https://doi.org/10.1007/s40430-020-02776-7

128. Liu, X., Zhou, Q., Zhao, J., Shen, H., Xiong, X.: Fault diagnosis of rotating machinery under noisy environment conditions based on a 1-D convolutional autoencoder and 1-D convolutional neural network. Sensors **19**(4), 972 (2019). https://doi.org/10.3390/s19040972

129. Zhang, C., Cheng, X., Liu, J., He, J., Liu, G.: Deep sparse autoencoder for feature extraction and diagnosis of locomotive adhesion status. J. Control Sci. Eng. **2018**, 1–9 (2018). https://doi.org/10.1155/2018/8676387

130. Zheng, Y., Wang, T., Xin, B., Xie, T., Wang, Y.: A sparse autoencoder and softmax regression based diagnosis method for the attachment on the blades of marine current turbine. Sensors **19**(4), 826 (2019). https://doi.org/10.3390/s19040826

131. Zhao, X., Jia, M., Liu, Z.: Semisupervised deep sparse auto-encoder with local and nonlocal information for intelligent fault diagnosis of rotating machinery. IEEE Trans. Instrum. Meas. **70**(3501413), 1–13 (2021). https://doi.org/10.1109/TIM.2020.3016045

132. Nguyen, C.D., Prosvirin, A.E., Kim, C.H., Kim, J.-M.: Construction of a sensitive and speed invariant gearbox fault diagnosis model using an incorporated utilizing adaptive noise

control and a stacked sparse autoencoder-based deep neural network. Sensors **21**(1), 18 (2021). https://doi.org/10.3390/s21010018

133. Qi, Y., Shen, C., Wang, D., Shi, J., Jiang, X., Zhu, Z.: Stacked sparse autoencoder-based deep network for fault diagnosis of rotating machinery. IEEE Access **5**, 15066–15079 (2017). https://doi.org/10.1109/ACCESS.2017.2728010

134. Sun, M., Wang, H., Liu, P., Huang, S., Fan, P.: A sparse stacked denoising autoencoder with optimized transfer learning applied to the fault diagnosis of rolling bearings. Measurement **146**, 305–314 (2019). https://doi.org/10.1016/j.measurement.2019.06.029

135. Saufi, S.R., Asrar, Z., MS Leong, MH Lim: Differential evolution optimization for resilient stacked sparse autoencoder and its applications on bearing fault diagnosis. Meas. Sci. Technol. **29**(12), 125002 (2018). https://doi.org/10.1088/1361-6501/aae5b2

136. Shi, P., Guo, X., Han, D., Fu, R.: A sparse auto-encoder method based on compressed sensing and wavelet packet energy entropy for rolling bearing intelligent fault diagnosis. J. Mech. Sci. Technol. **34**(4), 1445–1458 (2020). https://doi.org/10.1007/s12206-020-0306-1

137. Sohaib, M., Kim, J.-M.: Reliable fault diagnosis of rotary machine bearings using a stacked sparse autoencoder-based deep neural network. Shock Vib. **2018**, 1–11 (2018). https://doi.org/10.1155/2018/2919637

138. Zhu, H., Cheng, J., Zhang, C., Wu, J., Shao, X.: Stacked pruning sparse denoising autoencoder based intelligent fault diagnosis of rolling bearings. Appl. Soft Comput. J. **88**, 106060 (2020). https://doi.org/10.1016/j.asoc.2019.106060

139. Yang, P., Wen, C., Geng, H., Liu, P.: Intelligent fault diagnosis method for blade damage of quad-rotor UAV based on stacked pruning sparse denoising autoencoder and convolutional neural network. Machines **9**(12), 360 (2021). https://doi.org/10.3390/machines9120360

140. Jia, F., Lei, Y., Guo, L., Lin, J., Xing, S.: A neural network constructed by deep learning technique and its application to intelligent fault diagnosis of machines. Neurocomputing **272**, 619–628 (2018). https://doi.org/10.1016/j.neucom.2017.07.032

141. Qi, Y., Shen, C., Zhu, J., Jiang, X., Shi, J., Zhu, Z.: A new deep fusion network for automatic mechanical fault feature learning. IEEE Access **7**, 152552–152563 (2019). https://doi.org/10.1109/ACCESS.2019.2948661

142. Yi, J., Fu, S., Cui, S., Zhao, C., : A deep contractive auto-encoding network for machinery fault diagnosis. In: ISCIT 2018 - 18th International Symposium on Communication and Information Technology, pp. 85–89 (2018). https://doi.org/10.1109/ISCIT.2018.8587983

143. Shen, C., Qi, Y., Wang, J., Cai, G., Zhu, Z.: An automatic and robust features learning method for rotating machinery fault diagnosis based on contractive autoencoder. Eng. Appl. Artif. Intell. **76**(8), 170–184 (2018). https://doi.org/10.1016/j.engappai.2018.09.010

144. Zhang, Y., Li, X., Gao, L., Chen, W., Li, P.: Ensemble deep contractive auto-encoders for intelligent fault diagnosis of machines under noisy environment. Knowledge-Based Syst. **196**, 105764 (2020). https://doi.org/10.1016/j.knosys.2020.105764

145. Martin, G.S., Droguett, E.L., Meruane, V., das Chagas Moura, M.: Deep variational auto-encoders: A promising tool for dimensionality reduction and ball bearing elements fault diagnosis. Struct. Heal. Monit. **18**(4), 1092–1128 (2019). https://doi.org/10.1177/147592 1718788299

146. Yan, X., Xu, Y., She, D., Zhang, W.: Reliable fault diagnosis of bearings using an optimized stacked variational denoising auto-encoder. Entropy **24**(36), 1–26 (2022). https://doi.org/10.3390/e24010036

147. Zhao, D., et al.: Enhanced data-driven fault diagnosis for machines with small and unbalanced data based on variational auto-encoder. Meas. Sci. Technol. **31**(3), 035004 (2020). https://doi.org/10.1088/1361-6501/ab55f8

148. Dixit, S., Verma, N.K.: Intelligent condition-based monitoring of rotary machines with few samples. IEEE Sens. J. **20**(23), 14337–14346 (2020). https://doi.org/10.1109/JSEN.2020.3008177

149. Wang, Y.-r, Sun, G.-d, Jin, Q.: Imbalanced sample fault diagnosis of rotating machinery using conditional variational auto-encoder generative adversarial network. Appl. Soft Comput. **92**, 106333 (2020). https://doi.org/10.1016/j.asoc.2020.106333

150. Zhao, G., Liu, X., Zhang, B., Liu, Y., Niu, G., Cong, H.: A novel approach for analog circuit fault diagnosis based on Deep Belief Network. Measurement **121**, 170–178 (2018). https://doi.org/10.1016/j.measurement.2018.02.044

151. Yan, X., Liu, Y., Jia, M.: Multiscale cascading deep belief network for fault identification of rotating machinery under various working conditions. Knowledge-Based Syst. **193**, 105484 (2020). https://doi.org/10.1016/j.knosys.2020.105484

152. Shang, Z., Liao, X., Geng, R., Gao, M., Liu, X.: Fault diagnosis method of rolling bearing based on deep belief network. J. Mech. Sci. Technol. **32**(11), 5139–5145 (2018). https://doi.org/10.1007/s12206-018-1012-0

153. Qin, Y., Wang, X., Zou, J.: The optimized deep belief networks with improved logistic sigmoid units and their application in fault diagnosis for planetary gearboxes of wind turbines. IEEE Trans. Ind. Electron. **66**, 3814–3824 (2019). https://doi.org/10.1109/TIE.2018.2856205

154. Yan, J., Hu, Y., Guo, C.: Rotor unbalance fault diagnosis using DBN based on multi-source heterogeneous information fusion. Procedia Manuf. **35**, 1184–1189 (2019). https://doi.org/10.1016/j.promfg.2019.06.075

155. Han, D., Guo, X., Shi, E.: An intelligent fault diagnosis method of variable condition gearbox based on improved DBN combined with WPEE and MPE. IEEE Access **8**, 131299–131309 (2020). https://doi.org/10.1109/ACCESS.2020.3008208

156. Ma, J., Li, S., Wang, X.: Condition monitoring of rolling bearing based on multi-order FRFT and SSA-DBN. Symmetry (Basel) **14**(2), 320 (2022). https://doi.org/10.3390/sym14020320

157. Zhang, Y., Zhang, Y., Wen, L., Cui, Z., He, Y., Liu, G.: Power grid fault diagnosis based on improved deep belief network. J. Phys. Conf. Ser. **1585**(1), 012021 (2020). https://doi.org/10.1088/1742-6596/1585/1/012021

158. Yu, B.: A deep belief network and dempster-shafer theory multiclassifier for reliability of wind turbine system. IOP Conf. Ser. Mater. Sci. Eng. **1043**, 032057 (2021). https://doi.org/10.1088/1757-899X/1043/3/032057

159. Shi, P., Xue, P., Liu, A., Han, D.: A novel rotating machinery fault diagnosis method based on adaptive deep belief network structure and dynamic learning rate under variable working conditions. IEEE Access **9**, 44569–44579 (2021). https://doi.org/10.1109/ACCESS.2021.3066594

160. Niu, G., Wang, X., Golda, M., Mastro, S., Zhang, B.: An optimized adaptive PReLU-DBN for rolling element bearing fault diagnosis. Neurocomputing **445**, 26–34 (2021). https://doi.org/10.1016/j.neucom.2021.02.078

161. Fan, X., Fang, Y.J., Wang, D., Liang, J.Q., Tsui, K.L.: Combining DBN and FCM for fault diagnosis of roller element bearings without using data labels. Shock Vib. **2018**, 1–12 (2018). https://doi.org/10.1155/2018/3059230

162. Gao, S., Xu, L., Zhang, Y.: Rolling bearing fault diagnosis based on intelligent optimized self-adaptive deep belief network. Meas. Sci. Technol. **31**, 055009 (2020). https://doi.org/10.1088/1361-6501/ab50f0

163. Kamada, S., Ichimura, T., Hara, A., Mackin, K.J.: Adaptive structure learning method of deep belief network using neuron generation–annihilation and layer generation. Neural Comput. Appl. **31**(11), 8035–8049 (2018). https://doi.org/10.1007/s00521-018-3622-y

164. Shen, C., Xie, J., Wang, D., Jiang, X., Shi, J., Zhu, Z.: Improved hierarchical adaptive deep belief network for bearing fault diagnosis. Appl. Sci. **9**(16), 3374 (2019). https://doi.org/10.3390/app9163374

Finite Time Trajectory Tracking of a Mobile Robot Using Cascaded Terminal Sliding Mode Control Under the Presence of Random Gaussian Disturbance

Adisu Safo Bosera[1]([⊠]), Ayodeji Olalekan Salau[2], Asrat Gedefa Yadessa[1], and Kaheli Anteneh Jembere[1]

[1] Department of Electrical and Computer Engineering, Mettu University, Mettu, Ethiopia
addisuusaafoo@gmail.com

[2] Department of Electrical/Electronics and Computer Engineering, Afe Babalola University, Ado-Ekiti, Nigeria

Abstract. In this paper, dynamic modeling of a differential drive mobile robot (DDMR) using Langrage formulation and terminal sliding mode trajectory tracking control is presented. The proposed controller is a cascaded controller designed to improve the dynamic response of the system, i.e. kinematic and dynamic problems, asymptotical convergence, and chattering problem using terminal sliding mode control (TSMC). The terminal sliding mode control provides faster convergence and higher-precision control than the conventional linear hyperplane sliding control which guarantees the asymptotic stability. This is due to fact that the terminal sliding mode control system guarantees a finite time convergence to the sliding phase. The entire control design consists of an outer loop kinematics control and inner loop speed control system. Here, outer kinematic control system provides an appropriate velocity control input for the inner loop angular velocity control of each wheel. An angular and linear velocity control input is designed in order to make angular and posture error to converge to zero in a finite time based on global fast terminal sliding mode control (GFTSMC). Then, the inner loop GFTSMC of the robot is designed to ensure that the tracking error between the actual and desired angular velocity of each wheels converges to zero in a finite time. Both the inner and outer closed loop controllers achieve path following in a finite time and avoids high frequency switching in the closed loop such that the overall dynamic response of the system is improved using the cascaded control technique and the stability of each controller was checked using lyapunov criteria. Generally, the proposed control system shows the performance and effectiveness of the proposed method compared to conventional SMC, and the simulation results indicate good convergence and robustness of the system for circular trajectories under both model uncertainty and random Gaussian disturbances using GFTSMC.

Keywords: Differential drive mobile robot · Langrage formulation approach · TSMC · Finite time · Trajectory tracking

© ICST Institute for Computer Sciences, Social Informatics and Telecommunications Engineering 2023
Published by Springer Nature Switzerland AG 2023. All Rights Reserved
B. H. Woldegiorgis et al. (Eds.): ICAST 2022, LNICST 455, pp. 63–78, 2023.
https://doi.org/10.1007/978-3-031-28725-1_5

1 Introduction

Wheeled mobile robotics is one of the most developed fields in robotics, having many uses in different fields like entertainment, military, and exploration. Mobile robots have caught the interest of many researchers in robotic control because of their complications in control schemes and practice applications [1]. Non-holonomic constraint is a feature of a wheeled mobile robot. The problem of tracking control of a mobile robot has been addressed by a number of control approaches [2, 3]. Two models (kinematic and dynamic) are used to categorize research approaches in two axes and the tracking control problem as a kinematic or dynamic challenge [4]. In [5], a navigation system slide mode control was proposed for a mobile robot, but in the paper, the influence of dynamics of the mobile robot was not well thought-out and only steering system (kinematic problem) controller was designed. A back stepping controller for a Wheeled Mobile Robot was presented by the authors in [6]. The kinematic control unit's job was to provide velocity outputs to the robot, which assisted to keep the posture mistakes to a minimum. The response, on the other hand, was slow. Due to the large amount of difficult computations, the control has a reduced reaction time [7]. Authors in [8] designed a cascaded control system that has a master feedback kinematic and slave feedback dynamics based PID with the translational velocity kept constant and only the robot's rotational velocity controlled, i.e., the robot's pose is controlled by varying the rotational velocity alone, resulting in a low response time. In [9], a traditional sliding mode control (SMC) system which used Gao's reaching method was implemented for a mobile robot, while a fuzzy kinematic controller with dynamic proportional integral control was designed for a mobile robot in [10]. The results show that the controller has finite time convergence problems.

In this paper, a controller is proposed to eliminate the chattering occurrence and asymptotical convergence in traditional sliding mode control systems by using cascaded robust finite time global fast terminal SMC for trajectory tracking of a differential drive mobile robot. Furthermore, accuracy and effectiveness of the proposed control system is evaluated using different performance index such as ITSE, IAE and ITAE and compared to conventional SMC.

This remaining sections of the paper are structured as follows. Section 2 describes the kinematic model of the differential robot and proposed Langrage formulation. Section 3 describes both the GFTSMC and SMC controller design. Section 4 presents the analysis and simulation results, and Sect. 5 concludes the paper.

2 Kinematic and Dynamic Model of DDMR

In this section, both kinematic model and dynamic model of differential drive mobile robot was derived.

2.1 Kinematic model of DDMR

We consider the following a differential drive mobile robot which has two wheels and one castor wheel at front. Geometrical parameters of the differential drive mobile robot

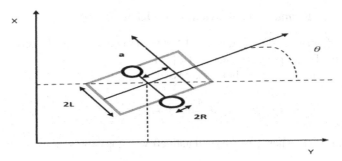

Fig. 1. Posture description of DDMRM

is specified such that the radius of R is placed with a distance L from the center of the mobile robot which illustrated in Fig. 1.

In Fig. 1, let's assume that the robot can never slide which makes the robots movement along y axis always zeros, i.e. $\dot{Y}_r = 0$. Since, the robot is subjected to Nonholonomic constraints, namely: pure rolling constraints and no side move (forward and backward) motion only. A two different coordinate system (frames) must be defined to characterize the WMR's position in its environment such as world coordinate and robot coordinate system.

The linear velocity of the robot in its robot coordinate system frame is:

$$\dot{X}_r = \frac{R\dot{\Phi}_r}{2} + \frac{R\dot{\Phi}_l}{2} \tag{1}$$

Given the speeds $\dot{\Phi}_r$ and $\dot{\Phi}_l$ of the right and left wheels, respectively, the rotational velocity is expressed as:

$$\dot{\theta} = \frac{R\dot{\Phi}_r}{2} + \frac{R\dot{\Phi}_l}{2} \tag{2}$$

The differential drive mobile robot's velocities can be obtained also in the inertial frame (world frame) with respect to the robot coordinate system as follows using the rotational matrix along z axis:

$$\begin{pmatrix} \dot{x}_I \\ \dot{y}_I \\ \dot{\theta}_I \end{pmatrix} = \begin{pmatrix} \frac{R\cos\theta}{2} & \frac{R\cos\theta}{2} \\ \frac{R\sin\theta}{2} & \frac{R\sin\theta}{2} \\ \frac{R}{2L} & -\frac{R}{2L} \end{pmatrix} \begin{pmatrix} \dot{\Phi}_r \\ \dot{\Phi}_l \end{pmatrix} \tag{3}$$

The relationship of linear and angular velocities to angular velocities of the wheels are given by Eqs. (1) and (2).

$$\begin{pmatrix} v \\ \omega \end{pmatrix} = M \begin{bmatrix} \omega_r \\ \omega_l \end{bmatrix} \tag{4}$$

where, $M = \begin{bmatrix} \frac{R}{2} & \frac{R}{2} \\ \frac{R}{2L} & -\frac{R}{2L} \end{bmatrix}$ which forward kinematics of DDMR.

2.2 Langrange Formulation of Dynamic Model of DDMR

We begin by describing the Lagrangain (L) of the mobile robot as the difference of its kinetic energy (T) and potential energy (U) [11].

The kinetic energy of the robot is expressed as:

$$T_c = \frac{1}{2}m_c v_c^2 + \frac{1}{2}I_c \dot{\theta}^2 \tag{5}$$

While the kinetic energy of the right and left wheel is expressed as:

$$T_{wr} = \frac{1}{2}m_w v_{wr}^2 + \frac{1}{2}I_m \dot{\theta}^2 + \frac{1}{2}I_w \dot{\Phi}_r^2 \tag{6}$$

$$T_{wl} = \frac{1}{2}m_w v_{wl}^2 + \frac{1}{2}I_m \dot{\theta}^2 + \frac{1}{2}I_w \dot{\Phi}_l^2 \tag{7}$$

where, m_c is the mass of the DDMR, m_w is the mass of each driving wheel (with actuator), I_c is the moment of inertia of the DDMR about the vertical axis through the center of mass, I_w is the moment of inertia of each driving wheel with a motor about the wheel axis, and I_m is the moment of inertia of each driving wheel with a motor about the wheel diameter. All velocities are first articulated as a function of the generalized coordinates using the general velocity equation in the inertial/world frame.

$$v_i^2 = x_i^2 + y_i^2 \tag{8}$$

The kinetic energy of the system is then written conveniently by

$$T(q, q') = \frac{1}{2}m\left(x_i^2 + y_i^2\right) + m_c a\dot{\theta}\left(\cos\theta \dot{y}_I - \sin\theta \dot{x}_I\right) + \frac{1}{2}I\dot{\theta}^2 + \frac{1}{2}I_w(\dot{\Phi}_r^2 + \dot{\Phi}_l^2) \tag{9}$$

where, the following new parameters are introduced as follows, $m = m_c + 2m_w$, and $I = 2m_w D^2 + m_c a^2 + 2I_m + I_c$ and the potential energy of the DDMR is considered to be zero, because the DDMR is travelling in the x_i- y_i plane.

The dynamic equations are simplified to the following form using state space representation, we get

$$\left(I_w + \frac{R^2(mL^2 + I)}{4L^2}\right)\dot{\omega}_r + \left(\frac{r^2(mL^2 - I)}{4L^2}\right)\dot{\omega}_l + \frac{R}{L}\left(\frac{R^2 m_c a}{2L}\right)(\omega_r - \omega_l)\omega_l = \tau_r \tag{10}$$

$$\left(I_w + \frac{R^2(mL^2 + I)}{4L^2}\right)\dot{\omega}_l + \left(\frac{r^2(mL^2 - I)}{4L^2}\right)\dot{\omega}_r - \frac{R}{L}\left(\frac{R^2 m_c a}{2L}\right)(\omega_r - \omega_l)\omega_r = \tau_l \tag{11}$$

where, the angular velocities of right and left wheel (ω_r, ω_l) of the DDMR and the driving motor torques $((\tau_r, \tau_l)$ are right and left wheel torque respectively.

3 Controller Design

To have improved motion over all control performance speed control system must employed because, due to additional load, surface, dynamics of mobile robot and among others perfect velocity tracking impossible. In this case the controller structure should be split into two stages as shown Fig. 2 [13].

Inner loop used for control of angular velocity of each wheel's, depending on the mobile robot dynamics. Whereas, the outer loop position control system is used to control both the translational and rotational positions of the mobile robot steering system. The outer loop finds suitable or desired velocity control inputs, which stabilize the kinematic closed loop control.

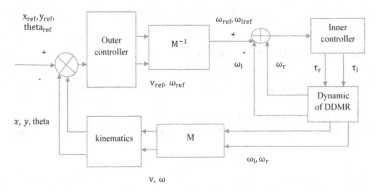

Fig. 2. Entire closed loop control system for DDMR.

3.1 Kinematic Outer Loop Global Fast Terminal Sliding Mode Controller Design

This control method is specially implemented to the kinematic differential drive mobile robot systems which are controlled by linear and angular velocity inputs.

The posture error $p_e = (x_e, y_e, \theta_e)^T$ can be expressed as using from Fig. 3, [12].

$$\begin{pmatrix} x_e \\ y_e \\ \theta_e \end{pmatrix} = \begin{pmatrix} \cos\theta & \sin\theta & 0 \\ -\sin\theta & \cos\theta & 0 \\ 0 & 0 & 1 \end{pmatrix} \begin{pmatrix} x_d - x \\ y_d - y \\ \theta_d - \theta \end{pmatrix} \quad (12)$$

The sliding surface is chosen as:

$$s = \dot{\theta}_e + \alpha\theta_e + \beta\theta_e^{\frac{q}{p}} \quad (13)$$

rearranging Eq. (13), we get

$$\dot{\theta}_e = \omega_d - \omega = -\alpha\theta_e - \beta\theta_e^{\frac{q}{p}} \quad (14)$$

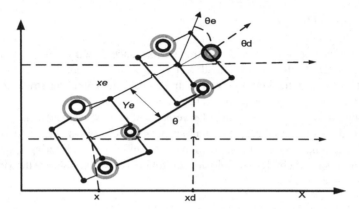

Fig. 3. Posture error description.

Then, the control law can be obtained as [14]:

$$\omega_d - \omega = -\alpha\theta_e - \beta\theta_e^{\frac{q}{p}} = \omega = \omega_d + \alpha\theta_e + \beta\theta_e^{\frac{q}{p}} \tag{15}$$

When, θ_e converges to zero, $\omega_d = \omega$.

The error model becomes to be as follows taking derivative of Eq. (12), we get

$$x_e' = y_e\omega_d - v + v_d \tag{16}$$

$$y_e' = -x_e\omega_d \tag{17}$$

By setting,

$$z = x_e - y_e \tag{18}$$

The control law can be found from the global fast terminal sliding mode as [14]:

$$\dot{z} + \alpha z + \beta z^{p/q} = 0 \tag{19}$$

So, the result can be obtained as:

$$\dot{z} = -\alpha z - \beta z^{p/q} \tag{20}$$

By using the Eqs. (16), (17), and (20), we have

$$\dot{z} = \dot{x}_e - \dot{y}_e = y\omega_d - v + v_d + x\omega_d \tag{21}$$

The control law can be obtained as:

$$\dot{z} = \dot{x}_e - \dot{y}_e = y_e\omega_d - v + v_d + x_e\omega_d \tag{22}$$

$$\dot{z} = -\alpha z - \beta z^{p/q} \tag{23}$$

The control law can be obtained as:

$$\begin{pmatrix} v \\ \omega \end{pmatrix} = \begin{pmatrix} y_e\omega_d + v_d + x_e\omega_d + \alpha z + \beta z^{p/q} \\ \omega_d + \alpha\theta_e + \beta\theta_e^{\frac{p}{q}} \end{pmatrix} \tag{24}$$

Analysis of stability of the control system is performed by selecting the lyapunov function as follows

$$v_1 = 0.5\theta_e^2 \tag{25}$$

By differentiating the Eq. (25) is given as:

$$\dot{v}_1 = \theta_e\dot{\theta}_e = -\alpha\theta_e^2 - \beta\theta_e^{p/q+1} < 0 \tag{26}$$

It was observed that $v_1 > 0$ and $\dot{v}_1 < 0$, thus, the error state variables, θ_e and $\dot{\theta}_e$ are stabilized at the equilibrium point.

The posture error control system stability is proved using the lyapunov function as follows

$$v_2 = 0.5z^2 \tag{27}$$

By differentiating the Eq. (27) is given as:

$$\dot{v}_2 z\dot{z} = -\alpha z^2 - \beta z^{p/q+1} < 0 \tag{28}$$

It can be seen that, $v_2 > 0$ and $\dot{v}_2 < 0$, thus, the error state variables, z and $\dot{z}$ are stabilized at the equilibrium point.

3.2 Inner Loop Speed Control System Using Global Terminal Sliding Mode Controller Design

Considering the linear and the terminal sliding surfaces, a new global fast terminal sliding surface is proposed given by Eq. (29).

$$s = \dot{x} + \alpha x + \beta x^{\frac{q}{p}} \tag{29}$$

where, $x \in R$, $\alpha, \beta > 0$ and $p > q$ are positive odd numbers [14].

First by describing the tracking error as follows

$$\dot{e}_i = e_{i+1} \text{ and } e_i = x_{id} - x_i \tag{30}$$

where, e_i is tracking error.

By choosing the sliding surface for angular velocity of right and right wheel according to GFTSMC, we have,

$$\begin{cases} s_1 = \dot{e}_{1r} + \alpha e_{1r} + \beta e_{2r}^{\frac{q}{p}} \\ s_2 = \dot{e}_{11} + \alpha e_{11} + \beta e_{21}^{\frac{q}{p}} \end{cases} \tag{31}$$

where, s_1 and s_2 sliding surface & $e_{ll} = \omega_{ld} - \omega_l$, $e_{lr} = \omega_{rd} - \omega_r$ is error b/n desired and actual angular velocity.

The control law for first order given by,

$$s_i = 0 \tag{32}$$

Therefore, the dynamic equation in Eqs. (10) and (11) using the state space representations are given by Eqs. (33) and (34).

$$\begin{cases} \dot{x}_1 = (m_1\tau_r - m_2\tau_l - m_1 V(x_1 - x_2)x_1 - m_2 V(x_1 - x_2)x_2)\dfrac{1}{m_1^2 - m_2^2} \\ \dot{x}_2 = (m_2\tau_r - m_1\tau_l - m_2 V(x_1 - x_2)x_1 - m_1 V(x_1 - x_2)x_2)\dfrac{1}{m_2^2 - m_1^2} \end{cases} \tag{33}$$

where, $m_1 = I_w + \dfrac{R^2(mL^2 + I)}{4L^2}$, $m_2 = \dfrac{R^2(mL^2 - I)}{4L^2}$, $V = \dfrac{R}{L}\left(\dfrac{R^2 m_c a}{2L}\right)$, $x_1 = \omega_r$ and $x_2 = \omega_l$.

The linear and angular velocity speed control system using GFTSMC is given by

$$\begin{cases} u_1 = (m_1^2 - m_2^2)\left(\alpha e_1 + \beta e_2^{\frac{q}{p}} + \dot{\omega}_{rd}\right) + m_1 V(x_1 - x_2)x_1 + m_2 V(x_1 - x_2)x_2 \\ u_2 = (m_2^2 - m_1^2)\left(\alpha e_1 + \beta e_2^{\frac{q}{p}} + \dot{\omega}_d\right) + m_2 V(x_1 - x_2)x_1 + m_1 V(x_1 - x_2)x_2 \end{cases} \tag{34}$$

where, $u_1 = m_1\tau_r - m_2\tau_l$ and $u_2 = m_2\tau_r - m_1\tau_l$.

In a finite amount of time, we can bring the system state to a condition of equilibrium t_s given by such that the initial state $x(0) \neq 0$ attains at $x = 0$,

$$t_s = \frac{p}{\alpha(p - q)} \ln \frac{\alpha x(0)^{(p-q)/p} + \beta}{\beta} \tag{35}$$

By designing α, β, p, q [15].

3.3 Inner Loop Speed Control System Using Conventional Sliding Mode Control

In conventional sliding mode control, sliding surface is given by

$$s_i = \left(\lambda + \frac{d}{dt}\right)^{n-1} e_i \tag{36}$$

where, s_i is sliding surface, n is order of system and e_i tracking error [15].

Sliding surface is chosen as follow for angular velocity of right and left wheel:

$$\begin{cases} s_1 = e_{1r} \\ s_2 = e_{1l} \end{cases} \tag{37}$$

Taking the derivative of Eq. (37), we obtain

$$\begin{cases} \dot{s}_1 = \dot{e}_{1r} \\ \dot{s}_2 = \dot{e}_{1l} \end{cases} \tag{38}$$

where, $e_{1r} = \omega_{rd} - \omega_r$ and $\omega_{rl} - \omega_l$ is difference b/n actual and desired angular velocity of right and left wheel.

Then, the control law can be obtained as,

$$\begin{cases} \dot{e}_{1r} = \dot{\omega}_{rd} - \dot{\omega}_r = -\eta\,\text{sign}(s_1) \\ \dot{e}_{11} = \dot{\omega}_{ld} - \dot{\omega}_l = -\eta\,\text{sign}(s_2) \end{cases} \tag{39}$$

where, η is positive constant,

Therefore, the control law of speed control systems for inner control using traditional control system is given by Eq. (40).

$$\begin{cases} u_1 = \left(m_1^2 - m_2^2\right)(\eta\,\text{sign}(s_1) + \dot{\omega}_{rd}) + m_1 V(x_1 - x_2)x_1 + m_2 V(x_1 - x_2)x_2 \\ u_2 = \left(m_2^2 - m_1^2\right)(\eta\,\text{sign}(s_2) + \dot{\omega}_{ld}) + m_2 V(x_1 - x_2)x_1 + m_1 V(x_1 - x_2)x_2 \end{cases} \tag{40}$$

4 Simulation Results and Analysis

Ntrol parameters for the proposed system assuming $v_d = 2\,\text{m/s}$, and $\omega_d = 2\,\text{rad/s}$ are $\alpha = 10, \beta = 1, q = 3$, and $p = 5$ for outer GFTSMC, $\beta = 2, \eta = 50,000, q = 3$, and $p = 5$ for inner GFTSMC and for SMC, $\eta = 50,000$. The numerical parameter used for demonstration of the proposed control is listed in Table 1 for the DDMR [16].

Table 1. Numerical parameter value of DDMR used for simulation.

Parameter	Value and unit
The mass of the DDMR without the driving wheels and actuators [mc]	6 [kg]
The mass of each driving wheel (with actuator) [mw]	0.5 [kg]
The moment of inertia of the mobile robot about the vertical axis through the center of mass [Ic]	3 [kgm^2]
The moment of inertia of each driving wheel with a motor about the wheel axis [Iw]	0.01875 [kgm^2]
The moment of inertia of each driving wheel with a motor about the wheel diameter [Im]	0.5 [kgm^2]
Radius of mobile robot wheel [R]	0.05 [m]
Half of the distance between the wheels [L]	0.5 [m]
Distance between center of mass of robot to the cut axis point of robot [a]	0.1 [m]

In order to test the robustness of the inner loop GFTSMC and the SMC speed control system against random gaussian disturbance shown in Fig. 4 is applied after 5 s of simulation time.

The accuracy and effectiveness of both inner and outer loop controllers against model uncertainty and random external disturbance was tested using different performance

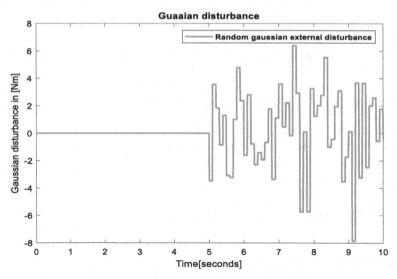

Fig. 4. Random Gaussian external disturbance.

Table 2. Performance index of outer loop controller.

Error along X, Y, and theta	GFTSMC-GTFSMC			GFTSMC-SMC		
	ITSE	IAE	ITAE	ITSE	IAE	ITAE
Theta	0.000686	0.05354	0.01489	0.007078	0.05745	0.02279
X axis	0.001383	0.07725	0.01114	0.1022	0.6755	0.3016
Y axis	0.01594	0.2469	0.03027	0.1242	0.7078	0.3148

Table 3. Performance index of inner loop controller.

Error along X, Y, and theta	GTFSMC			SMC		
	ITSE	IAE	ITAE	ITSE	IAE	ITAE
Right	0.1146	0.2323	1.712	4.081	2.512	12.53
X axis	0.1146	0.2296	1.712	4.06	2.639	12.53

index measurements such as integral absolute error (IAE), integral time square error (ITSE), and integral time absolute error (ITAE) as presented in Tables 2 and 3.

The results reveal that the along θ with external and uncertainty disturbance achieved the least error 0.000686 using GFTSMC (ITSE), while the largest error was achieved along θ with (0.05745) using SMC (IAE). Similarly, least error along X and Y axis 0.001383 and 0.01594 respectively using GFTSMC (ITSE). The largest error 0.6755 and 0.7078 achieved along X and Y axis respectively using SMC (IAE) as indicated

Table 2. The results show that the right and left wheel angular velocity achieved the least error of 0.1146 using GFTSMC (ITSE), while the Left and right wheel angular velocity achieved the largest error (12.53) with SMC (ITAE) as presented in Table 3. The performance index versus time of GFTSMC and conventional SMC illustrated in Fig. 5 and Fig. 6 respectively. It was observed from Fig. 5 that GFTSMC gives good convergence under model uncertainty and Gaussian external disturbances.

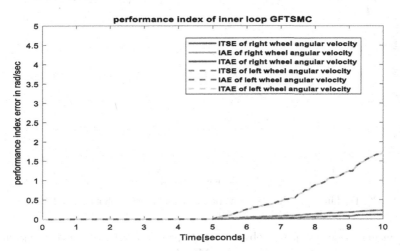

Fig. 5. Performance index of inner loop GFTSMC.

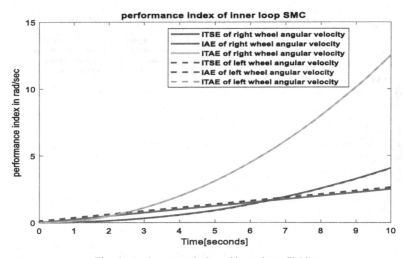

Fig. 6. Performance index of inner loop SMC.

From Fig. 7, an initial tracking error along X axis, Y axis, and rotation position starting from point $\left(1, 2, \frac{pi}{6}\right)$ approached to zero in finite time.

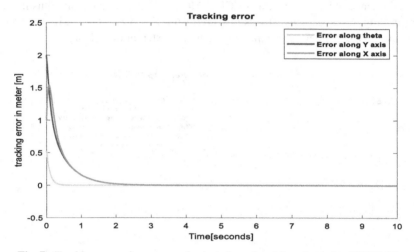

Fig. 7. Tracking error along x, y and orientation of mobile robot using GFTSMC.

From Figs. 8 and 9, it was observed that control signal input angular and linear velocity generated using outer loop GFTSMC changed in to angular velocity of each wheels using transformational matrix in Eq. (4).

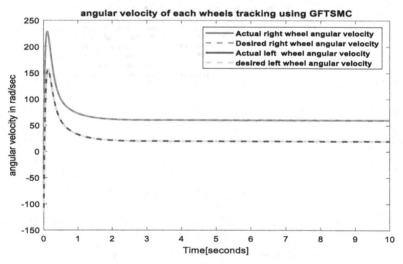

Fig. 8. Right and left wheel angular velocity speed tracking

The inner loop GFTSMC speed control system makes the actual linear and angular velocity to track the desired linear and angular velocity of 2 m/s and 2 rad/s respectively.

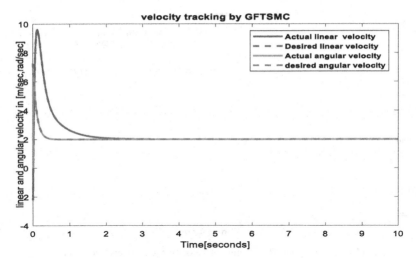

Fig. 9. Angular and linear velocity control signals

XY 2D movement of the mobile robot shown in Fig. 10. The robot starts from the initial point (1 m, −1 m) and moves to the desired circular trajectory radius of 1 m.

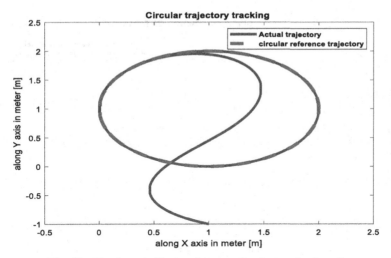

Fig. 10. Circular trajectory tracking starting from point $(1, -1)$

Similarly, the Fig. 11 shows the inner loop GFTSMC speed control system performance which shows the error between the angular velocity of each wheel to zero at a finite time. In this case, random Gaussian disturbance was applied after 5 s for both wheels and a model uncertainty of $\rho = 0.1 * \sin(20 * t)$ a was taken to test the robustness of the controller, but the inner loop controller is suitable for both disturbances inputs.

In traditional SMC, sliding mode parameters can be adjusted to get faster error convergence, however, this will in turn increase the control gain, which may cause severe

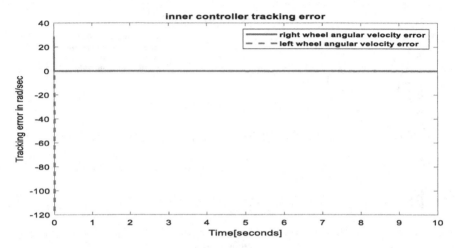

Fig. 11. Inner loop GFTSMC angular velocity each wheels tracking error

chattering of the sliding surface and, therefore, deteriorate the system performance as shown in Fig. 13. But, sliding surface GFTSMC smooth which means that it is free from chattering problems as shown in Fig. 12.

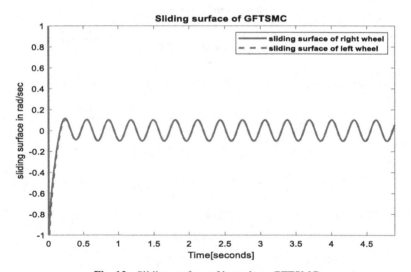

Fig. 12. Sliding surface of inner loop GFTSMC

As observed in Fig. 14, the mobile robot approaches the desired trajectory starting from (0, 0). Here, starting error along X, Y and theta will be (2, 1, pi/6).

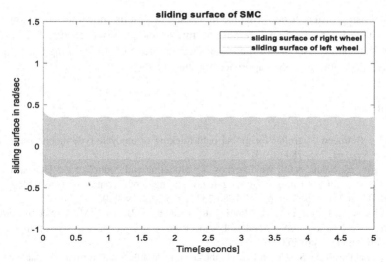

Fig. 13. Sliding surface of inner loop SMC

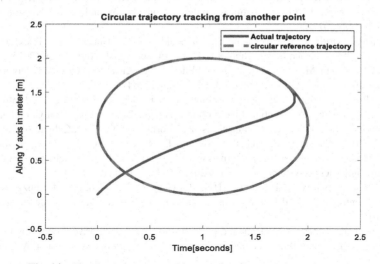

Fig. 14. Circular trajectory tracking starting from another point (0, 0)

5 Conclusion

In this paper, the dynamics of DDMR is obtained using Langrage formulation. The trajectory tracking control for the DDMR is designed to achieve the desired reference trajectory such that the angle error and posture error comes to zero with a finite time global fast terminal sliding mode control (GFTSMC). The GTSMC without chattering effect for kinematic outer closed loop and dynamic inner loop velocity controller was designed to achieve velocity tracking so that errors between the actual and desired velocity control input are reduced to zero at a finite time. The robustness of both controllers was tested using different performance indexes. The results show that both controllers are insensitive to any random external and model uncertainty disturbance. Moreover, the overall dynamic response of the system was improved by integrating the dynamics using cascaded motion control approach for the DDMR.

References

1. Do, K.: Bounded controller for global path tracking of unicycle type mobile robot. Rob.t Auton. Syst. **61**, 775–784 (2013)
2. Ashagrie, A., Salau, A.O., Weldcherkos, T.: Modeling and control of a 3-DOF articulated robotic manipulator using self-tuning fuzzy sliding mode controller. Cogent Eng. **8**(1), 1950105 (2021). https://doi.org/10.1080/23311916.2021.1950105
3. Martins, N.A., Bertol, D.W., De Pieri, E.R., Castelan, E.B., Dias, M.M.; Neural dynamics control of a Nonholonomic mobile robot incorporating the actuator dynamics. In: CIMCA, IAWTIC and ISE, pp. 563–568 (2008)
4. Taganamathar, T.d.a.S.: Control of Nonholonomic mobile robot formations back stepping kinematics in to dynamics. In IEEE, Singapore (2007)
5. Benaziza, W., Slimane, N., Mallem, A.: Mobile robot trajectory tracking using terminal sliding mode control. In: 6th International Conference on Systems and Control (ICSC), Banta, pp. 538–542 (2017)
6. Zidani, G., Drid, S., Chrifi-Alaoui, L., Benmakhlouf, A., Chaouch, S.: Back stepping controller for a wheeled mobile robot. In: IEEE, Sousse, Tunisia April 28–30 (2015)
7. Butt, C., Rahman, M.A.: Limitations of simplified fuzzy logic controller for IPM motor drive. In: Industry Applications Conference, **3**, 1891–1898 (2004)
8. Vinod Raj, N., Mathew, A.T.: Design, simulation and implementation of cascaded path tracking controller for a differential drive mobile robot. In: 2015 International Conference on Advances in Computing, Communications and Informatics (ICACCI), Calicut, India (2015)
9. Salichs, M.A., Puente, E.A., Gachet, D., Moreno, L.: Robust controller for trajectory tracking of a mobile robot. In: 1st IEEE International Conference on Power Electronics. Intelligent Control and Energy Systems (ICPEICES-2016), Calicut, India (2016)
10. Ushikoshi, T., Peixoto, K.P., Souto, F.H.S., das Chagas, T.P., Schnitman, L.: Fuzzy maneuvering controller applied to a dynamic model of a differential drive mobile robot. IEEE (2018)
11. Dhaouadi, R., Hatab, A.A.: Dynamic modelling of differential-drive mobile robots using Lagrange and Newton-Euler methodologies: a unified Framework. Adv. Robot. Autom. **2**(2) (2013)
12. Kandalama, Y., Kimura, Y., Miyazaki, F., Noguchi, T.: A stable tracking control method for an autonomous mobile robot. In: Robotics and Automation, 1990. Proceedings, IEEE (1990)
13. Kolmanovsky, I., McClamroch, N.H.: Development in nonholonomic control problems. IEEE Control Systems 20–36 (1995)
14. Park, K.B., Tsuiji, T.: Terminal sliding mode control of second-order nonlinear uncertain systems. Int. J. Robust Nonlin. Control **9**(11), 769–780 (1999)
15. Liu, J., Wang, X.: Advanced Sliding Mode Control for Mechanical Systems. Springer, Beijing, China (2011). https://doi.org/10.1007/978-3-642-20907-9
16. Sharma, A., Panwar, V.: Control of mobile robot for trajectory tracking by sliding mode control technique. In: International Conference on Electrical, Electronics, and Optimization Techniques (ICEEOT) - 2016, Greater Noida, India (2016)

Artificial Intelligence-Based Breast and Cervical Cancer Diagnosis and Management System

Elbetel Taye Zewde[1,2] (ID), Mizanu Zelalem Degu[2,3],
and Gizeaddis Lamesgin Simegn[1,2(✉)] (ID)

[1] Biomedical Imaging Unit, School of Biomedical Engineering, Jimma Institute of Technology,
Jimma University, Jimma, Ethiopia
gizeaddis.lamesgin@ju.edu.et
[2] AI and Biomedical Imaging Research Unit, Jimma Institute of Technology, Jimma University,
Jimma, Ethiopia
[3] Faculty of Computing and Informatics, Jimma Institute of Technology, Jimma University,
Jimma, Ethiopia

Abstract. Breast cancer and cervical cancer are two of the most common and deadly malignancies in women. Early diagnosis and treatment can save lives and improve quality of life. However, there is a shortage of pathologists and physicians in most developing countries, including Ethiopia, preventing many breast and cervical cancer patients from early cancer screening. Many women, particularly in low resource settings, have limited access to early diagnosis of breast and cervical cancer and receive poor treatment which in turn increases the morbidity and mortality due to these cancers. In this paper, an integrated intelligent decision support system is proposed for the diagnosis and management of breast and cervical cancer using multimodal im-age data. The system includes breast cancer type, sub-type and grade classification, cervix type (transformation zone) detection and classification, pap smear image classification, and histopathology-based cervical cancer type classification. In addition, patient registration, data retrieval, and storage as well as cancer statistical analysis mechanisms are integrated into the proposed system. A ResNet152 deep learning model was used for classification tasks and satisfactory results were achieved when testing the model. The developed system was deployed to an offline web page which has added the advantage of storing the digital medical images and the labeled results for future use by the physicians or other researchers.

Keywords: Breast cancer · Cervical cancer · Decision support system · Screening · Histopathological images · Cancer management

1 Introduction

The world health organization (WHO) estimated that there were 9.6 million cancer-related deaths and 18.1 million new cases worldwide in 2018 [1]. Breast cancer and

B. H. Woldegiorgis et al. (Eds.): ICAST 2022, LNICST 455, pp. 79–94, 2023.
https://doi.org/10.1007/978-3-031-28725-1_6

cervical cancer, which are the second and fourth most common cancer types in women globally, are the most common and lethal of all known cancer kinds [1].

According to statistics provided by the Addis Ababa City Cancer Registry Quadrennial Report of 2012–2015 investigation, women made up two-thirds of the city's cancer burden, with breast cancer accounting for the majority of that burden. Cervical cancer was the next most common cancer to affect the city's female population [2].

The two most prevalent cancers, breast cancer and cervical cancer, are also the two that kill the most women from cancer worldwide. Through improvements in early diagnosis methods and prescreening procedures, illness survival can be increased. A clinical examination, imaging-based screening, and pathological evaluation (a biopsy test) are often used diagnostic techniques.

Unfortunately, most underdeveloped nations [3–5] including Ethiopia have a physician and pathologist scarcity, which makes it difficult for many women with breast cancer and cervical cancer to receive an early diagnosis. Due to this, many women, especially those living in rural regions, miss out on the opportunity to receive an early diagnosis of breast cancer and cervical cancer [6]. Because of these illnesses, the death rate of mothers has increased. In addition, pathologists must review a lot of biopsy samples each day, which exhausts them. Moreover, the complexity of the cancer cells and their subjective decision, which is dependent on expert's performance, may lead to misdiagnosis of the subtypes and grade of the diseases from biopsy tests, which is essential to understand the biological characteristics and clinical behavior of the cancer cells.

Even though, Pap smear test, Biopsy test and colposcopy image tests are the common screening and diagnosing techniques for cervical cancer and breast cancer [7–9] they are sometimes prone to misdiagnosis. That is, their result can be either over interpreted or under interpreted. Over interpreted means that women without cancer are exposed to potentially harmful treatments and unnecessary expenses. On the other hand, inaccurate interpretation (under interpretation) of biopsy result could prevent women from getting the treatment early causing the cancer to grow more to invasive stage.

Now a days, machine learning and deep learning techniques are applied in different medical image and signal analysis works [10–18] in order to automate diagnosis systems and help physicians in getting support for making accurate decision. Artificial intelligence's (AI) application in the analysis of medical images can help increase diagnosis accuracy and reduce subjective variability and misdiagnosis rate by reducing the work load of pathologists.

In this project, an integrated breast cancer and cervical cancer diagnosis decision support system is developed in the form of web application.

2   Methodology

2.1   System Development and Technologies Used

When properly developed, clinical decision support systems can greatly improve the quality of treatment by delivering more accurate diagnoses, fewer mistakes, reduced costs, and more patient and provider satisfaction [19].

An automatic breast cancer and cervical cancer diagnosing system is developed and tested. The essence of this project is to develop an integrated web application which can be installed in any desktop or personal computer of users. It is designed as a decision support system for physicians. Furthermore, the system has many added features which can help researchers to acquire important information like statistical data of the prevalence of the diseases and organized labeled digital images for future studies can be acquired from the systems database.

The developed offline web application has the ability of classifying digital histopathological image of breast cancer in to binary (benign and malignant) classes and has the ability of classifying the given image into further eight subtypes (adenosis tumor, fibroadenoma tumor, phyllodus tumor and tubularadenoma tumor subtypes under the benign class and Ductal carcinoma, Lobular carcinoma, Mucinious Carcinoma and Pappilary Carcinoma as a subtypes under the malignant cancer types), and classifies digital histopathological image of cervical cancer in to precqancerous cells, squamous carcinoma and adenocarcinoma. Furthermore, the developed system is capable of classifying pap smear images as normal and abnormal cells. Furthermore, using a colposcopy image as input, physicians can use the developed system to classify cervix type as type 1, type 2 and type 3. This information will help physicians to exactly know the transformation zone of the cervix.

To develop the overall system Django frame work was used. Django is an open source high-level python web framework that enables rapid development of secure and maintainable websites [20]. For front end development HTML, CSS, JavaScript, Bootstrap, and AJAX were used. For backend development python is used for storage SQLite database is used. Furthermore, the classification models were developed by using ResNet152 pretrained model. Finally, the Incremental Development Model was utilized to create the web application. This is a software development process in which a web application is developed, implemented, and tested gradually, with little adjustments made until the application is complete.

2.2 Image Classification System

2.2.1 Data

Microscopic histopathology images of breast cancer were collected from datasets provided by Jimma University Medical Center (JUMC), "break-his," [10] and "zendo" [11] repositories for the purposes of model training, validation, and testing. The images from JUMC were collected using an Opti-ka-vision camera attached to a simple light-microscope with four magnification powers (40X, 200X, and 400X), and a resolution of 2592 X 1936. The images were stained using the H&E staining process. Similar to how images from the break-his dataset were collected, different magnification factors were used to capture the images (40X, 100X, 200X, and 400X). The image frames were obtained from areas that had been affected by tumor growth.

A total of 915 Hematoxylin and Eosin (H&E) stained histopathology images from Jimma University Medical Center (JUMC) and St. Paul Hospital were obtained for the classification of cervical cancer. The OPTIKA light microscope with mounted digital camera, a smartphone camera, and a digital scanner was used to gather the data. 4x, 10x, 40x, and 100x magnification powers were used to capture microscopic images, with resolutions ranging from 419×407 to 2048×1536.

For the classification of cervix types, a total of 4005 pictures were gathered. After applying 5% acetic acid to the cervix, 133 colposcope images from Tercha General Hospital were obtained via a speculum (to reach the cervix) and 13MP and 18MP Tecno smartphone cameras. Three experienced gynecologists, including a senior medical doctor and an integrated emergency surgical officer (IESO), categorized the images after they had been gathered. The remaining pictures are from the Kaggle dataset, which is accessible to everyone [20]. Prior to data collection, Jimma Institute of Health's intuitional research review board granted ethical permission.

For testing and training, the Herlev Pap smear dataset (general public) was used. It includes 917 pictures of single cervical cells that have been classified and segmented using ground truth [16].

2.2.2   Classification Model

An image is recognized based on its visual information by the intricate image classification system used in computer vision. The accuracy of classification is primarily influenced by the characteristics of the dataset, the complexity of the analytical problem, and the efficiency of the classification method. Inference time, memory usage, computational complexity, recognition accuracy, and model complexity can all be used to assess how effective a classification method is. It is difficult to train a highly deep neural network because of the vanishing gradient problem. The gradient becomes exceedingly small as a result of repeated multiplication since it propagates back to earlier layers. Whenever the result of a network grows deeper, its performance also reaches a saturation point or soon drops [21, 22].

For the cancer classification tasks, the state-of-the-art pre-trained classification model, ResNet152, was used to classify all the multi class histopathology images, colposcopy images and Pap smear images. The model was trained with raw images acquired from the data repositories. Adam optimizer with a learning rate of 0.0001, maximum of 300 number of epochs, and loss function of sparse categorical cross entropy, were used for training. The models were validated using train-test split technique. Finally, new data were used to test the performance of the model for the different classification tasks.

2.3   Key Functions of the System

The web application provides several key functions including:

- **Doctor's login:** the doctor/ physician will be able to login after inputting the assigned username and password on the provided space.
- **Diagnosis:** the system is able to diagnose breast cancer and cervical cancer from histopathological images and screen cervical cancer from digital Pap smear images and colposcopy image.
- **Store patient information and data:** the system has the ability of saving all recorded and diagnosed cases for future use.
- **Statistics:** the system statistically provides the prevalence of the disease in the form of charts, so that researchers and health professionals can take necessary measures.
- **View and print results:** the doctor will be able to view all the diagnosed results and images and print out the results in case needed.

3 Result

3.1 Classification

For the BC binary classification, classification in to benign and malignant breast cancer types, better validation was achieved at the 120th epoch with a validation loss of 0.1284. at this epoch, 96.53% training accuracy and 94.62% validation accuracy. For the BC benign subtype classification, the lowest validation loss (0.4628) was achieved at the 170th epoch having a training accuracy of 82.14% and validation accuracy of 82.83%. For BC malignant subtype classification task, the lowest validation loss (0.1725) was achieved at the 276th epoch with a training accuracy of 97.7% and validation accuracy of 95.42%. Figure 1 indicates the model's accuracy and loss result obtained for BC

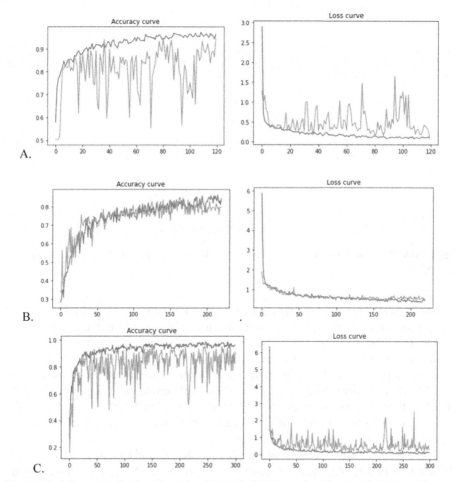

Fig. 1. Model accuracy (left) and loss plots (right) **A.** BC binary classification, **B.** Benign sub-type classification, **C.** BC Malignant subtype classification

histopathological image binary classification, BC benign subtype classification and BC malignant subtype classification.

Furthermore, the models were further evaluated using unseen test dataset. The test results for the trained models on the binary classification, benign subtype classification and malignant subtype classification are demonstrated below using the ROC plot in Fig. 2.

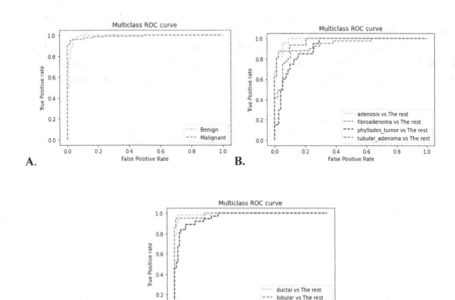

Fig. 2. ROC for BC **A.** Binary classification model, **B.** Binary subtype classification, **C.** Malignant subtype classification

For cervical cancer classification from histopathological image using ResNet152 the lowest validation loss (0.1515) was achieved at the 151th epoch with training accuracy 95.49% and validation accuracy of 96.17%. For cervix type classification the lowest validation loss of 0.7179 with training accuracy of 76.53% and validation accuracy of 67.91%. For pap smear classification the model achieved the lowest validation loss at the 173th epoch with a training accuracy of 86.1% and validation accuracy of 87.27%. Figure 3 shows training accuracy and loss results of ResNet152 model for classifying cervical cancer from histopathological images, cervix type classification task and pap smear image classification task respectively.

The trained models to perform cervix type classification and Pap smear digital image classification task were further evaluated using unseen test dataset. The test results for the models to perform the necessary classification are shown below using the ROC plot in Fig. 4.

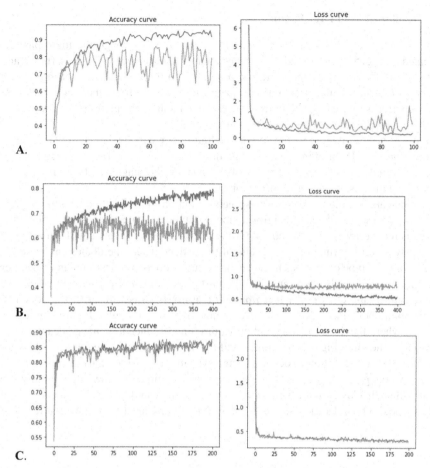

Fig. 3. Model accuracy (left) and loss plots (right) **A.** Cervical cancer histopathological image classification, **B.** Cervix type classification, **C.** Pap smear image classification

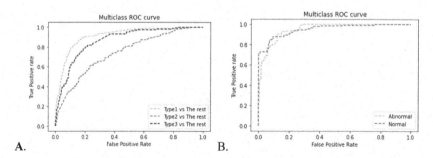

Fig. 4. ROC graph **A.** cervix type classification model, **B.** Pap smear image classification

3.2 Developed Web Page Result

The first interaction between the user and the web app is through the login page as indicated on Fig. 5. A registered doctor/physician who have been given a username and password by the admin of the system can login in to the system by inserting his/her unique ID /username and password. This page will help to restrict other people who are not given access to use the system from accessing patient's information on the web app. This will increase patient data confidentiality.

After a successful login by the doctor, the **homepage** will appear as shown in Fig. 6. This section provides brief information regarding Breast cancer and cervical cancer. This page also directs to different functions of the system like diagnosis, browsing patient records and images, statistical information etc.

As indicated in Fig. 7, **Diagnosis/screen** option on the left side of the home page will allow doctors to fill personal information about the patient and create new ID, if the patient is registering as a new patient. Or the doctor can use the patients ID to retrieve previously saved information and results of the patient. Then, the doctor can continue by selecting the type of analysis he/she is intending to do as demonstrated in Fig. 8. For example: if the doctor's intention is to do breast cancer diagnosis. Then he will choose **breast cancer** and click on "**select image for analysis**" this will direct the doctor to either capture new image or browse saved image from the computer. Figure 9 indicates the page where the doctor can upload or acquire image. Once the image is captured or browsed as shown in Fig. 10, the image will be displayed as indicated in Fig. 11. After that, by clicking the "**Start processing**", the system will start to analyze the class and subclass of the given image based on the deep learning model's knowledge. Finally, the result will be displayed in the form of percentage as indicated in Fig. 12. As the system is developed to be a decision support system for doctors, he/she will take the result of

Fig. 5. Login page of BCCDMS

the system as an input and the final decision on the diagnosis result will be made by the doctor. Finally, the result achieved with the detail patient's information can be saved and printed as indicated in Fig. 13.

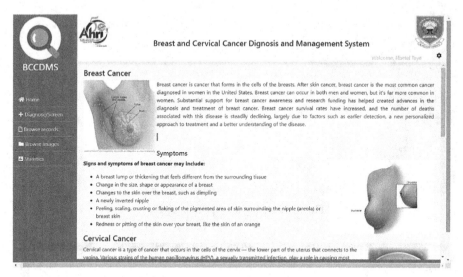

Fig. 6. Homepage of BCCDMS

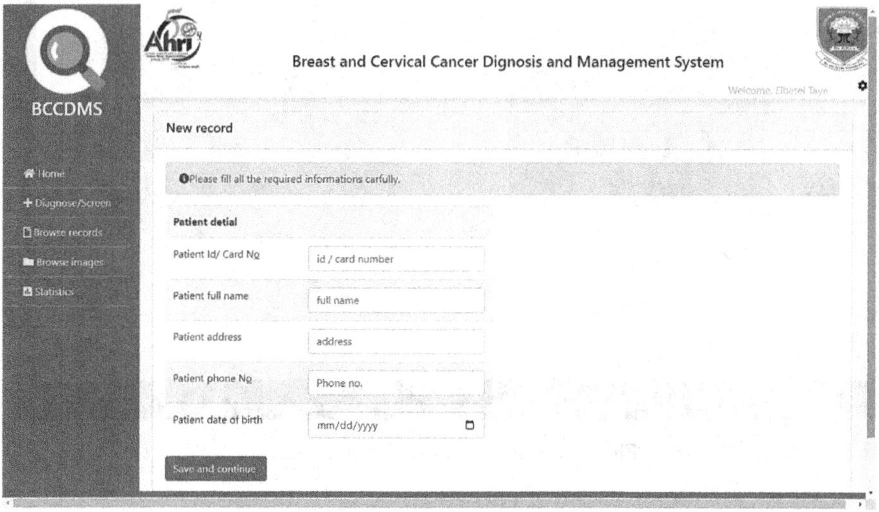

Fig. 7. Patient registration page

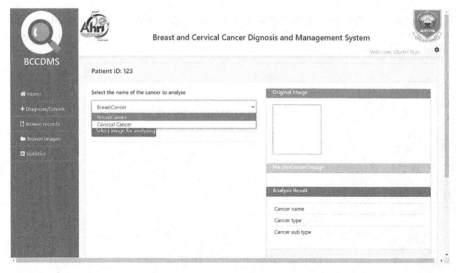

Fig. 8. Selecting the type of analysis

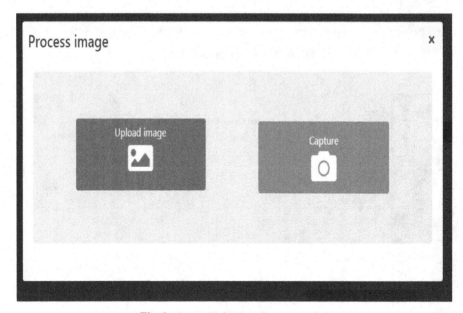

Fig. 9. Image Upload or Capture option

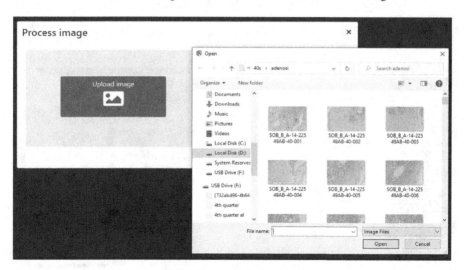

Fig. 10. Image Browsing

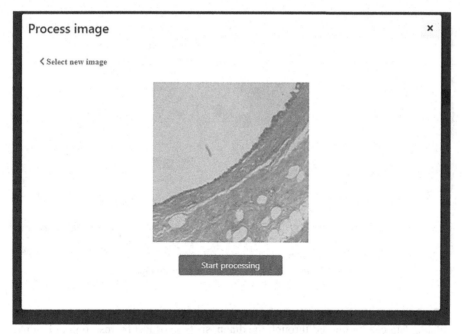

Fig. 11. Selected image processing

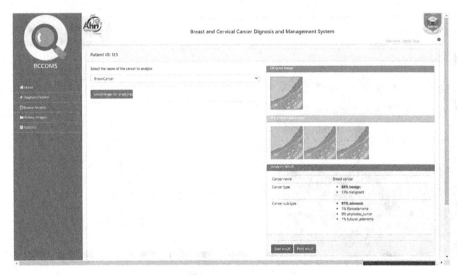

Fig. 12. Breast cancer analysis result

Fig. 13. Patient details and result printing Option

Furthermore, the system allows data retrieving through the **Browse data** option the available data can be filtered through the diagnosis type or just by inserting individual patients ID or name. This feature of the developed system will help researchers to find an organized digital dataset. Figure 14 shows each analyzed image saved under its labeled class/folder. Moreover, the system will help researchers and other concerned bodies to know how prevalent the diseases are in a given area. Figure 12 shows example of the chart indicating the frequency of the diseases from all up to date analyzed images (Fig. 15).

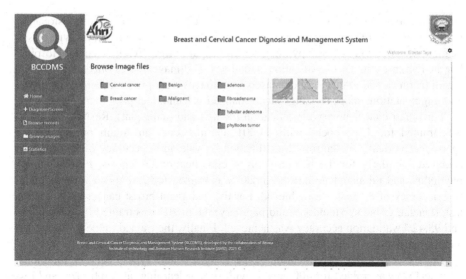

Fig. 14. List of images under their respective class/ folder

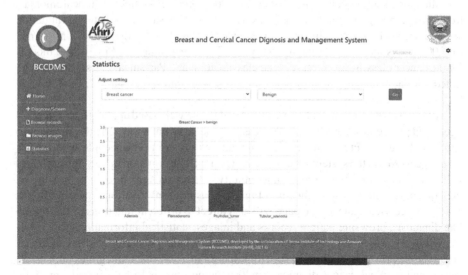

Fig. 15. Statistical representation of analyzed data

4   Discussion

In this paper, a web page was developed based on deep learning model in order to help pathologists and physicians in their decision making process when diagnosing breast cancer and cervical cancer.

For deep model training, validation and testing, histopathological images of breast cancer were obtained from BreakHis online dataset and locally acquired from Jimma

University Medical Center (JUMC). For the cervical cancer classification, histopathological images were acquired from JUMC, and St. Paul Hospital. For cervix type classification, 4,005 cervix colposcopy images were acquired from Tercha General Hospital. For the Pap smear image classification, a total of 1,417 images were collected from local health facilities and online public datasets. All images were pre-processed by applying data augmentation, image resizing and normalization techniques prior to model training.

For breast cancer binary classification (benign and malignant), ResNet152 model was trained for 120 epochs with 0.0001 learning rate, an Adam optimizer, and sparse_categorical_crossentropy loss function. A validation accuracy of 94.62% was achieved. Similarly for the benign subtype classification (Adenosis, Fibroadenoma, phyllodus, and tubularadenoma) the model was trained for 220 epochs and a validation accuracy of 82.83% was achieved. For the malignant breast cancer classification task (Ductal, Lobular, Mucinous, and papillary) the model was trained for 300 epochs and 95.42% validation accuracy was achieved. Finally, the trained models were tested and the satisfactory result was achieved as shown in the ROC in Fig. 2.

For the cervix type classification (Type 1, Type 2 and Type 3 cervix), the model (ResNet152) was trained for 400 epochs with similar learning rate, optimizer and loss function. A result of 76.53% validation accuracy was achieved. For the cervical cancer classification ('Squamous cell carcinoma', 'Precancer', 'Normal',' Adinocarcinoma') from histopathological image using ResNet152 having the same parameters a validation accuracy of 96.17% was achieved. For Pap smear classification the model was trained for 200 epochs and a validation accuracy of 87.27% was attained. Finally, the models for the cancer classification, cervix type classification and Pap smear classification were tested with new images and a satisfactory result was achieved as shown in the ROC in Fig. 4.

All the classification models were integrated in one web based application for pathologists and physicians to perform the necessary tasks in an easy way. The developed web app as indicated in Fig. 5 has a login page that will restrict other people who are not given access to use the system from accessing patient's information on the web app. This will help to increase patient data confidentiality. After a successful login by the doctor, the homepage will appear as shown in Fig. 6. It has brief information regarding Breast cancer and cervical cancer. The page will also direct to different functions of the system like diagnosis, browsing patient records and images, statistical information etc.

This work presents an integrated diagnosis and screening system which has an advantage over the previous literatures [10–18] proposed in a way that it integrates both breast cancer and cervical cancer diagnosis and screening options to help in reducing misdiagnosis rate, fatigue on doctors and to increase the reliability of test results on the diagnosis. Besides, the developed system, through its option to save the images, has the ability to store the analyzed images with their result (the system will save the images as labeled images). This is one of the major contributions offered by the system to the existing health care system and the research world. As a result, anyone who wants to retrieve previous data can easily have access to it as indicated in Fig. 14.

In summary, in this paper, a full-fledged, integrated, magnification power independent and robust system is proposed for cervical cancer screening and diagnosis system by automating both the pre-screening (cervix type classification) and cervical cancer type classification tasks. We acknowledge that, staging of cervical cancer has not been included in this work. Moreover, adding more data may improve the accuracies of the proposed systems.

5   Conclusion

The proposed intelligent web-based tool is a robust decision support system designed to diagnose and manage breast and cervical cancer using multimodal image data including histopathological images, pap smear digital images and colposcopy images by employing the state-of-the-art artificial intelligence technique. The developed system will help in the reduction of cancer misdiagnosis, minimize the work load of physicians and improve management of breast and cervical cancer diagnosis. Moreover, the system has added a unique feature of storing labeled images to solve the existing problem of lack of organized labeled digital medical data. This will allow researchers and other health care professionals to easily retrieve digital images for further analysis and research.

References

1. World Health Organization Releases Latest Global Cancer Data - Cancer Health. https://www.cancerhealth.com/article/world-health-organization-releases-latest-global-cancer-data. Accessed 15 Dec 2019
2. Timotewos, G., et al.: First data from a population based cancer registry in Ethiopia. Cancer Epidemiol. **53**, 93–98 (2018)
3. Naicker, S., Plange-Rhule, .J, Tutt, R.C., Eastwood, J.B.: Shortage of healthcare workers in developing countries Africa. Ethn. Dis. **19**(1 Suppl 1), S1-60-4 (2009 Spring). PMID: 19484878
4. Legason, I.D., et al.: A protocol to clinically evaluate liquid biopsies as a tool to speed up diagnosis of children and young adults with aggressive infection-related lymphoma in East Africa "(AI-REAL)." BMC Cancer **22**(1), 1–9 (2022)
5. Kinfu, Y., Dal Poz, M.R., Mercer, H., Evans, D.B.: The health worker shortage in Africa: are enough physicians and nurses being trained? (2009)
6. Kisiangani, J., et al.: Determinants of breast cancer early detection for cues to expanded control and care: the lived experiences among women from Western Kenya. BMC Womens Health. **18**(1), 81 (2018). https://doi.org/10.1186/s12905-018-0571-7
7. Sornapudi, S., et al.: Automated cervical digitized histology whole-slide image analysis toolbox. J. Pathol. Inform. **12**(1), 26 (2021)
8. Safaeian, M., Solomon, D., Castle, P.E.: Cervical cancer prevention—cervical screening: science in evolution. Obstet. Gynecol. Clin. North Am. **34**(4), 739–760 (2007). https://doi.org/10.1016/j.ogc.2007.09.004
9. Veta, M., Pluim, J.P., Van Diest, P.J., Viergever, M.A.: Breast cancer histopathology image analysis: a review. IEEE Trans. Biomed. Eng. **61**(5), 1400–1411 (2014)
10. Spanhol, F.A., et al.: A dataset for breast cancer histopathological image classification. IEEE Trans. Biomed. Eng. **63**(7), 1455–1462 (2015)
11. Dimitropoulos, K., et al.: Grading of invasive breast carcinoma through Grassmannian VLAD encoding. PLoS ONE **12**(9), e0185110 (2017)
12. He, K., et al.: Deep residual learning for image recognition. In 2016 IEEE Conference on Computer Vision and Pattern Recognition (CVPR) (2016)

13. Alyafeai, Z., Ghouti, L.: A fully-automated deep learning pipeline for cervical cancer classification. Exper. Syst. Appl. **141**, 112951 (2020). https://doi.org/10.1016/j.eswa.2019.112951

14. Guo, P., et al.: Nuclei-based features for uterine cervical cancer histology image analysis with fusion- based classification. IEEE J Biomed Heal. Inf. **20**(6), 1595–1607 (2015). https://doi.org/10.1109/JBHI.2015.2483318

15. Zewde, E.T., Simegn, G.L.: Automatic diagnosis of breast cancer from histopathological images using deep learning technique. In: Berihun, M.L. (ed.) ICAST 2021. LNICSSITE, vol. 411, pp. 619–634. Springer, Cham (2022). https://doi.org/10.1007/978-3-030-93709-6_42

16. Almubarak, H.A., et al.: A hybrid deep learning and handcrafted feature approach for cervical cancer digital histology image classification. Int. J. Healthc. Inf. Syst. Inform. **14**(2), 66–87 (2019). https://doi.org/10.4018/IJHISI.2019040105

17. Tian, Y., et al.: Computer-aided detection of squamous carcinoma of the cervix in whole slide images. Comput. Vis. Pattern Recognit. https://arxiv.org/abs/1905.10959

18. Wei, L., Gan, Q., Ji, T.: Cervical cancer histology image identification method based on texture and lesion area features. Comput. Assist. Surg. **22**(sup1), 186–199 (2017). https://doi.org/10.1080/24699322.2017.1389397

19. Zewdie, E.T., Tessema, A.W., Simegn, G.L.: Classification of breast cancer types, sub-types and grade from histopathological images using deep learning technique. Heal. Technol. **11**(6), 1277–1290 (2021). https://doi.org/10.1007/s12553-021-00592-0

20. Django, https://developer.mozilla.org/en-US/docs/Learn/Server-side/Django/Introduction Last Accessed: 15–5–2022

21. Cervical cancer dataset. https://www.kaggle.com/code/prakharpipersania/cervical-cancer. Accessed 17 Feb 2022

22. Gao, Z., Wang, L., Zhou, L., Zhang, J.: HEp-2 cell image classification with deep convolutional neural networks. IEEE J. Biomed. Health Inform. **21**(2), 416 (2017)

Numerical Simulation and Optimization of a Locally Built Midibus Structure in Quasi-static and Rollover Condition

Hailemichael Solomon Addisu[1](✉) [iD], Ermias Gebrekidan Koricho[2],
and Adino Amare Kassie[1]

[1] School of Mechanical and Industrial Engineering, Institute of Technology, Dire Dawa
University, Dire Dawa, Ethiopia
hailasolomon15@gmail.com
[2] Department of Mechanical Engineering, Addis Ababa Science and Technology University,
Addis Ababa, Ethiopia

Abstract. Rollover crashworthiness concerns the ability of a vehicle's structural system and components to absorb energies with complete protection of occupants in dynamic (rollover) crash scenarios. First, this study aims to analyze a locally built midibus structure in rollover crashes using numerical investigation (LS-DYNA) as stated by United Nations Regulation 66 (UNECE R66). Also, this study considered the quasi-static simulation to determine the energy absorbing and load-deformation behavior of the midibus frame sections. Then, the two alternatives in design optimization were presented via reinforcement design and numerical optimization (Successive Response Surface Method in LS-OPT) to improve the strength and weight of the midibus structure. As a rollover simulation result, the maximum deformation of the baseline structure occurred at pillar A and three bays. As a result, the baseline midibus structure failed the standard requirement and has unacceptable strength in both quasi-static and rollover simulation. Moreover, related to the baseline model, the structure's weight of the reinforced Model was effectively reduced by 5.2%. However, an optimized model (using the Successive Response Surface Method) has reduced the weight of the reinforced model by 5.6%. Lastly, the Energy Absorption and Specific Energy Absorption of the baseline and the two alternative models were evaluated and compared.

Keywords: Crashworthiness · Deformation · FE methods · Midibus · Reinforcement · Rollover

1 Introduction

Locally built buses (midibuses) and public transport vehicles are frequently used in Ethiopia. Moreover, most buses are locally manufactured in Ethiopia from ISUZU N-Series truck chassis with accessible materials. However, these locally built midibuses are not analyzed and tested using numerical or experimental approaches. Also, for approval,

© ICST Institute for Computer Sciences, Social Informatics and Telecommunications Engineering 2023
Published by Springer Nature Switzerland AG 2023. All Rights Reserved
B. H. Woldegiorgis et al. (Eds.): ICAST 2022, LNICST 455, pp. 95–126, 2023.
https://doi.org/10.1007/978-3-031-28725-1_7

most midibuses are expected to conform to the set parameters, usually spatial (length, width, and height), seating, and weight measurement. However, this approval technique leads to low strength and overheavy bus structure [1]. Moreover, rollover crashes frequently happen due to the reasons of rolling down into a cliff, collision with vehicles and rotating sideways by obstacles (ditch, kerb, or objects) [2–4]. Most rollover crashes happen on the road of curved (tangent) sections in Ethiopia due to the pedestrian's priority and fast-moving (high speed) crashes [5]. Additionally, bus rollover accidents occur under the circumstance of dropping down into a cliff in this country. Consequently, These accidents led to 50% of fatalities and 50% of passengers' injuries in 2004 [6]. Mainly, the bus rollover accidents obtained serious structural deformation and severe injuries & fatalities to the passengers [2, 3]. The public transport vehicles (buses) involved high numbers of fatalities and injuries during crashes. In Ethiopia, fatalities by bus crash involved nearly 35.42% (1,324 road traffic deaths) in 2018. (UN ECE 2020) reported that in Ethiopia, the crash tendency increased on average by 9% from 2010–2018 [7]. Specifically, rollover crashes consist of 17.34% fatalities & 17.17% of injuries within six years (2005–2011) [5].

Crashworthiness is the ability to absorb vehicle crashes and protect the occupants in survival space [8, 9]. Structural crashworthiness is concerned with designing a vehicle's structural system and components, which requires absorbing the dynamic energies and loads and energies in dynamic case (collision (impact) occasion) [10, 11]. Therefore, after the design of the bus, the numerical simulation by FE Method is a better approach to visualize the strength and decrease the development time of the bus before manufacturing and testing. In a quasi-static analysis, (Micu et al., 2014; Nurhadi & Zain, 2010) [12, 13] presents a study of quasi-static loading test on bus body sections by regulation of UNECE R66 via both FEM simulation (ANSYS) and experimental test. Likewise, (Nor & Baharin, 2014) [14] studied a quasi-static simulation test by ANSYS Explicit Dynamic Analysis using the standard of UN ECE R66. According to (Mahajan et al., 2003; Na et al., 2014) [15, 16], the bus structure resistance of rollover by experimental setup for the rollover test of bus section and numerical model with FEM (LS-DYNA solver). Moreover, (Bai et al., 2019; Phadatare, 2017; Rogov Petr Sergeevich & Orlov Lev Nikolaevich, 2015; Thosare & Patil, 2017; Zhou et al., 2019) [2, 17–20] studied the rollover of a bus frame (structure) analyzes by the experimental method and detailed FEM method. Also, (Wang, Pan, Zhang, & Cui, 2015) [32] discussed and analyzed the system's energy dissipation and the effects of energy-absorbing of the main structure in the rollover crash process using experiment tests and FE model. Again, (Yang & Deng, 2015) [21] analyzed and studied the structural optimization and lightweight of the bus body skeleton using the numerical Model by Hyperworks.

M. K. Mohd Nor & M. Z. Dol Baharin, 2014) [14] studied the numerical simulation (ANSYS) of a heavy vehicle bus structure according to FMVSS 220 & ECE R66 for quasi-static and rollover analysis. Lastly, (Karliński et al., 2014) [4] focused on the strength of the Volkswagen LT vans bus structure according to ECE R66 using a numerical model & simulation using FEM. (Korta & Uhl, 2013) [22] analyzed and studied multi-material optimization of bus structure using Genetic Algorithm (GA) optimization and numerical simulation for different materials. Similarly, (Reyes-ruiz et al.,

2013) [23] optimized and analyzed the passenger bus frame by using the design concept, numerical analysis, and simulation for different materials with FE Method on the parameters of strength (torsion, bending), dynamics (vibration response) constants and the thickness of bus frame structure. Then again, (Hu et al., 2012; Li et al., 2012a; Yusof & Afripin, 2013) [24–26] studied that bus superstructures undergo rollover events through experimental tests and numerical simulation with FEM. Lastly, (Su et al., 2011; Yusof et al., 2012) [3, 27] discovered the bus frame and validated it through a single box model experiment and FE analysis for the masses effects and bus structure strength.

Consequently, (Bojanowski & Kulak, 2011; Tech & Iturrioz, 2009) [28, 29] studied the bus structure using numerical simulation with FEM by LS-DYNA, Multi-Objective (MO) Optimization, and optimization by super-beam elements by plastic hinge lateral-base union of two samples. On the other hand, (Matolcsy 1997) [30] focuses on severe conditions and parameters in rollover accidents of bus structures through statistical and qualitative data analysis using standards and general testing methods. (R. P. Sergeevich & O. L. Nikolaevich, 2015) [20] focused on verifying and analyzing the bus body structure components in the rollover test by experimental and numerical crashworthiness investigations of the bus structure for each element. According to (Friedman et al., 2006) [31], the composite roof structures in transit buses are implemented and analyzed. (Lan et al., 2004; Cho Chung Liang & Le, 2009; Lin & Nian, 2006) [32–34] studied the design and analysis of the bus frame structure during a rollover by numerical simulation of bus structure sections and optimized through LS-OPT (successive_respond_surface_method (SRSM)). (Rahman, 2011) [35] The numerical modeling and analysis of the middle section of the bus structure are only designed and analyzed. According to (Park et al., 2006) [36], the beam's analytical and numerical Model and non-linear spring elements of the bus structure in a rollover are discussed. (Tech et al., 2007) [8] studied experimental and theoretical predictions of the collapse of basic bus structure with FEM (LS-DYNA) for plastic hinges by a super-beam element of the bus section in a rollover. (C. C. Liang & Nam, 2010) [37] studied the bus rollover protection using the numerical simulation using FEM (ANSYS) according to ECE R66 & FMVSS 220. (Subic & He, 1997) [38] discussed the experimental and analytical modal analysis of the bus roll-cage structure by an Alternative Research Approach. And also (Bojanowski et al., 2011; Valladares et al., 2010) [39, 40] studied the experimental setup and numerical Model of a paratransit bus for roof crush and rollover.

Accordingly, this study mainly focuses on the rollover crashworthiness analysis and optimization of the locally built midibus structure with the FE Approach according to UNECE R66 Standard. Foremost, the existing midibus structure is carefully studied using FE Method (LS-DYNA) according to the testing standard of UNECE R66. This approach visualizes which structure components lead to low strength and overweight. And then, two techniques of structural optimization were developed, structural modification (reinforcement) and numerical optimization (SRSM) in LS-OPT. Lastly, the comparison of the three models (baseline, reinforced (Model – I), and optimized (Model – II) model) was measured according to the structural strength and weight.

2 Methodology

In Ethiopia, the size of the midibuses depends on the chassis model types and specifications used by local manufacturers. Nevertheless, local manufacturers commonly used
ISUZU NPR 71K chassis to manufacture a midibus in Addis Ababa, Ethiopia [1]. Hence,
this study focuses on the locally built midibus (using ISUZU NPR 71K chassis model),
which has a twenty-nine (28 + 1) passenger capacity. And also, the mass of the unladen
kerb (M_k) and Gross Vehicle Weight (GVW) are 4500 kg and 7350 kg, respectively.
The existing (baseline) model of a locally built midibus structure has six parts: front,
roof, rear, floor, left, and right frame, as shown in Fig. 1. Therefore, the FE Model development consists of quasi-static and rollover crash analyses of the midibus structure.
LS-DYNA is the best and most efficient for explicit dynamic analysis because of using a
return mapping algorithm and the central difference method to avoid expensive numerical iteration and matrix inversion [41]. Accordingly, the rollover crash FE analysis with
the quasi-static FE analysis is developed using the explicit code of LS-DYNA R11.0 as
stated by the standard of UNECE R66.

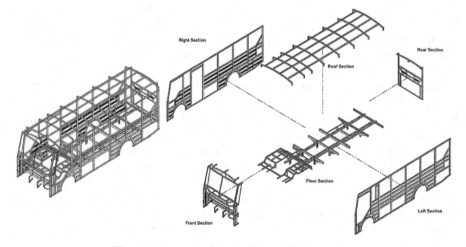

Fig. 1. Main parts of the baseline midibus structure

The material used for all parts (sections) of the bus frame is conventional structural
steel. Moreover, the material properties of conventional structural (CS) steel, such as
*density (kg/mm³), Yield Strength (MPa), Ultimate Tensile Strength (MPa), Elongation
(%), and Young's Modulus (MPa)* are *7850, 260, 360, 30, 210,* respectively. Consequently, the material input data of the FE Simulation needs the effective stress versus
the effective plastic strain curve, as shown in Fig. 2.

The input data of the material in numerical simulation (LS-DYNA) needs the effective
stress versus the effective plastic strain curve. The true stress (σ_t) and the true plastic
strain (ε_t) determined by:

$$\sigma_t = \sigma_{eng}(1 + \varepsilon_{eng}) \tag{1}$$

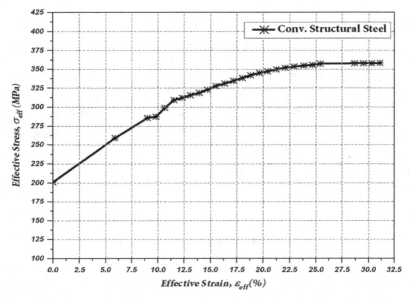

Fig. 2. Effective stress vs plastic strain curve for conventional structural steel

$$\varepsilon_t = \ln[1 + \varepsilon_{eng}] \qquad (2)$$

where: σ_{eng} - the engineering stress and ε_{eng} - the engineering strain.

2.1 Quasi-static and Rollover Simulation Using ECE R66

Procedures for the Quasi-static Analysis
The quasi-static simulation uses to check whether the bus sections and their bays withstand the rollover crash or not [42, 43]. Hence, the body sections and their bays evaluate whether they failed or passed the quasi-static loading test. Once studying structure roof-crush and buckling behavior, the LS-DYNA explicit time integration is the best to bring reliable results [44]. Thus, the quasi-static simulation procedures have been done according to UNECE R66 using explicit code in LS-DYNA Software to study and analyze the bus structure energy absorbing and load-deformation behavior. This paper develops the bus superstructure model as a shell element for fast and high computational simulation. The selected Belytschko-Lin-Tsay shell element is defaulted to calculate the shell element formulation in LS-DYNA R11.0. Moreover, this shell element has high computational efficiency compared to others [42, 45]. Furthermore, the residual space and rigid plate (impactor) model are developed for the quasi-static simulation, as shown in Fig. 3.

The FE model for quasi-static simulation is developed as a shell element (quadrilateral and triangular), as shown in Fig. 3. Also, due to the variation of frame size, the maximum and minimum element sizes of the bus body frame are 10 mm & 2.5 mm. Thus, the total quasi-static FE model consisted of 885,722 shell elements with 905,923

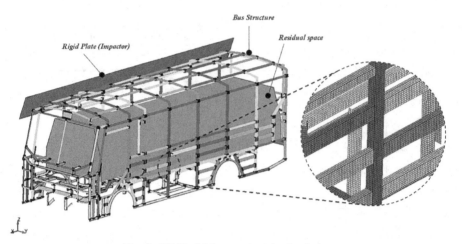

Fig. 3. FE Model for quasi-static simulation

nodes with an average element size of 15 mm (see Table 1). Hence, the weld formulation between all frame parts is modeled as spot welds (rigid nodes) through the 'CONSTRAINED_SPOTWELD' card without defining the failure force [26, 42, 46–48]. Experimental material properties of structural steel were executed into the FE models formulation. The PIECEWISE_LINEAR_PLASTICITY (MAT_24) material definition was implemented [42]. A rigid plate (the impactor) is developed by the material model of Rigid (MAT_20). A null (MAT_09) material is used to represent occupant space.

Table 1. Statistics of the existing (baseline) FE Model for the quasi-static analysis

Description	Bus structure	Residual Space & Impactor	Entire Model
Material type	Pieces-wise Linear Plasticity (MAT_24)	Rigid (MAT_20) & Null (MAT_09)	-
Min. element size (mm)	2.5	-	2.5
Max. element size (mm)	10	15	15
Number of parts	355	2	357
Number of nodes	572,578	333,345	905,923
Number of elements	553,040	332,682	885,722

Loading and Boundary Conditions. During the simulation, the rigid rectangular plate (impactor) touches the roof frames at an angle between the vehicle's longitudinal vertical center plane (VLCP) and the load direction [43]. A quasi-static load is distributed on the cant rail section with a rectangular plate (impactor). Thus, the quasi-static loading rate is applied quasi-equally for the 500 mm displacement with a short incremental time of 2.25 s. However, the computational time of the simulation takes 1.75 s to reach the passenger compartment. The external load card 'LOAD_BODY_Z' is applied as the gravity of the bus structure.

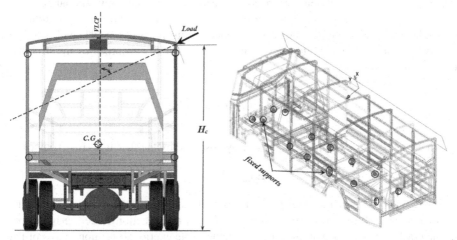

Fig. 4. Loading direction (left) and fixed supports (right) in the quasi-static simulation.

The fixed supports are causing no effect on the structure's deformation. These supports are applied to the underfloor structure section (see Fig. 4 (right)). The angle between the load direction & the vehicle's longitudinal vertical center plane (VLCP), (α) is determined by

$$\alpha = \frac{\pi}{2} - \sin^{-1}\left(\frac{800}{H_c}\right) \tag{3}$$

where: H_c- height of the vehicle's cant rail (in mm) from the horizontal plane (see Fig. 4 (left)).

The penalty-based contacts are contacts algorithms used in crash simulation [47]. The contact algorithm of 'AUTOMATIC_SINGLE_SURFACE' is appropriate for self-contacting cases. This contact type was used to state the contact between all bus section components. The contact 'AUTOMATIC_SURFACE_TO_SURFACE' is defined as the contact relationship between the bus section and a rigid plate (impactor). The static and dynamic friction coefficient for steel-to-steel contact is a value of 0.15 and default, respectively [49].

Evaluation Criteria for Quasi-static Simulation Results. According to the UNECE R66 standard, the minimum energy absorbed by the structure (body sections) (Emin) is equals to the sum of the energy of the ith bay and calculated by:

$$E_{min} = \sum_i^s E_i = E_T \frac{\sum_i^s m_i}{M} \tag{4}$$

Then, the total absorbed Energy (E_T) by the vehicle is calculated by:

$$E_T = 0.75 \, Mg \, \Delta h \tag{5}$$

where: $M(M_k)$ – the unladen kerb mass of the vehicle, g – the gravitational constant (9.81 m^2/s), Δh – the vertical distance of the vehicle center of gravity in the rollover test, - the energy absorbed by the "i^{th}" bay, and m_i – the mass of the "i^{th}" bay.

By substituting Eq. (5) into Eq. (4), the minimum energy absorbed by the structure (E_{min}) is determined by:

$$E_{min} = 0.75 \, g \, \Delta h \sum_i^s m_i \tag{6}$$

In the quasi-static loading test, the energy absorbed by the structure ($E_{st,a}$) passes if:

$$E_{st,a} \geq E_{min} \tag{7}$$

Otherwise, the structure fails the tests, even if only one of the bays is touched the residual space. Moreover, energy absorption ($E_{st,a}$) and reaction force are indicators of the crashworthiness capability of the structure. The energy absorption ($E_{st,a}$) of the structure can be determined by integrating of the load-displacement curve. Furthermore, it can be formulated by:

$$E_{st,a} = \int_0^l P d\delta \tag{8}$$

where: P – applied load (reaction force), l – length of the crushed structure, and δ – displacement.

The specific energy absorption (SEA) is also vital to the design of structural parts that involve the reduction of weight of the structure. Therefore, the Specific Energy Absorption capacity define as the energy absorbed per the mass of the structure [50] and determined by

$$SEA = \frac{E_{st,a}}{m} \tag{9}$$

where: m – mass of the structure.

Procedures for Rollover Analysis.
According to the UNECE R66 standard, the rollover test of a vehicle is used to evaluate the crashworthiness capability and occupant safety during rollover crashes. The rollover crashworthiness via finite element analysis (FEA) is extensively done due to the experimental test's long time and extreme cost [2]. In the bus rollover crash, passenger safety is affected by structure, seats, and seat belt strength [51]. In this study, the bus structure and seat frames are the main components of the bus to study the strength and crashworthiness behavior using LS-DYNA with the explicit time integration code. Moreover, the tare-weight load (Unladen Kerb Mass (M_k)) of the locally built midibus (NPR 71K

Table 2. Components of Unladen Kerb Mass (M_k) of NPR 71K Chassis midibus

Items	Qty	Total mass (kg)
Seat (each 8.5 kg)	29 + 1	255
Self-weight of the structure	–	577
Chassis body with fuel tank, engine, and battery, wheel, axle (NPR 71K)	–	3668
Unladen Kerb Mass (M_k)		4500

Chassis midibus) is 4500 kg (4.5 tons). Table 2. Illustrates the quantities and masses of components in the case of tare weight loading.

In this study, the bus skin, glasses, and other sensitive parts of the vehicle are not modelled due to the difficulty of modelling and simulation. However, the structure and seat are developed by direct measurement from local manufacturers (bodybuilders). First, the chassis components are modeled using the ISUZU N-series body builder manual and guide [52–54]. However, the structure and seat are developed by measuring and observing the bus construction from the available local manufacturers (bodybuilders). Then, the bus structure, chassis, seat frame, tilt platform, and other assembled components were imported to the LS-Prepost as an Initial Graphics Exchange Specification (IGES) file format to develop a finite element mesh.

As shown in Fig. 5, the main parts of the bus structure, such as the cant-rail, window rail, waist rail, A & B pillar, vertical pillars, and skirt rail, are mentioned. Moreover, the structures' eight bays (B1–B8) are also arranged.

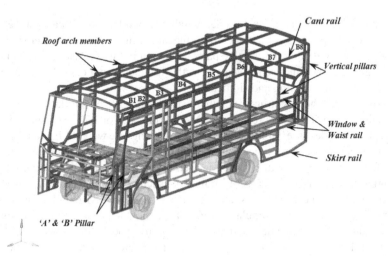

Fig. 5. Components of existing bus body structure

To decrease the computational time of the simulation, the entire rollover model is developed as a Belytschko-Lin-Tsay shell element. However, the thickness of the shell element is defined by its section properties and values. Hence, the weld connection between all structural parts with seats frame and rail is modeled as spot welds (rigid nodes) using a card 'CONSTRAINED_SPOTWELD' without defining the failure forces and coefficients [26, 42, 46–48]. The keyword 'CONSTRAINED_RIGID_BODIES' defines the assembly between the rigid parts of the chassis by assigning one of them as a master part and merging the others [55].

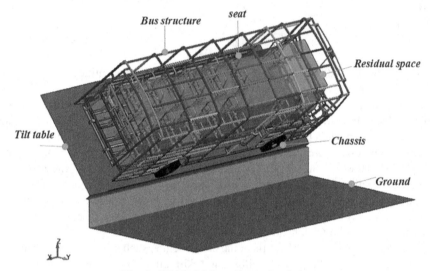

Fig. 6. FE Model for rollover simulation

The bus structure, chassis, seat frame, tilt table, ground, residual space, and element mass of other components are considered for the tare-weight vehicle rollover simulation without bus skin and other deformable parts, as shown in Fig. 6. Also, in this study, the bus skin, glasses, and other sensitive parts of the vehicle are not considered because of their difficulties in modeling and simulation.

In this section, the bus structure element size is constructed similarly to the quasi-static analysis case. The entire rollover FE Model consisted of 1,771,305 shell elements with 1,790,729 nodes (see Table 3.). The overall existing tare weight FE Model contains 412 parts. Hourglass control type (type 4, 5) defines ranges of coefficient between 0.03 to 0.05 for parts of the structure, which is used to reduce the response of non-physical stiffening [56]. The shell elements of FE models, such as fully integrated (type 16) and Belytschko-Tsay (type 2) elements, are the most accurate and have high computational efficiency in crashworthiness simulation [46]. Thus, this study also conducts Belytschko-Tsay (default type) shell element and hourglass control type 4 with a coefficient of 0.05 for all FE rollover models. The computational time decrease when the entire rollover model is developed as a Belytschko-Lin-Tsay shell element.

Table 4 describes the mesh quality of the rollover FE model. The numbers of quadrilateral and triangular elements are 1,757,153(99.2%) and 14,149(0.799%), respectively. Hence, the weld connection between all structural parts with seats

Table 3. Statistics of baseline FE model for rollover simulation

Parameters	Bus structure and seat frame	Tilt table, residual space & ground	Chassis body	Entire model
Min. element size (mm)	2.5	–	–	2.5
Max. element size (mm)	10	15	30	30
Number of Parts	391	3	18	412
Number of Nodes	904,442	818,935	80,076	1,790,729
Number of Elements	885,012	805,934	88,718	1,771,305

frame and rail is modeled as spot welds (rigid nodes) using a card 'CON-STRAINED_SPOTWELD' without defining the failure force [26, 42, 46–48]. The keyword 'CONSTRAINED_RIGID_BODIES' defines the assembly between the chassis' rigid parts by assigning them as a master part and merging the others [55]. The material properties of the rollover structure are similar, as mentioned in the quasi-static analysis. The material model used for all structural frames and seat rail is executed by using PIECEWISE_LINEAR_PLASTICITY (MAT_24) material definition [4, 34, 42, 49]. In most rollover impacts, the chassis parts are in motion but not directly affected by the crashes [57]. Hence, the chassis, tilt table, and ground are developed by rigid material models (MAT_20). A Null (MAT_09) material is used as a symbolic representation for occupant space [58].

Table 4. Summary of shell element quality report for rollover FE model

Criteria	Allowable (threshold) value
Min side length (mm)	3
Max side length (mm)	30
Aspect ratio	10
Warpage	10
Min quad. angle (deg)	45
Max quad. angle (deg)	135
Min tria. angle (deg)	30
Max tria. angle (deg)	120
Taper	0.7
Skew (deg)	45
Jacobian	0.6
#Quads (%): 1,757,153 (99.2%), #Trias (%): 14,149 (0.799%)	

In the rollover simulation, 'ELEMENT_MASS_NODE_SET' is used to lump the mass of passengers, luggage, battery, and engine [46, 59, 60]. In this study, the mass element of the passengers is assigned to the seat frame node. And also, the mass of luggage, battery, fuel tank, engine, and other miscellaneous parts are equally distributed on the chassis and structural components of the bus. During the rollover simulation, the complete vehicle is initially tilted in an unstable (equilibrium) position and a ditch of 800 mm [43]. The angle of the tilt table shall be greater than 35° and the trial & error simulation is needed to obtain the minimum tilt table angle that is used to trigger the vehicle to tip over [48, 61]. Hence, the minimum angle of the tilt table is 49° in an unstable equilibrium position. The bus's initial angular velocity should not exceed 0.0875 rad/s (5 deg/s) [43]. The external load card 'LOAD_BODY_Z' is applied as the gravity of the bus [58]. Furthermore, A card 'INITIAL_VELOCITY_GENERATION' defines the initial angular velocity of the bus [62].

The contact algorithm 'AUTOMATIC_ SURFACE_TO_SURFACE' defines the contact relationship between the structure & ground. Moreover, the coefficient of friction for steel-to-concrete contact is 0.65 [17, 42, 63, 64]. The contact definition between tilt table & tires (rubber-to-steel) is developed using 'AUTOMATIC_NODE_TO_SURFACE' [36]. In this contact, the value of the coefficient of friction is 0.7 [42]. The self-contact between bus body frames (bus section components) and seats (steel to steel) define using 'AUTO-MATIC_SINGLE_TO_SURFACE', with its coefficient of friction is 0.15 [42]. Moreover, The Switch 'DEFORMABLE_TO_RIGID_AUTOMATIC' was highly recommended for flexible switch activation and easy use application. This switch is commonly used for a component to switch deformable as well as rigid automatically by the change of contact surface force or rigid wall force [17, 65, 66] hence, the rollover simulation of a vehicle needs to change from deformable material to rigid material or vice versa using a 'DEFORMABLE_TO_RIGID_AUTOMATIC' switch. The first switch activates all model parts, from deformable to rigid. The second switch activates only the structure and seat with seat rail to deformable material. The two switches are paired and related to each other to switch back and forth using a contact force automatically.

The overall rollover simulation (Elapsed) time took over two days (48 h) by using the Intel CORE (R) i7-7700HQ CPU @ 2.80 GHz processor, depending on the initial and finish contact time. For accurate energy distribution with stable models during simulation, the option of mass-scaling was considered to regulate the time step (DT2MS) [57]. During the tare–weight rollover case, the time step is reduced from $1.00e-5$ to $1.00e-6$ s between simulation time of 1.62–2.25 s. This rollover procedure is used to minimize computational time [42]. However, the computational time of the simulation takes 2.25 s until the vehicle reaches motionless.

FE Verification of Energy Balance in Rollover simulation. The numerical simulation error assesses by the task of the model verification of the FE model [46]. In the rollover simulation, checking the energy balance is the main factor in assessing the solution's errors. this verification guidelines presents the concept of energy conservation laws [42, 46, 67]. The applied total energy of the structure can be determined as stated by ECE

R66 [43]:

$$E_{Total} = 0.75 \, Mg \, \Delta h \text{ or}$$

$$E_{Total} = 0.75 \, Mg \left(\sqrt{\left(\frac{W}{2}\right)^2 + H_0^2} - \frac{W}{2H}\sqrt{H^2 - (800)^2} + \frac{(800 * H_0)}{H} \right) \quad (10)$$

where: W – the overall width of the bus, I – the height of the bus, and H_o – the height of the center of gravity.

The LS-DYNA total energy (E_{Total}) is the sum of six components such as current internal energy (E_I), current kinetic energy (E_K), current sliding energy (E_{Sli}), current hourglass energy (E_{HG}), current system damping energy (E_D), and current stonewall/rigid wall energy (E_{RW}) and also must be equal with the sum of initial total energy ($E_{Total,o}$) & work done by external loads (W_{Ext}) [68, 69]:

$$E_{Total} = E_K + E_I + E_{Sli} + E_{RW} + E_{Damp} + E_{HG} = E_{Total,0} + W_{Ext} \quad (11)$$

$$E_{Total,0} = E_{K_0} + E_{I_0} \quad (12)$$

where: E_{I_0} – the initial internal energy & E_{K_0} – the initial kinetic energy

Angular Deformation Index (DI_θ) is a quantitative measure used to evaluate a margin of safety and a deformation extent in rollover simulation [42, 46]. Thus, this index also measures the structure's strength during rollover crashes. Figure 7 shows the angle between wall & floor ("3), angle of waist rails (θ_2), and angle between roof & wall (θ_1) with a measured angular deformation. The Angular Deformation Index (DI_θ) can be determined as [46]:

$$DI_\theta = \frac{l}{400} \tan(\Delta\theta_3) + \frac{(1250 - l)}{400} \tan(\Delta\theta_2) \quad (13)$$

where: $\Delta\theta_3$ – the angle changes between wall & floor, $\Delta\theta_2$ – the angle changes of waist rails and l – the distance from the floor to waist rails.

2.2 Structural Reinforcement and Optimization of the Structure

Optimization is the development of objective functions with constraints to maximizing or minimizing the desired value [70]. In the computer-based optimization process, the simulation requires a high number of iterations, the performance of computers, and large numbers of design parameters [46]. However, reinforcement optimization is a technique to strengthen the vehicle's body [1, 71]. Therefore, this study's primary goal is to specify the two alternative design solutions. However, this process was done by use of practical reinforcement (model – I (RD)) and numerical optimization technique using the Successive Response Surface Method (SRSM) in LS-OPT (Model – II (SRSM)) for dynamic (rollover) conditions.

Structural Reinforcement Design.
Most local bus manufacturer experiences the design of the structure's construction

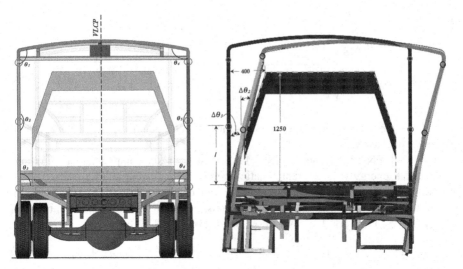

Fig. 7. The concept of deformation angle and angular deformation index [46]

through the try and error methods, this approach outcomes in low structural strength and safety of the passengers during accidents. The existing bus structure needs a structural design analysis and improvement to maximize the strength and minimize the structure's weight. First, the existing bus structure carefully examines using the manufacturer constraint and the dynamic rollover simulation to select the component's layout and cross-sections that might be changed. When these components lead to low strength, the change in in the cross-section and layout of structural components were measured. As shown in Fig. 8, the most local bus bodybuilder used the wall support members as support and cover of the walls. However, Due to high number of wall support, the weight of the bus structure was increased. Even though, these components do not strengthen the bus structure in rollover conditions due to their arrangements and cross-sections.

In this study, the design modification (reinforcement) conducts the addition and replacement of the structural components. First, the layout of the structure and the shape of the cross-section in the structural elements are changed. Next, the supports were added to assemble the section of the structure. As a result, the rectangular (lying orientation) profile has a much higher energy absorption for oblique impact load directions than the square cross-section [72]. The other important consideration for the rollover case is the strength of the floor-wall connection. Hence, the supporting member attaches at the corner of the floor and sidewall section on the front, medium, and rear portions to reduce the structural deformation.

Accordingly, the baseline design defects improve through additional support and change the layout and cross-section of the structure, as shown in Fig. 9. Therefore, the new modified (reinforced) midibus structure has been considered the following structural design modifications:

- Roof arc members to strengthen the roof -parts using rectangular cross-section (RHS)

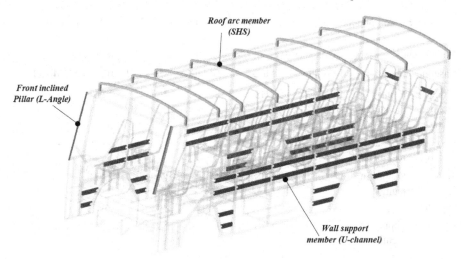

Fig. 8. Selected Configurations to improve the baseline design

- Inclined connecting elements (Supporting members – #1) at the left & right sections of frames
- Supporting members from #2 to #4 developed to build a connection between roof and sides pillars with rear pillars, Support for extended floor section & chassis, and Support for connection side walls and floor during the rollover, respectively
- Inclined front (A) pillar at the front section of the structure (square cross-section (SHS)) to strengthen the front pillar

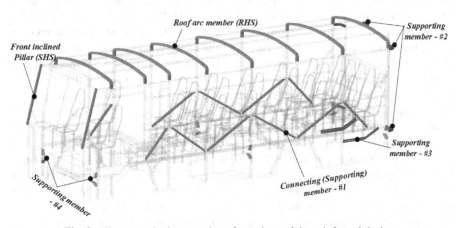

Fig. 9. Illustrates the improved configurations of the reinforced design

The comparison of masses of the components of baseline and reinforced configurations is described, as shown in Table 5. In sum, using this reinforcement design, the baseline model decreases its weight from 577.13 kg to 547.15 kg (29.98 kg).

Table 5. Components with its mass of baseline and reinforced (Model - I) configuration

Component		Cross-section		Quantity		Mass (kg)	
Baseline	Model - I	Baseline	Model - I	Baseline	Model - I	Baseline	Model - I
Front inclined pillar	Front inclined pillar	L-angle	SHS	2	2	2.26	4.53
Roof arc member	Roof arc member	SHS	RHS	8	8	36.06	44.74
Wall support member	Supporting member – #1	U – channel	SHS	44	11	77.11	19.7
–	Supporting member - #2	–	SHS	–	5	–	2.58
–	Supporting member - #3	–	L-Angle	–	2	–	7.86
–	Supporting member - #4	–	RHS	–	12	–	6.12
Total mass of the parts						115.43	85.53

Optimization Method via Successive Response Surface Method (SRSM)

Optimization methods using Response Surface Method (SRSM) are commonly used in design optimization of crashworthiness [34, 42, 73, 74]. Moreover, the main aim of this optimization is to maximize the reinforced structure's strength within a maintained level of weight during the rollover case. However, the analysis was carried out using LS – DYNA and LS-OPT (successive response surface method (SRSM)). Furthermore, the optimization task was done to identify the significance of the section of the reinforced structure in response to the loading experienced in ECE – R66 quasi-static testing procedures. From Eq. 6, the minimum requirement of absorbed energy of the reinforced design (Model – I) was equal to 4.86 kJ. Moreover, it is a fact that the highest value of energy absorption indicates a better strength of the structure. Consequently, Multi-objective optimization was done by reducing the total mass of the selected parts (objective function) and keeping the structure's energy absorption above 4.86 kJ (constraint). Thus, the energy absorption capacity should be greater than or equal to 4.86 kJ because the energy absorption capability is one of the parameters that indicates the structure's strength level. The energy absorption response was specified on the history output by integrating the structure's reaction force vs displacement. The interactive approach used to classify a single best compromise solution by objective and constraint function [70]:

$$\text{minimize } f_1(t)$$
$$\text{subject to } f_2(t) \geq c \qquad (14)$$
$$\text{and } t \in \Omega$$

where: Ω – sets of the lower and the upper bounds on inputs variable ($t_L \leq t \leq t_U$) and c – the bounds of the outputs ($c_{min} \leq c \leq c_{max}$).

And the only difference was that the constraint of this optimization is absorbed energy response. Furthermore, it can be expressed as:

Minimize the total mass of selected parts

Subjected to: Absorbed Energy(E_{st}) ≥ 4.86 kJ

Figure 10 displays the selected parts of the structure for optimization tasks. In the design of experiments (DOE), The design variables are the thickness of the parts. First, these parts are grouped based on their assembly section. Then, the initial thickness values for these parts and the manufacturable lower and upper bounds (ranges) in the experiment design are listed, as shown in Table 6.

Table 6. The selected part's thickness and their range of values from reinforced structure

Sections	Symbols	Cross-section	Initial thickness (mm)	Ranges of mfg. values (mm)
Roof	troof	RHS & SHS	1.5	1, 1.5, 2, 2.5, 3
Rails	trail	SHS	1.5	1, 1.5, 2, 2.5, 3
Pillars	tplr	RHS	1.5	1, 1.5, 2, 2.5, 3

Figure 11 displays the optimization process to minimize the total mass of the structural member in the three thickness parameters (trail, troofa, and tplr) using LS-OPT. First, the three parameters are assigned for all selected parts in the Finite Element (FE) quasi-static model using LS – DYNA.

Moreover, it is imported to the optimization tool of LS-OPT. Then, linear ordered polynomial sampling is selected to identify the seven simulation points using D – optimal selection using LS-OPT. Next, the response of mass parts, internal energy, reaction force, and absorbed energy was formulated. Moreover, the output history of reaction force, displacement, and force vs. displacement plot were defined. Lastly, the objective and constraint of the optimization process was developed using a Genetic Algorithm for two iterations. The overall optimization iteration was performed within the linear response surface approximation. These approaches requested seven jobs per iteration. This means the optimization runs 14 + 1 jobs for all quasi-static simulations. Thus, the optimization approach using successive response surface methods took over ten days (238 h) for all simulations by Intel® Core™ i7-7700HQ CPU @ 2.80 GHz processor.

Study of Sensitivity via ANOVA and Global (GSA/Sobol)

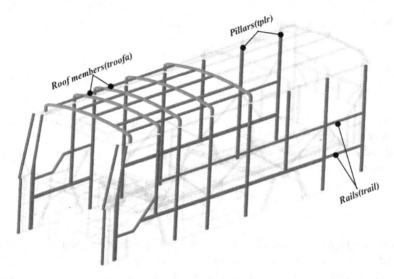

Fig. 10. Selected parts of the reinforced structure for quasi-static case optimization.

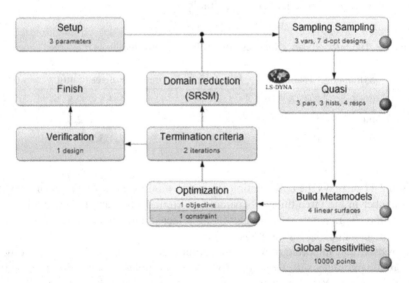

Fig. 11. Flow chart of the structural optimization of the reinforced structure in quasi-static simulation via LS-OPT

Figures 12 and 13 shows sensitivity analysis results for the total mass of the parts and responses absorbed energy by Global Sensitivity Analysis (GSA) with Sobol's and Analysis of Variance (ANOVA) approach. The trails and troofa are the minor influence variables for the mass parts. However, the thickness of the vertical pillars (tplr) is the essential variable, and it has a 95% confidence interval of the response function. Moreover, the roof members also do not influence the response of absorbed energy, as shown in Fig. 13.

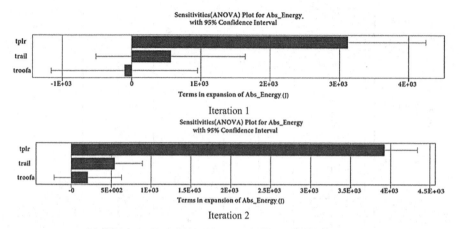

Iteration 1

Iteration 2

Fig. 12. Sensitivity (ANOVA) plot for the response of absorbed energy for all iterations

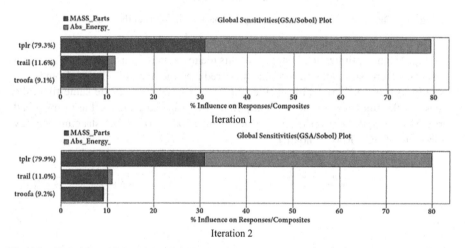

Iteration 1

Iteration 2

Fig. 13. Global Sensitivity (GSA/Sobol) plot for the response of mass of the parts and absorbed energy for all iterations

Figure 14 displays the tradeoff between the absorbed energy and total mass of the selected parts in all iterations. As shown in the first iteration of the design space, responses and design points have resulted in a most extensive and widely spread. Thus, the more focused design points are identified to obtain a new optimum result. During metamodel adequacy, when determining the best quality of the fit for responses, the error of root mean square (RMS) and the square (R^2) should be small and higher, respectively [46, 75]. Thus, the perfect fit ($R^2 = 1$) was found in both iterations for the mass parts response.

However, the R^2 of the absorbed energy response in the first and second iterations equals 0.987 and 0.999, respectively. This result denotes an almost perfect fit for all iteration.

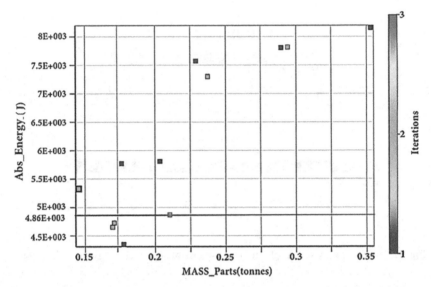

Fig. 14. The response of absorbed energy vs mass of the parts at the design variables

In Fig. 15, the optimization history presents the response of mass parts and absorbed energy. The decreasing rail and roof thickness reduced the mass of all parts from 0.176 tons (176 kg) to 0.145 tons (145 kg), equal to 17.6% (31 kg). At the same time, the increase of the thickness in the vertical pillars (*trail*) increases the value of absorbed energy. Moreover, it improves the energy absorbing capacity of the bus structure (energy requirement of ECE R66 Standard).

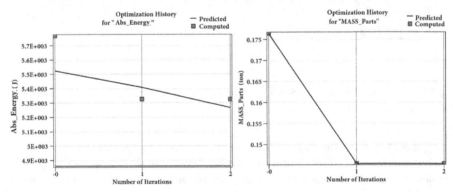

Fig. 15. Optimization history of the absorbed energy of the models (left) and the mass of the parts (right)

3   Result and Discussion

The reaction force and absorbed energy parameters were determined during a quasi-static simulation. The contour of stress specifies the maximum stress present at the contact area of impactor (rigid plate) loading, the joint areas of the vertical pillar-roof, and the vertical pillar-floor section, as shown in Fig. 16. A vertical pillar is highly intruded into residual space (RS) from the baseline simulation, as shown in Fig. 16 b). During the quasi-static simulation, the impactor load gradually increased until it touched the residual space. Perhaps, At the displacement of 375 mm, the simulation was motionless. Although the peak reaction forces for each model change differently based on the load resistance capacity. Figure 17(a) displays the reaction force developed on the baseline model, Model – I, and Model – II are 14.5 kN, 17.7 kN, and 16.3 kN. In addition, only the reinforced force converged after the displacement of 280 mm.

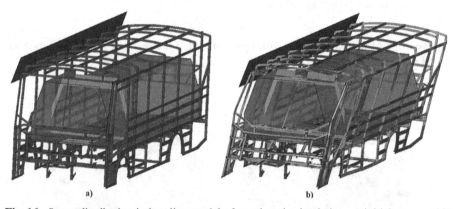

Fig. 16. Stress distribution in baseline model of quasi-static simulation: a) initial stage and b) final deformed condition

Figure 17(b) describes the minimum requirement of absorbed energies and energy absorbed by the structure (E_{st}) of the three structure models. The Absorbed energy of the baseline model, Model – I (RD), and Model – II (SRSM) are 4.43 kJ and 5.67 kJ, and 5.32 kJ, respectively. Accordingly, the energy-absorbing capability of the bus structure ($E_{st,ext}$) is lower than the minimum value of energy absorbed by the existing bus structure (E_{min}). This result also implies that the existing (existing) bus structure fails the tests cause one of the bays is touched the residual space. However, the reinforced energy absorption capacity ($E_{st,RD}$) is greater than the minimum requirement energy-absorbing structure ($E_{min,RD}$). Therefore, it was found that the reinforced structure passed the tests.

Moreover, model – II (SRSM) also passed the standard requirement. Notably, the reinforced Model (Model – I) is stronger than the baseline and Model – II (SRSM) models. This important finding is undoubtedly an equivalent method for the rollover test, as stated by UNECE R66.

Furthermore, all the models for the quasi-static analysis, shown in Fig. 18, follow the same deformation pattern when the cages' deformation is reached in a residual space.

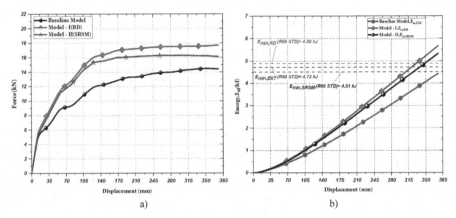

Fig. 17. Quasi-static simulation results for all models; a) Force versus displacement curves and b) Energy Absorption vs displacement curves

However, one of the baseline structural parts intruded the residual space, as shown in Fig. 18 a).

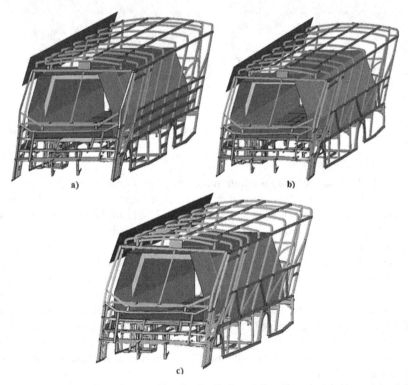

Fig. 18. Deformed models after quasi-static simulation: a) baseline model; b) model – I (RD); and c) model – II (SRSM)

The comparison of baseline and two alternative solutions using specific energy absorption capacity vs mass of the models were explained, as shown in Fig. 19. Thus, The Specific energy absorption of the Baseline model, Model – I, and Model – II are 7.68 J/kg, 10.36 J/kg, and 10.31 J/kg, respectively. This result shows that Model – I has a better energy absorption capacity than others. However, Model – II is more effective for better energy absorption capacity and less structure weight than others.

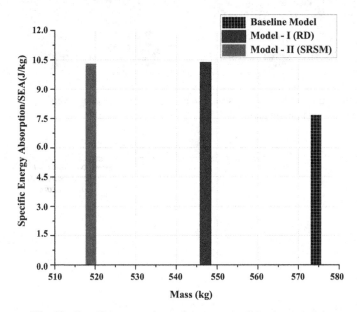

Fig. 19. Specific energy absorption vs mass of the three models

In a baseline (existing) tare–weight rollover case, the value of the internal energy of structure and seats are estimated. Significantly, the internal energy of the structure's main components and bus sections compared to visualize their capacity throughout the simulation. During the tare–weight rollover case, the maximum deformation of the existing structure and seat frame was located at pillar A and bays (B1 – B3), as shown in Fig. 20. Additionally, the high deformation of seats presents at the first and second seats of the passenger.

In the tare-weight rollover, the structure and seat's internal energies are 23.3 kJ and 0.85 kJ, respectively, as shown in Fig. 21. This result shows the internal energy of the structure is higher than the internal energy of the seat frames. Moreover, this finding shows that when the absorbed energy of the frame is higher, the structural part of the midibus is the most significant for the strength of the rollover case. A similar conclusion was reached by (Cezary Bojanowski, 2009), [46], where the author showed the effect of skin parts on the strength of a bus in a rollover crash test.

Although, the maximum internal energy for six sections of the structure is displayed, as shown in Fig. 22 (left). Accordingly, the minimum and maximum internal energy of 2.25 kJ and 6.56 kJ were obtained at the roof and right section of the bus structure because

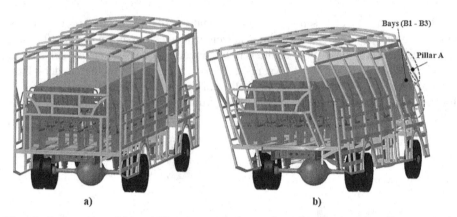

Fig. 20. Deformation of the existing structure and seats frame in tare – weight rollover case: a) initial phase and b) maximum deformation

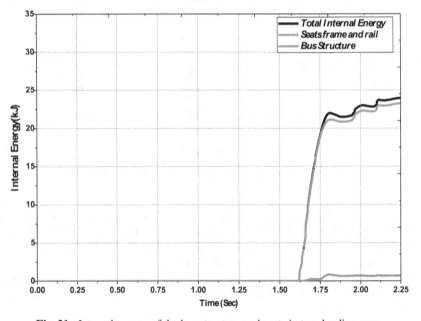

Fig. 21. Internal energy of the bus structure and seats in tare-loading cases

they are highly deformed until the rollover crash is stopped, leading to more energy absorption. Furthermore, the roof and front section have the lowest energy absorbing capacity because the rollover crash is not directly affected.

The sum of each component's internal energy contributes to the overall energy capability of the structure after the crash. In addition, Fig. 22 (right) shows the internal energy of each component. The absorbed energy of the components differs depending on the deformation throughout the crash. Hence, the internal energy of the A & B pillar, roof

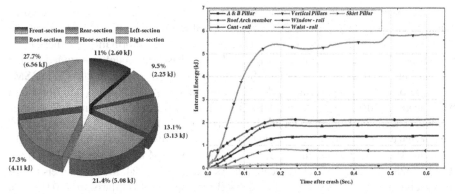

Fig. 22. Internal energy of the sections (left) and the components (right) of the baseline model in tare-weight rollover case

arc member, vertical pillars, window-rail, waist rail, and skirt pillar are 1.42 kJ, 2.15 kJ, 1.90 kJ, 5.85 kJ, 0.15 kJ, 0.83 kJ, and 0.19 kJ, respectively. This result shows that the lower and greater internal energy occurred at the window-rail and vertical pillar. Due to the high impact on the roof arc members and vertical pillar, the absorbed energy of these components is high compared to other components.

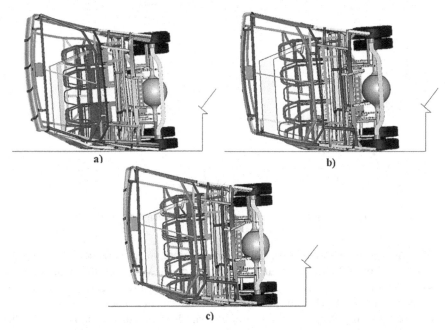

Fig. 23. Comparison of deformation in the tare-weight rollover: (a) Baseline model; (b) Model – I (RD); and (c) Model – II (SRSM)

In this section, angular deformation index and internal energy parameters are estimated to compare the crashworthiness capacity of the baseline model and alternative solutions during the tare-weight rollover scenario. The final deformations of these three models are displayed, as shown in Fig. 23. First, a baseline model is intruded on the residual space and is highly deformed (See Fig. 23 (a). Furthermore, model – II (SRSM) is less deformed than a baseline model. However, Model – I (RD) is less deformed than baseline and Model – II. Next, a model – I (RD) is less deformed than all models, and the residual space is far from the Model's frame, as shown in Fig. 23(b). Figure 24 displays the internal energy of the baseline model and two alternative designs in the tare – weight scenario. Thus, the internal energy of the baseline model, Model – I, and Model – II are 24.15 kJ, 28.3 kJ, and 25.7 kJ, respectively. As shown in Fig. 24, the internal energy of the baseline model, Model – I, and model – II are converged similarly from 0.0–0.2 s. The above results show that model – II (SRSM) has less weight with enough energy absorbing capacity than the baseline model.

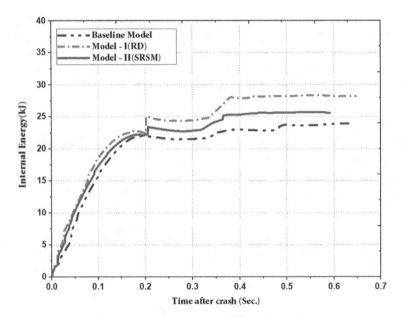

Fig. 24. Comparison of Internal energy between three models

The Angular deformation index (DI_θ) is another parameter to identify the crashworthiness capability of the structure during a rollover crash. Moreover, the angular deformation index of the structure response was measured based on angles between nodes with time in LS-DYNA. The rating of the angular deformation index (DI_θ) indicates the strength of structure during structural deformation. Therefore, this section evaluates the angular deformation index to compare the three models in tare-weight rollover simulation. The maximum DI_θ of the baseline model, Model – I and Model – II are 1.07, 0.70, and 0.78, respectively, as shown in Table 7. These maximum deformation indexes are located at pillar A with bays (B1–B3) for all models in the tare-weight rollover scenario except Model – III (bays (B4–B8)). As depicted in Table 7, the change of plastic hinge angles and the maximum angular deformation index at the pillar and bays of the structure are described in the tare-weight scenarios for all three models.

Table 7. Comparison of deformation index among three models in tare-weight scenario

Change of angles (deg) and DI	Pillar A & bays (B1–B3)			Bays (B4–B8)		
	Baseline	M- I (RD)	M-II (SRSM)	Baseline	M- I (RD)	M-II (SRSM)
$\Delta\theta_1$	−32.9	−19.6	−22.7	5.8	2.2	2.6
$\Delta\theta_2$	2.6	8.4	12.0	0.6	0.5	0.5
$\Delta\theta_3$	17.5	7.6	6.8	−17.8	−8.4	−10.7
$\Delta\theta_4$	26.4	18.2	21.9	18.8	−12.3	−20.4
$\Delta\theta_5$	6.5	9.4	13.1	21.4	11.9	21.5
$\Delta\theta_6$	−10.9	−3.9	−2.8	5.9	3.2	3.1
DI_θ	1.07	0.70	0.78	0.69	0.40	0.59

Figure 25 compares the deformation index vs time curves of the baseline model and alternative models in the tare – weight rollover case. In this scenario, the baseline structure has an unacceptable and poor strength. This strength describes that the entire structure failed due to high deformation at pillar A and bays (B1-B3). These results imply that the total structural strength of the baseline model is weak to survive in this rollover crash case. Moreover, both Model – I (RD) and Model – II (SRSM) have acceptable strength in tare–weight cases.

4   Conclusion

This research paper also facilitates the rollover crashworthiness capability of the structure in quasi-static loading and tare-weight rollover scenarios. In addition, the two alternative designs for rollover were conducted using reinforcement design and optimization by successive response surface method using LS-OPT for all quasi-static and rollover analyses. From both analysis and optimization results, the following conclusions are mentioned:

- The energy absorption of the existing (baseline) model is less than the minimum requirement energy by ECE R66 in quasi-static simulation, which means the existing bus structure fails the rollover tests due to one of the bays is touched the residual space. However, all design alternative solutions pass the test because of fulfilling the requirement of the standard.
- During the tare-weight rollover case, the baseline structure and seat frame contain 96.5% and 3.5% of internal energy, respectively. The entire structure failed due to

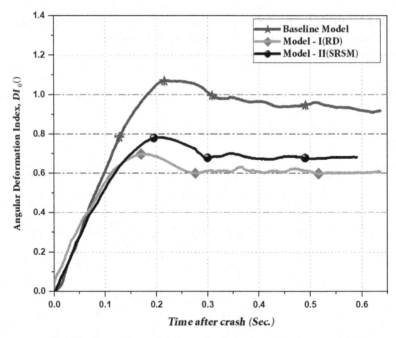

Fig. 25. Comparison of deformation index between three models

high deformation at pillar A and bays (B1–B3). Moreover, the lower and greater internal energy occurred at the skirt and vertical pillars. As per the tare – weight rollover analysis and optimization results of baseline and two alternative models, the internal energy and angular deformation index (DI_θ) are the main parameters to identify the crashworthiness capacity.

Generally, it can be determined that the baseline model, Model – I (RD), and Model – II (SRSM) have unacceptable, acceptable, and intermediate strengths, respectively. Therefore, the first design approach is that the reinforced design experiences sufficient strength by adding the support and change of cross-section on the front pillars. Moreover, a model – II (SRSM) has less weight with adequate energy absorbing capacity than a reinforced model by varying the thickness of pillars, windows, and waist rails. Moreover, a model – II (SRSM) has less weight with adequate energy absorbing capacity than a baseline model.

References

1. Addisu, H.S., Koricho, E.G.: Structural weight and stiffness optimization of a midibus using the reinforcement and response surface optimization (RSO) method in static condition. Model. Simul. Eng. **2022**, 1–15 (2022)
2. Bai, J., Meng, G., Zuo, W.: Rollover crashworthiness analysis and optimization of bus frame for conceptual design. J. Mech. Sci. Technol. **33**(7), 3363–3373 (2019)

3. Bin Yusof, M., Amirul, M., Bin, A.: Effect of mass on bus superstructure strength having rollover crash. Int. Sci. Index **6**(8), 1443–1449 (2012)
4. Karliński, J., Ptak, M., Działak, P., Rusiński, E.: Strength analysis of bus superstructure according to Regulation No. 66 of UN/ECE. Arch. Civil Mech. Eng. **14**(3), 342–353 (2013)
5. Tulu, G.S., Washington, S., King, M.J.: Characteristics of police-reported road traffic crashes in ethiopia over a six year period. In: Proceedings of the 2013 Australasian Road Safety Research, pp. 1–13 (2013)
6. UNECE 2004: Statistics about rollover accident of buses – VI. Hungary (2004)
7. UNECE 2020: Road Safety Performance Review – Ethiopia. United Nations Economic Commission for Africa & Europe, Geneva **20** (2020)
8. Tech, T.W., Iturrioz, I., De Meira Júnior, A.D.: Numerical simulation of bus rollover. SAE Tech. Pap. (2007)
9. Iozsa, M. D., Micu, D. A. N. A., Frățilă, G.: Influence of crash box on automotive crashworthiness. Recent Adv. Civ. Eng. Mech. 49–54 (2014)
10. Series, N.A.S.I., Base, N.: Crashworthiness of Transportation Systems: Structural Impact and Occupant Protection (1997)
11. Cezary, B. Jerry, W., Leslaw, K., Jerzy, K.: Florida standard for crashworthiness and safety evaluation of paratransit buses. Transit Off. Florida Dep. Transp. 1–14 (2009)
12. Micu, D.A., Iozsa, D., Stan, C.: Quasi-static simulation approaches on rollover impact of a bus structure. In: WSEAS, ACMOS, pp. 81–86 (2014)
13. Nurhadi, I., Zain, R.: Development of computer based procedure for quantitative evaluation of bus superstructure in type approval. J. KONES **17**(2), 371–378 (2010)
14. Mohd Nor, M.K., Dol Baharin, M.Z.: Rollover analysis of heavy vehicle bus. Appl. Mech. Mater. **660**, 633–636 (2014)
15. Na, J., Wang, T., Xu, Z.: Research on a one-step fast simulation algorithm for bus rollover collision based on total strain theory. Int. J. Crashworthiness **19**(3), 275–287 (2014)
16. Mahajan, R.S., Daphal, P.N., Athavale, S.M.: Study and analysis of rollover resistance of bus body structure by non-linear FEM technique and experimental method. In: SAE Tech. Pap., pp. 205–209 (2003)
17. Zhou, W., Kuznectov, A., Wu, C.Q., Telichev, I.: A comparative numerical study of motor-coach rollover resistance under ECE R66 and proposed NHTSA regulation conditions. Int. J. Crashworthiness **25**(2), 1–16 (2019)
18. Phadatare, V.D.: Performance improvement of bus structure for rollover analysis using FEA and validation of roll bar. IOSR J. Mech. Civ. Eng. **17**(10), 16–19 (2017)
19. Thosare, A., Patil, S.B.: Rollover analysis of bus body to meet ais-052 regulations and optimization of the body. Int. Eng. Res. J. **3**, 96-102 (2017)
20. Rogov, P.S., Orlov, L.N.: Verification of computer simulation results of bus body section rollover. J. Traffic Transp. Eng. **3**(2), 118–127 (2015)
21. Yang, L., Deng, S.: Structural local analysis and optimization of bus body skeleton. In: 5th International Conference Civil Engineering Transportation no. ICCET, pp. 1975–1979 (2015)
22. Korta, J., Uhl, T.: Multi-material design optimization of a bus body structure. J. KONES. Powertrain Transp. **20**(1), 139–146 (2013)
23. Reyes-ruiz, C., Cervantes, O.R., Prado, A.O., Ramirez, E.: Analysis and optimization of a passenger bus frame through finite element software. In: 2013 SIMULIA Community Conference (2013)
24. Bin Yusof, M., Bin Afripin, M.A.A.: Effect of beam profile size on bus superstructure strength having rollover crash. Appl. Mech. Mater. **372**, 620–629 (2013)
25. Hu, H., Yang, C.L., Wang, J.: Development and validation of finite element model for the welded structure of transit bus. Int. J. Heavy Veh. Syst. **19**(4), 371–388 (2012)
26. Li, Y., Lan, F., Chen, J.: Experimental and numerical study of rollover crashworthiness of a coach body section. SAE Int., vol. 8 (2012)

27. Su, R., Gui, L., Fan, Z.: Multi-objective optimization for bus body with strength and rollover safety constraints based on surrogate models. Struct. Multidiscip. Optim. **44**(3), 431–441 (2011)
28. Bojanowski, C., Kulak, R.F.: Multi-objective optimisation and sensitivity analysis of a paratransit bus structure for rollover and side impact tests. Int. J. Crashworthiness **16**(6), 665–673 (2011)
29. Tech, T.W., Iturrioz, I.: Structural optimization of a bus in rollover conditions. SAE Tech. Pap. (2009)
30. Matolcsy, M.: The severity of bus rollover accidents. Crashworthiness Transp. Syst. 07 (1997)
31. Friedman, K., Hutchinson, J., Weerth, E., Mihora, D.: Implementation of composite roof structures in transit buses to increase rollover roof strength and reduce the likelihood of rollover. Int. J. Crashworthiness **11**(6), 593–596 (2006)
32. Lin, Y.C., Nian, H.C.: Structural design optimization of the body section using the finite element method. SAE Tech. Pap. (2006)
33. Lan, F., Chen, J., Lin, J.: Comparative analysis for bus side structures and lightweight optimization. Proc. Inst. Mech. Eng. Part D J. Automob. Eng. **218**(10), 1067–1075 (2004)
34. Liang, C.C., Le, G.N.: Bus rollover crashworthiness under European standard: An optimal analysis of superstructure strength using successive response surface method. Int. J. Crashworthiness **14**(6), 623–639 (2009)
35. Rahman, M.K.: Body section analysis in bus rollover simulation. J. East. Asia Soc. Transp. Stud. **9**, 1967–1981 (2011)
36. Park, S.J., Yoo, W.S., Kwon, Y.J.: Rollover analysis of a bus using beam and nonlinear spring elements. WSEAS Trans. Math. **5**(5), 526–531 (2006)
37. Liang, C.C., Nam, L.G.: Comparative analysis of bus rollover protection under existing standards. WIT Trans. Built Environ. **113**, 41–53 (2010)
38. Subic, A., He, J.: Improving bus rollover design through modal analysis. Int. J. Crashworthiness **2**(2), 139–152 (1997)
39. Bojanowski, C., Gepner, B., Kwasniewski, L., Rawl, C., Wekezer, J.: Roof Crush Resistance and Rollover Strength of a Paratransit Bus. In: 8th European LS-DYNA® Users Conference, vol. 66, no. May 2011, pp. 1–13 (2011)
40. Valladares, D., Miralbes, R., Castejon, L.: Development of a numerical technique for bus rollover test simulation by the F.E.M. In: WCE 2010 - World Congress on Engineering 2010, vol. 2, pp. 1361–1365 (2010)
41. Liu, Y.: ANSYS and LS-DYNA used for structural analysis. Int. J. Comput. Aided Eng. Technol. **1**(1), 31–44 (2008)
42. Gepner, B.D.: Rollover Procedures for Crashworthiness Assessment of Paratransit Bus Structures. Florida State University (2014)
43. UNECE R66: Uniform technical prescriptions concerning the approval of large passenger vehicles with regard to the strength of their superstructure. Geneva (2006)
44. Schweizerhof, K., Walz, M., Rust, W.J.H., Franz, U.: Quasi-static analyses using explicit time integration - applications of LS-DYNA. In: 2nd Eur. LS-DYNA Conference, no. May 2018, pp. 1–18 (1999)
45. Gertsch, J., Shim, T.: Interpretation of roll plane stability models. Int. J. Veh. Des. **46**(1), 72–93 (2008)
46. Bojanowski, C.: Verification , Validation and Optimization of Finite Element Model of Bus Structure for Rollover Test. Florida State University (2009)
47. Kwa, L., Wekezer, J.W., Gepner, B., Siervogel, J.: Development of simplified safety assessment procedure for paratransit buses. Comput. Methods Mech., no. 028818 (2011)
48. Wekezer, J.W., Cichocki, K.: Structural response of paratransit buses in rollover accidents. Int. J. Crashworthiness **12**(3), 217–225 (2007)

49. Guler, M.A., Elitok, K., Bayram, B., Stelzmann, U.: The influence of seat structure and passenger weight on the rollover crashworthiness of an intercity coach. Int. J. Crashworthiness **12**(6), 567–580 (2007)
50. Boria, S.: Lightweight Design and Crash Analysis of Composites. Elsevier Ltd. (2016)
51. Wicaksono, S., Rizka Faisal Rahman, M., Mihradi, S., Nurhadi, I.: Finite element analysis of bus rollover test in accordance with UN ECE R66 standard. J. Eng. Technol. Sci. **49**(6), 799–810 (2017)
52. Isuzu Motors Limited. Isuzu N-Series Body Builders Guide (2014)
53. General Motors Isuzu Commercial Truck and American Isuzu motors Inc. Isuzu Body Builder's Guide (2003)
54. Isuzu Motors Inc.: Isuzu N-Series Body Builder Guides. www.isuzutruckservice.com%0ADownload (2016)
55. Bitzenbauer, J., Franz, U., Schweizerhof, K.: Deformable Rigid Bodies in LS-DYNA with Applications – Merits and Limits (2005)
56. LSTC 2021: Hourglass-Welcome to the LS-DYNA support site. Livermore Software Technology Corporation (LSTC). https://www.dynasupport.com/howtos/element/hourglass. Accessed 19 May 2021
57. Seyedi, M., Jung, S., Wekezer, J.: A comprehensive assessment of bus rollover crashes : integration of multibody dynamic and finite element simulation methods. Int. J. Crashworthiness **27**(3), 1–16 (2020)
58. Wang, Q., Zhou, W., Telichev, I., Wu, C.Q.: Load transfer analysis of a bus bay section under standard rollover test using U*M index. Int. J. Automot. Technol. **19**(4), 705–716 (2018)
59. Elseufy, S.M., Mawsouf, N.M., Ahmad, A.: Safety evaluation of buses during rollover. J. Manag. Eng. Integr. **6**(March), 102–108 (2013)
60. Chirwa, E.C., Li, H., Qian, P.: Modelling a 32-seat bus and virtual testing for R66 compliance. Int. J. Crashworthiness **20**(2), 200–209 (2015)
61. Kwasniewski, L., Bojanowski, C., Siervogel, J., Wekezer, J.W., Cichocki, K.: Crash and safety assessment program for paratransit buses. Int. J. Impact Eng. **36**(2), 235–242 (2009)
62. Livermore Software Technology Corporation, Keyword User's Manual Vol II, vol. I, no. May (2007)
63. Rabbat, B.G., Russell, H.G.: Friction coefficient of steel on concrete or grout. J. Struct. Eng. **111**(3), 505–515 (1985)
64. Gleba, M.: Effect of Friction on Vehicle Crashworthiness during Rollover. Florida State University Libraries (2015)
65. Zhu, L.: Development of guidelines for deformable and rigid switch in Ls-Dyna simulation. University of Nebraska (2009)
66. Zhou, W., Kuznetcov, A., Telichev, I., Wu, C.: Deformable-rigid switch in computational simulation of bus rollover test. Int. Union Theor. Appl. Mech., no. August, pp. 3–4 (2016)
67. Hamid, I.A., Kamarudin, K.A., Osman, M.R., Abidin, A.N.S.Z., Zulkipli, Z.H.: Finite element bus rollover test verification. J. Soc. Automot. Eng. Malaysia **3**(4), 57–63 (2019)
68. LSTC 2021: Total energy-Welcome to the LS-DYNA support site. Livermore Software Technology Corporation (LSTC). https://www.dynasupport.com/howtos/general/total-energy. Accessed 19 May 2021
69. LSTC 2021: Energy data-Welcome to the LS-DYNA support site. Livermore Software Technology Corporation (LSTC). https://www.dynasupport.com/tutorial/ls-dyna-users-guide/energy-data. Accessed 19 May 2021
70. Belegundu, A.D., Chandrupatla, T.R.: Optimization Concepts and Applications in Engineering, 3rd ed. Cambridge University Press (2019)
71. Vanderplaats, G.N.: Structural optimization for statics, dynamics and beyond. J. Brazilian Soc. Mech. Sci. Eng. **28**(3), 316–322 (2006)

72. Witteman, W.J.: Improved Vehicle Crashworthiness Design by Control of the Energy Absorption for Different Collision Situations, Thesis, 1999, Eindhoven University of Technology, ISBN 90-386-0880-2, no. 1999 (1999)

73. Kurtaran, H., Eskandarian, A., Marzougui, D., Bedewi, N.E.: Crashworthiness design optimization using successive response surface approximations. Comput. Mech. 29(4–5), 409–421 (2002)

74. Esfahlani, S.S., Shirvani, H., Nwaubani, S., Shirvani, A., Mebrahtu, H.: Comparative study of honeycomb optimization using Kriging and radial basis function. Theor. Appl. Mech. Lett. 3(3), 031002 (2013)

75. Stander, N., Roux, W., Goel, T., Eggleston, T., Craig, K.: LS – OPT User's Manual: A Design Optimization and Probabilistic Analysis Tool. Livermore Softw. Technol. Corp., no. February (2012)

Transfer Learning with Pre-trained CNNs for Breast Cancer Stage Identification

Tesfahunegn Minwuyelet Mengistu$^{(\boxtimes)}$ ⓘ, Birtukan Shegaw Arega, and Birhanu Hailu Belay

Bahir Dar University, Bahir Dar, Ethiopia
tesfahunegn9@gmail.com

Abstract. Breast cancer stage identification is an important prerequisite for early treatment to increase the chance of survival, and predict the recurrence of cancer. Research works done so far were mainly focused on the classification of breast cancer types while many of them are neglecting to stage of breast cancer. Obtaining an adequate labeled breast cancer image dataset for training machine learning algorithms is challenging. In this paper, we propose a pre-trained Convolutional Neural Network (Pretrained-CNN) model for Breast Cancer Stage Identification. The proposed method is designed by leveraging transfer learning techniques. Further, the performance of the pre-trained model is compared with CNN-based models that are trained from scratch. The performance of the proposed model is tested using a publicly available breast cancer-image dataset taken and achieved a promising result with an overall classification accuracy of 90%

Keywords: Breast cancer · Pre-trained model · CNN · Segmentation · Transfer learning

1 Introduction

In many clinical practices, the detection and identification of diseases including breast cancer diagnostic and medical image interpretation have been made by the expertise of individual clinicians and/or physicians (Chen et al. 2022; Birtukan et al. 2020). The medical decision mainly relies on the physician's knowledge, and experience which results in large variability in interpreting medical images (Chen et al. 2022). To address such problems various types of research proposed machine learning-based models for interpreting and analysis of medical images (Birtukan et al. 2020). Artificial Intelligence (AI)-based models have been widely applied in medical image processing including diagnosis and staging and detection of breast cancer from digital mammography (Li et al. 2020; Dabeer et al. 2019). These tools help to detect the suspicious region in mammogram images and classify the suspicious regions into different classes. Compared to the manual detection of breast cancer, such AI-assisted systems improve the accuracy of diagnosis and detection of the stage of breast cancer (Dembrower et al. 2020; Tahmooresi et al.

B. H. Woldegiorgis et al. (Eds.): ICAST 2022, LNICST 455, pp. 127–136, 2023.
https://doi.org/10.1007/978-3-031-28725-1_8

2018). Furthermore, as a simultaneous assistant to a radiologist, the use of these AI-enabled cancer detectors can identify other cancers (Kim 2022; Schaffter et al. n.d.; McKinney et al. 2020).

Breast cancer is one of the most prominent cancer types and is the leading cause of death in women (Birtukan et al. 2020; Boughorbel et al. 2016). Generally, breast cancer is classified into five stages (Birtukan et al. 2020). In clinical settings, the stages of breast cancer are usually expressed on a scale of 0 through IV where stage 0 denotes non-invasive cancer and stage IV denotes invasive cancers (Birtukan et al. 2020; Boughorbel et al. 2016) (Breastcancer.org. 2018). Sample images of the five breast cancer images taken from (Birtukan et al. 2020) are illustrated in Fig. 1.

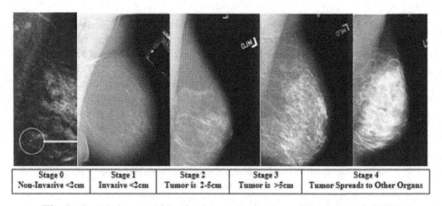

Fig. 1. Sample images of the five breast cancer stages (Birtukan et al. 2020)

Advances in deep learning and its remarkable success in clinical applications has attracted the attention of researcher in medical domains (Boughorbel et al. 2016; Goswami 2018; Saha 2015). Despite the success, the lack of a labeled image dataset has been the major bottleneck in developing a robust deep learning-based model for medical image analysis (Chen et al. 2022; Birtukan et al. 2020). This is also true in breast cancer detection and identification. Therefore, various researchers (Breastcancer.org. 2018; Nadig 2017; McCowan et al. 2007) use CNN and handcrafted feature-assisted classical machine learning algorithms for breast cancer detection and staging. Since these handcrafted features consist of limited information about the image, the recognition performance of the models was very limited (Birtukan et al. 2020).

Nowadays, to overcome the shortage of training datasets, researchers employ transfer learning techniques whereby models are first trained on a problem and one or more layers from the trained model are then used in a second new model of a related problem (Li et al. 2020; Kim 2022). As a continuation of our previous work (Birtukan et al. 2020), which employed CNNs and traditional machine learning algorithms as feature extractors and classifiers respectively, this paper presents a model with pre-trained CNN backbones by leveraging transfer learning techniques for Breast cancer staging.

The rest of the paper is organized as follows: Sect. 2 reviews the relevant methods and related works. The proposed methods and training settings are described in Sect. 3.

Section 4 presents all experiments and results obtained from the experiments. Finally, Sect. 5, presents conclusions and future research directions.

2 Related Works

The existing breast cancer detection and staging models can utilize either handcrafted or automatic features. Methods belonging to the first category were mainly applied before the introduction of deep learning and follows step-wise routines. In contrast, the second approach integrated the feature extraction and classification steps and trained from end to end. Therefore, in this section, we review the research trends in breast cancer detection and the existing state-of-the-art techniques that are applied for medical image analysis where labeled images are very limited.

A k-Means based Gaussian Mixture Model (GMM) is proposed to detect and classify breast cancer as benign or malignant (Dheeba 2019). The authors developed a model by following three steps. First, they found a region of interest by using the K-means-based GMM segmentation technique, and then they applied texture feature extraction and optimization of features of the Region of Interest (ROI) by using a Genetic Algorithm (GA). Finally, classified the detected abnormality as benign or malignant. To develop and evaluate the model they used the publicly available Mammographic Image Analysis Society (MIAS) dataset and got an accuracy of 95.8%. They classified images as benign or malignant but they did not work on identifying stages of detected malignant breast cancer.

Researchers in (Shen 2017) also developed a deep learning algorithm that can detect breast cancer on screening mammograms by using an end-to-end training approach. They used a combined dataset of publicly available Digital databases for Screening mammograms (CBIS-DDSM) and a private dataset of Full-Field Digital Mammography (FFDM) from the INbreast database. They achieved 86.1% of sensitivity and 80.1% of specificity in CBIS-DDSM and 86.7% of sensitivity and 96.1% of specificity in FFDM of the INbreast dataset. The mammograms are classified as cancer or normal. It did not identify whether the cancerous are benign or malignant and at which stage the cancerous case is reached.

Another CNN-based approach for cancer diagnosis on the histopathological image was also proposed in (Dabeer et al. 2019). They used datasets from the BreakHis database and reported 99.86% of accuracy. They introduced deep learning architecture for breast cancer detection as benign and malignant but they did not work on which stage malignant breast cancer reach that is used to start treatment, estimate recurrence, and increase survival from cancer.

The breast cancer histopathology image classification scheme was also proposed by researchers in (Zhu et al. 2019) by assembling multiple compact CNNs. They used two breast cancer datasets these are the BreakHis database which contains 7909 images taken from the breast tissue biopsy side and the BreAst Cancer Histology (BACH) which contains 400 breast histology images and achieved 84.4% accuracy. They classified breast cancer histopathological images based on multiple compact CNN as cancer or not cancer.

CNN was used by researchers for breast cancer screening as a multi-view deep CNN (Geras et al. 2017). They used a mammography-based breast cancer screening

Breast Imaging Reporting and Database System (BI-RADS) dataset having 886,000 images to study the impact of training set size and image size on CNN cancer prediction accuracy. They focused on identifying the impact of the training set size and image size on the prediction of cancer accuracy. They did not consider how can identify and classify stages of malignant breast cancer since identifying stages of breast cancer. A similar work (Dalmış et al. 2018) also proposed a CNN-based model for breast cancer screening from a Magnetic Resonance Image (MRI). 385 MRI scans, containing 161 malignant lesions.

Nine-layer CNN with the parametric rectified linear unit and rank-based stochastic pooling is also employed for abnormal breast identification. Researchers focused on how to select the optimal number of convolution layers and the effect of data augmentation on breast cancer detection. They were using a mini-MIAS database that contains 209 normal breast images and 113 abnormal breast images. They achieved results over 100 test sets with 94.0% of accuracy, 94.5% of precision, 93.4% of sensitivity, and 94.6 specificity by combining ReLU and rank-based stochastic pooling. They did not consider the classification of stages of abnormal breast images.

Researchers in (Wang et al. 2014) used convolutional neural network features by combining them with handcrafted features for Mitosis detection in breast cancer pathology images. They used the public ICPR12 mitosis dataset that has 226 mitoses and 15 testing HPFs and got an F-measure of 0.7345. They considered mitosis count for grading of breast cancer they did not work on the identification of stages of breast cancer necessary to start treatment and for better treatment suggestions.

CNN was used for automated breast ultrasound lesions detection (Yap et al. 2018). They used two different datasets of US images that were obtained from US systems. Dataset A contains 306 images with 246 benign and 60 malignant cases. Dataset B contains 163 images with 110 benign and 53 malignant cases. The proposed model detects lesions either benign or malignant.

Researchers in (Nadig 2017) proposed stage-specific predictive models for breast cancer survivability by using three different machine learning methods (naïve Bayes, Logistic regression, and decision tree). And compared their accuracy to predict survivability. They used a publicly available SEER dataset. Unlike image-based staging, they considered text-based stage information as a factor. The other method for the prediction of breast cancer using big data analytics by using the K-nearest neighbor algorithm was also proposed in (Shailaja et al. 2018). The Wisconsin breast cancer dataset taken from the UCI machine learning repository that contains 699 instances with 11 attributes are classified as either benign or malignant by using KNN.

Though there are limited labeled breast cancer training datasets, nowadays, the detection of breast cancer for early diagnosis from image data is becoming common and many researchers reported promising results employing Deep convolutional neural networks (Breastcancer.org 2018). Hence, in this paper, we propose a method that can overcome the issue of labeled training image scarcity with a great emphasis on the identification of stages of breast cancer. The following sections give a detailed overview of the proposed breast cancer stinging model.

3   Materials and Methods

In this section, we describe the breast cancer dataset employed in model development and elaborate on the details of the proposed breast cancer staging model architecture. To develop the breast cancer stages identification models we follow experimental research which consists of dataset preparation, model training, and evaluation.

3.1   Dataset

In the era of deep learning, the shortage of training datasets is one of the bottlenecks that limit the development of robust models for disease diagnosis in general and specifically for breast cancer staging. Few datasets have been used in various works on breast cancer detection (Dheeba 2019). Many of the datasets used in the literature consist of very limited images that are prepared for a very specific use case. Recently, an updated and organized breast cancer staging dataset has introduced by (Birtukan et al. 2020) where the images were collected from various sources including the Curated Breast Imaging Subset of DDSM (CBIS-DDSM) (Lee et al. 2017) and it is publicly available in Kaggle at [https://www.kaggle.com/datasets/tesfahunegn/breast-cancer-stage-identi fications]. This dataset consists of 1469 Images in MINIST files that are split as training and testing data. The details of the dataset used in this experiment are summarized in Table 1.

Table 1. Number of breast cancer images in each stage

No	Classes	Image format	Number of images
1	Stage 0	Png	395
2	Stage 1	Png	495
3	Stage 2	png	310
4	Stage 3	Png	173
5	Stage 4	Png	96
Total			1469

Sample breast cancer images are illustrated in Figs. 2 and 3.

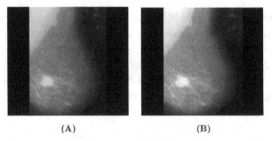

(A)   (B)

Fig. 2. Sample enhanced breast cancer image (Birtukan et al. 2020). (A) Original image, (B) image after filtering and enhancement

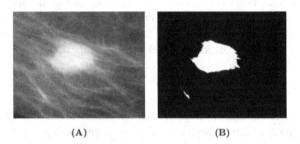

(A)   (B)

Fig. 3. Example of ROI segmentation (Birtukan et al. 2020)

3.2   The Proposed Model Architecture

Considering the nature and size of the dataset, we propose two experimental setups. The first approach employs a CNN-based architecture which is trained from scratch while the second breast cancer stage identification approach uses pre-trained CNN models as a backbone. The two proposed approached and the overall model architectures are depicted in Figs. 4 and 5 respectively.

The network architecture in the first approach consists of the feature learner that takes the segmented image of the breast cancer and encodes it to a high-level feature vector representation and the classification layers (which consists of a fully connected layer with softmax) which are responsible to produce the probability distribution of breast cancer stages over a given input feature vectors. The architecture of the CNN-based model that is trained from the scratch is illustrated in Fig. 4.

This CNN-based model consists of seven convolution layers and the number of filters that are used to control the depth of the output volume was 64, 128, and 256 filters also a 3×3 filter size at a single layer have used. We applied the ReLU activation function, a pooling size of two (2×2) after the two consecutive convolution layers, dropout layers with dropping probability (p) = 0.25, 0.4, and 0.5 at each fully connected layer, batch normalization, Adam optimizer as an optimization function, a learning rate of 0.003, categorical cross-entropy as a loss function. The model was for 25 epochs with a batch size of 32. Unlike our previous work (Birtukan et al. 2020), in this paper, we haven't applied data augmentation.

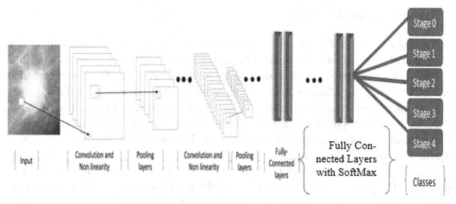

Fig. 4. Proposed CNN model architecture

The second approach employed a pre-trained CNN in which the model reuses the already learning method for other tasks. This approach is commonly called transfer learning. In this experimental setup, the pre-trained VGG16 model trained with ImageNet recognition tasks is applied for breast cancer stage identification where the number of class labels is set to five. To demonstrate the effect of the depth of fine-tuned layers on recognition performance, we fine-tuned some of the specific layers (up to 3 layers), while the remaining layers are set to freeze. Since the pre-training VGG net input size was 224 × 224, we changed the input tensor size of the breast cancer image to a similar size of 224 × 224. During loading the pre-trained VGG16 model, we don't load the fully connected layers; thus, we add custom fully connected (FC) layers which will be trained together. The fine-tuned and frozen layers of the pre-trained VGG net model are illustrated in Fig. 5.

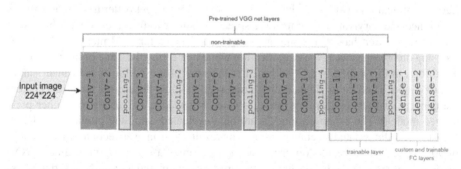

Fig. 5. Pre-trained VGG net setups: all frozen layers except the last three and custom fully connected layers

In both experimental setups, the whole dataset is split as 80/20 for training and testing respectively and after that again we classified the remaining training data as 80/20 for the training and validation phase respectively that is we used 60% for training the model, 20% for validating the model and to remove bias to training dataset and 20% for testing and evaluating the model.

4   Experimental Results

Experiments were conducted using the dataset (Birtukan et al. 2020), which is a freely available breast cancer dataset. The model architecture and experimental setups described in Sect. 3 are implemented using Keras Application Program Interface (API) with a Tensor Flow as a backend. To select a suitable model parameter, different values of these parameters were considered and tuned during model training. In addition to dropouts, we also employ early stopping to avoid over-fitting The best results recorded during experimentation are reported in Table 2. The performance of both models is measured using an accuracy metric.

We compared the accuracy of the CNN-based model trained from the scratch with a pre-trained CNN model, and our pre-trained CNN model outperforms the classification accuracy of the model that was trained from scratch by a large margin.

Table 2. Summary of comparison of models based on testing accuracy

Dataset used	Model	Testing accuracy
Full mammogram images	CNN-SoftMax	39%
	Pretrained-CNN model	42%
Segmented ROI images	CNN- SoftMax	84%
	Pretrained-CNN model	90%

5   Conclusion

Early identification of stages of breast cancer allows one to get better treatment, expect recurrence and survival, and control the spread of cancer to the other part of the body. Extensive research has been done on breast cancer detection, and many of them are focusing on breast cancer type classification while neglecting the stage of breast cancer. Therefore, in this paper, two CNN-based models are proposed for breast cancer stage identification. The first model is trained from scratch while the second model leverages the concept of transfer learning through which knowledge is reused from other pre-trained models. The proposed models are then evaluated using the publicly available breast cancer dataset and it achieves promising results with an over-recognition accuracy of 90% Based on the results observed during experimentation, using pre-trained CNN models gives significant discrimination performance compared to the CN-based models that are trained from scratch. To improve the performance of the proposed model, as part of future work, other pre-trained CNN models could be explored and investigated. In addition, instead of developing independent models, we plan to develop a multi-tasking deep learning model that can learn the breast cancer type and stage simultaneously.

Author Contributions. Conceptualization, T.M. and B.S; Methodology, T.M. and B.S.; Writing—original draft preparation, T.M. and B.S.; Writing—review and editing, B.H. All authors have read and agreed to the published version of the manuscript.

References

Birtukan, S., et al.: Breast cancer Stage identification using Machine learning. Univerity of Gondar, Gondar, Ethiopia (2020)

Breastcancer.org.: Breast cancer stages: 0 through iv. (2018). Retrieved from http://www.breast cancer.org/symptoms/diagnosis/stagi

Zhu, C., et al.: Breast cancer histopathology image classification through assembling multiple compact CNNs. BMC Med. Inform. Decis. Mak. **19**(1), 1–17 (2019). https://doi.org/10.1186/s12911-019-0913-x

Dheeba, S.S.: A research on detection and classification of breast cancer using k-means gmm & CNN algorithms. Int. J. Eng. Adv. Technol. **8**(6), 501–509 (2019). https://doi.org/10.35940/ijeat.F1102.0886S19

Goswami, T.: Impact of deep learning in image processing and computer vision. In: Anguera, J., Satapathy, S.C., Bhateja, V., Sunitha, K.V.N. (eds.) Microelectronics, Electromagnetics and Telecommunications. LNEE, vol. 471, pp. 475–485. Springer, Singapore (2018). https://doi.org/10.1007/978-981-10-7329-8_48

Wang, H., et al.: Mitosis detection in breast cancer pathology images by combining handcrafted and convolutional neural network features. J. Med. Imaging **1**(3), 034003 (2014). https://doi.org/10.1117/1.jmi.1.3.034003

McCowan, I.A., et al.: Collection of cancer stage data by classifying free-text medical reports. J. Am. Med. Informatics Assoc. **14**, 736–745 (2007). https://doi.org/10.1197/jamia.M2130

Dembrower, K., et al.: Effect of artificial intelligence-based triaging of breast cancer screening mammograms on cancer detection and radiologist workload: a retrospective simulation study. Lancet Digit. Health **2**(9), e468–e474 (2020). https://doi.org/10.1016/S2589-7500(20)30185-0

Geras, K.J., et al.: High-resolution breast cancer screening with multi-view deep convolutional neural networks. arXiv, 1–9

Shailaja, K., et al.: Prediction of breast cancer using big data analytics. Int. J. Eng. Technol **7**, 223–226 (2018). https://doi.org/10.14419/ijet.v7i4.6.20480

Kim, H., e.: Transfer learning for medical image classification: a literature review. BMC Med. Imag. (2022). https://doi.org/10.1186/s12880-022-00793-7

Yap, M.H., et al.: Automated breast ultrasound lesions detection using convolutional neural networks. IEEE J. Biomed. Heal. Inform. **22**(4), 1218–1226 (2018). https://doi.org/10.1109/JBHI.2017.2731873

Tahmooresi, M., et al.: Early detection of breast cancer using machine learning techniques. J. Telecommun. Electron. Comput. Eng. **10**, 21–27 (2018)

Dalmış, M.U., et al.: Fully automated detection of breast cancer in screening MRI using convolutional neural networks. J. Med. Imaging **5**(1), 1 (2018). https://doi.org/10.1117/1.jmi.5.1.014502

Nadig, R.J.: Stage-specific predictive models for breast cancer survivability. Int. J. Med. Inform. **97**, 304–311 (2017). https://doi.org/10.1016/j.ijmedinf.2016.11.001

Lee, R.S., et al.: A curated mammography data set for use in computer-aided detection and diagnosis research. (2017). https://doi.org/10.1038/sdata.2017.177

Boughorbel, S., et al.: Model comparison for breast cancer prognosis based on clinical data. PLoS ONE **11**, 1–15 (2016). https://doi.org/10.1371/journal.pone.0146413

McKinney, S.M., et al.: International evaluation of an AI system for breast cancer screening. Nature **577**, 89–94 (2020). https://doi.org/10.1038/s41586-019-1799-6

Saha, S.: A Comprehensive Guide to Convolutional Neural Networks — the ELI5 way. Towards Data Science, pp. 1–19 (2015). Retrieved from https://towardsdatascience.com/a-comprehensive-guide-to-convolutional-neuralnetworks-the-eli5-way-3bd2b1164a53

Shen, L.: End-to-end training for whole image breast cancer diagnosis using an all convolutional design. arXiv, pp. 1–12 (2017). https://doi.org/10.1038/s41598-019-48995-4

Dabeer, S., et al.: Cancer diagnosis in the histopathological image: CNN based approach. Inform. Med. Unlocked **16**. https://doi.org/10.1016/j.imu.2019.100231

Schaffter, T., et al.: Evaluation of combined artificial intelligence and radiologist assessment to interpret screening Mammograms. JAMA Netw. Open. **3**(3), e200265 (n.d.). https://doi.org/10.1001/jamanetworkopen.2020.0265

Li, X., et al.: Transfer learning in computer vision tasks: remember where you come from. Image Vision Comput. **93** (2020)

Chen, X., et al.: Recent advances and clinical applications of deep learning in medical image analysis. Med. Image Anal. **79**. (2022)doi:https://doi.org/10.1016/j.media.2022.102444

Fuzzy Sliding Mode Controller Based Trajectory Tracking Control of Free Flying Space Robot Manipulator System

Esubalew W. Shibabw$^{(\boxtimes)}$ and Gerbaw Y. Tamiru

Bahirdar Institute of Technology, Bahirdar University, Bahir Dar, Ethiopia
esubalewwalle378@gmail.com

Abstract. A free flying space robot manipulator system (FFSRMS) is made up of a six degrees of freedom (DOF) spacecraft and n degree of freedom manipulator mounted on the base spacecraft. The space robot manipulator is used to perform a variety of tasks in space or on orbit service (OOS), such as assembling and repairing spacecraft, refueling satellites in orbit, and removing space debris. Unlike ground-based robot manipulators, a space robot manipulator has no fixed base. As a result, the base and manipulator are strongly coupled. Furthermore, the system operates in unstructured and zero gravity environment. The kinematics and dynamics of a FFSRMS has been developed based on chaseles theorem and Euler-Lagrangian equation of motion respectively. The stability of the system and the convergence of the tracking errors to the origin has been checked using Lyapunov stability criterion. This work used a fuzzy sliding mode controller (FSMC) to examine a space robot manipulator trajectory tracking control capability in joint space and robustness. The proposed controller has been simulated in MATLAB/Simulink considering external disturbance and parametric variation and compared with the sliding mode controller to evaluate how effective it is. The results reveal that the proposed controller is robust and has good trajectory tracking capability with a reduced Integral Time Absolute Error (ITAE) and it eliminate the chattering effect. The comparative study shows that FSMC has reduced ITAE than SMC.

Keywords: FFSRMS · FSMC · OOS · Robust · Trajectory tracking

1 Introduction

Space robotic manipulator technology is the new frontier field of study which attracts the attention of many countries to take space power and a subsequent study has been conducted over the past 40 years. Space robotic manipulator plays a significant role for servicing a spacecraft after they are deployed on orbit (OOS) that would prolong the life span of a spacecraft and would be too risky and time consuming if performed by an astronaut, such as assembling,

© ICST Institute for Computer Sciences, Social Informatics and Telecommunications Engineering 2023
Published by Springer Nature Switzerland AG 2023. All Rights Reserved
B. H. Woldegiorgis et al. (Eds.): ICAST 2022, LNICST 455, pp. 137–158, 2023.
https://doi.org/10.1007/978-3-031-28725-1_9

constructing or repairing a space station in orbit [1], on orbit refueling [2], space debris removal [3, 4] and extravehicular activity(EVA) support [1].

The historical development and advancement of space robotic manipulator was presented in [5]. Shuttle Remote Manipulator System (SRMS) also known as canadarm1 was the first manipulator mounted on space shuttle orbiter on second mission of the space shuttle (STS-2) to capture Hubble Space Telescope (HST) and to positioning astronauts during extravehicular activity (EVA) in 1981. There are some other robotic manipulator mounted on international space station (ISS) such as; Space Station Remote Manipulator System (SSRMS) also known as canadarm2, the Japanese Experiment Module Remote Manipulator System (JEMRMS), European Robotic Arm (ERA), Robonaut1 and Robonaut2 [8].

In order to capture a target spacecraft, a series of stages must be taken. The first step is planning of how the robot arm should capture the target spacecraft by observing the motion and collecting information about the physical property of the target spacecraft using different sensors like vision, radar sensors. The second stage is controlling the base spacecraft to move to the capturing location (for free flying case) such that, the manipulator is ready to grasp the target satellite. The third step is the actual capturing of the target space robot by the end effector. Lastly, the captured spacecraft is stabilized along with the capturing satellite manipulator by considering the servicing and target spacecraft as one system, this is called post capturing stage [8].

An accurate and precise trajectory tracking of space robot is required when the space robot manipulator is commanded to capture some target spacecraft or object. However, unlike ground-based robot manipulator, space robot has no fixed base, which makes the system dynamics strongly coupled and difficult to formulate both the kinematic and dynamic motion of the space robot manipulator system accurately and to design the controller. In addition, the space robot is working in zero gravity and unstructured environment. Moreover, to cope up external disturbances and parametric variations due to, thrust firing and additional payload carried by the manipulator, a robust controller is required. Many scholars have proposed different path planning algorithm and controller scheme for space robot manipulator system. Researchers in [10–13] presented adaptive controller as a feasible solution to overcome the following two problems: The dynamic equation of the system cannot linearly parameterized and the uncertainty in kinematic mapping from inertial space to joint space when the base is free floating. Gu You-Liang and Yangsheng Xu [10] proposed a normal augmentation approach to adaptive control by modeling the entire free floating space robot system as extended robot. The simulation result shows that the proposed controller asymptotically stabilized the space robot manipulator system to track a given trajectory in Cartesian space. However, the proposed approach requires large computation and it requires measurement of spacecraft accelerations. To alleviate the drawback in [10], Parlaktuna et al. [11] proposed an adaptive controller based on an extended robot approach and an on-line adaptive estimation law for an unknown parameter along with a computed torque controller. Simulation of a two link planner space robot system is conducted using MATLAB and the simulation result shows that the proposed controller has a good position trajectory tracking performance. However, the algorithm that the authors follow to eliminates the measurement of acceleration requires, a large computation.

In [12], the parameterization problem of dynamic equation and adaptive control of free floating space robot are discussed. Abiko et al. [13] proposed adaptive control of fully free floating space robot with dynamic and kinematic model uncertainty. The authors discussed adaptive control for a torque controlled space robot without acceleration measurement. Limin Xie et al. [14] proposed a robust fuzzy slide mode controller for controlling the free floating space robot manipulator system, so that it can correctly track the required trajectory while also suppressing vibration induced by the flexible joints and flexible link. The simulation, however, is conducted in two dimension with two-link manipulator. Meanwhile, the error convergence speed is also slow. Yicheng Liu et al. [15] proposed trajectory tracking for a dual arm free floating space robot with a class of general non-singular predefined time terminal sliding mode. Simulation using MATLAB has been conducted to check the performance of the control and the simulation result shows that the joint angles successfully track the given trajectory. However terminal slide mode controller has chattering problem for bounded random-changing external disturbances. In order to solve this problem, the authors adopt a continuous function scheme at the cost of high gain.

Xin Zhang et al. [16] proposed adaptive robust decoupling control of multi-arm space robots using time delay estimation technique. The paper is focused on decoupling and counteracting the coupling between the base and the manipulator. The simulation is conducted using MATLAB for decoupling control of multi-arm space robot and they have made comparative analysis of computed torque control (CTC) based SMC and time delay estimation (TDE) based SMC and the result demonstrates that the TDE based SMC with a reduced time delay length still achieve a guaranteed control performance. However, the measure taken to reduce the effect of chattering problem of SMC controller is replacing the switching control action by saturation control action operated based on the boundary layer width. Meanwhile, the authors select the boundary layer width randomly (they do not use optimal selection algorithm.) because, if the boundary layer width variable is selected as large the control input will be smooth, but the robustness of the controller degrade, which may causes a steady-state error. Shiyuan Jia and Jin J. Shan [17], presented a trajectory tracking control of space manipulator in the presence of actuator uncertainties, such as actuator fault, actuator saturation and the bias control torque. In addition they have considered parametric uncertainty and external disturbance. In order to resolve the problem mentioned, the authors applies a continuous integral slide mode controller for trajectory tracking control of the space robot manipulator.

This paper focused on trajectory planning and tracking control of free flying space robot manipulator and stabilization of base spacecraft position and orientation using fuzzy slide mode controller. The attitude of base spacecraft is represented using Euler angles. The fifth order polynomial trajectory planning technique is adopted for the manipulator joint and spacecraft attitude desired trajectory. A slide mode controller is a robust controller that can compensate external disturbance and parameter uncertainty change in the system while maintaining the manipulator tracks it's given trajectory with a minimum error and fast convergence. However, SMC suffers a chattering problem due to a high frequency switching control law, in order to alleviate this problem, we adopt a fuzzy logic controller.

The main contributions of this paper are:

1. A fuzzy sliding mode controller is designed for trajectory tracking control of a free flying space robot manipulator system that allows the manipulator to track the desired trajectory with a minimum error and that can reject an external disturbances and parameter uncertainties while the base spacecraft positions and orientations are stabilized in a desired position with a minimum errors.
2. Most of the above mentioned research works such as [14,16,17] analyzed the performance of their controller in 2D plane. However, for real space mission a 3D motion analysis and response is required. So, we are conducted a 3D simulation of a 9 DOF (Six for base positions and attitudes, three for the manipulator joint angles) free flying space robot.
3. As most studies point out, controlling all of the $6+n$ DOF without decoupling the system dynamics is difficult, especially for a free-flying space robot, where the base spacecraft and the manipulator are strongly coupled. This paper, however, uses no decoupling strategies.

2   Modeling of Space Robot Manipulator System

2.1   Assumptions

The formulation of kinematics and dynamic motion of a free flying space robot manipulator system requires the following assumptions.

1. The space robot manipulator system is assumed to be rigid and works in micro-gravity or no gravity environment ($g = 0$).
2. The orbital mechanics is not considered (the body frame is assumed to be aligned with the orbiting frame).
3. As explained in [19], there are several external disturbance torques acting on spacecraft in orbit. In this paper all the disturbances considered are assumed to be bounded.

2.2   Kinematics of Space Robot Manipulator System

A spacecraft equipped with a single n rigid links connected by a revolute joints is shown in Fig. 1. The base spacecraft is defined as link 0, with joint 0 coinciding at the space robot base center of mass. The end- effector is located at a virtual

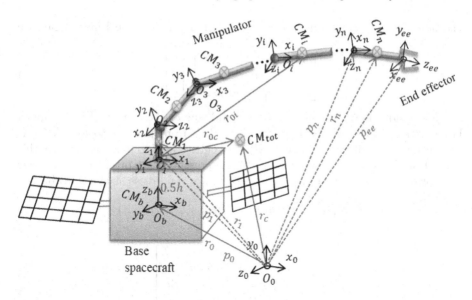

Fig. 1. Geometry of a free flying space robot manipulator system

joint $n+1$. For ground base robot manipulator system, the relationship between these two spaces is non-linear. However, as mentioned in [20], the position and orientation of the end effector of a free flying space robot manipulator may not have a closed form solution, since the motion of the end effector depends on the past history of the manipulator motion. This makes difficult to formulate the inverse kinematics. Thus, the kinematics is formulated at velocity level.

To develop the kinematics of FFSRMS, we use four coordinate systems; An inertial coordinate system with origin Ω, body coordinate system located at the center of mass of the base spacecraft with origin Q_b, a set of $n+1$ joint fixed coordinate systems J_i with origin, $Q_i (i = 0 : n+1)$, and a set of n link fixed coordinate systems, L_i located at the center of mass of each link. The orientation of the base spacecraft with respect to the inertial frame is expressed in Euler angle specified by a rotation matrix in Eq. (1). In order to be free of singularity we would prefer to choose roll-pitch-yaw (ϕ, θ, ψ) rotation sequence [22].

$$R_{0b} = R_\phi R_\theta R_\psi = \begin{bmatrix} 1 & 0 & 0 \\ 0 & c\phi & -s\phi \\ 0 & s\phi & c\phi \end{bmatrix} \begin{bmatrix} c\theta & 0 & s\theta \\ 0 & 1 & 0 \\ -s\theta & 0 & c\theta \end{bmatrix} \begin{bmatrix} c\psi & -s\psi & 0 \\ s\psi & c\psi & 0 \\ 0 & 0 & 1 \end{bmatrix}$$

$$R_{0b} = \begin{bmatrix} c\psi c\theta & c\psi s\theta s\phi - s\psi c\phi & s\psi s\phi + c\psi s\theta c\phi \\ s\psi c\psi & s\psi s\theta s\phi + c\psi c\phi & s\psi s\theta c\phi - c\psi s\phi \\ -s\theta & s\theta s\phi & s\theta c\phi \end{bmatrix} \tag{1}$$

The body fixed angular velocity $(\omega_x, \omega_y, \omega_z)$ is related to the Euler angle rate by;

$$
\begin{bmatrix} \omega_x \\ \omega_y \\ \omega_z \end{bmatrix} = \begin{bmatrix} 1 & 0 & s\theta \\ 0 & c\phi & s\phi c\theta \\ 0 & -s\phi & c\phi s\theta \end{bmatrix} \begin{bmatrix} \dot{\phi} \\ \dot{\theta} \\ \dot{\psi} \end{bmatrix} \tag{2}
$$

where, the notation c and s is for cosine and sine of the respected Euler angle. The rotation matrix of the manipulator, $^iR_{-1}$ from joint J_i to joint J_{i-1} is built using Denavit-Hartenberg (DH) convention as specified in [23]. In this convention the transformation matrix, T_i is represented as a product of four basic transformation with four DH parameters. θ_i, l_i, α_i, and d_i as shown in Table 1

$$
T_i = (rot_{z,\theta_i})(trans_z, d_i)(trans_x, l_i)(rot_{x,\alpha_i})
$$

$$
T_i = \begin{bmatrix} c\theta_i & -s\theta_i c\alpha & s\theta_i s\alpha & l_i c\theta_i \\ s\theta_i & c\theta_i c\alpha & -c\theta_i s\alpha & l_i s\theta_i \\ 0 & s\alpha & c\alpha & d_i \\ 0 & 0 & 0 & 1 \end{bmatrix} \tag{3}
$$

Consider Fig. 1, the position of the end-effector p_e, is given by;

Table 1. DH parameters and their geometric meaning.

DH parameters	Name	Geometric meaning
θ_i	Joint angle	Rotation from ^xJ_i to $^xJ_{i+1}$ about ^zJ_i
l_i	Link length	Distance along the common normal between ^zJ_i and $^zJ_{i+1}$
α_i	Link offset	Rotation from ^zJ_i to $^zJ_{i+1}$ about $^xJ_{i+1}$
d_i	Link twist	Distance between O_i and $^xJ_{i+1}$ along ^zJ_i

$$
p_e = r_0 + b_0 + \sum_{i=1}^{n} l_i \tag{4}
$$

Where, r_0, b_0, l_i are the position of center of mass of base spacecraft, the position of first joint coordinate system with respect to spacecraft center of mass and length of link i respectively.

By derivating Eq. (4) and substituting Eq. (8) and (9) on it, the linear and angular velocity of the end effector is given as;

$$
\begin{bmatrix} v_e \\ \omega_e \end{bmatrix} = \begin{bmatrix} v_b \\ \omega_b \end{bmatrix} + \begin{bmatrix} \omega_b \times (p_e - r_0) + \sum_{j=1}^{n}(k_j \times (p_e - p_j))\dot{\theta}_j \\ \sum_{j=1}^{n}\left(k_j \dot{\theta}_j\right) \end{bmatrix} \tag{5}
$$

Where, v_b and ω_b are the linear and angular velocity of the base spacecraft respectively, k_j is he rotation axis of the ith revolute joint usually defined as the z axis in the corresponding link frame, $\dot{\theta}_j$ denotes the angular velocity of ith rotational joint. Equation (5) can be expressed in canonical form as:

$$\begin{bmatrix} v_e \\ \omega_e \end{bmatrix} = \begin{bmatrix} J_s\dot{\phi} + J_m\dot{\theta} \end{bmatrix} \tag{6}$$

Where, $\dot{\phi} = [v_b, \omega_b]^T$ is the generalized linear and angular velocity vector of the base spacecraft, $\dot{\theta} = [\dot{\theta}_1, \dot{\theta}_2 \dots \dot{\theta}_n]$ are joint angular velocities. $J_s \in R^{6\times6}$ is the Jacobian of the base spacecraft, $J_m \in R^{6\times n}$ is the standard Jacobian of a ground base robot manipulator.

$$J_s = \begin{bmatrix} I_{3,3} & -(p_e - r_0)^\times \\ 0_{3,3} & I_{3,3} \end{bmatrix}, J_m = \begin{bmatrix} k_1 \times (p_e - p_1) \dots k_i \times (p_e - p_i) & 0_{3,n-i} \\ k_1 \dots k_i & 0_{3,n-i} \end{bmatrix}$$

The notation $r^\times$ is used to denote the skew symmetric matrix associated with the vector $r = [r_x, r_y, r_z]^T$ can be written in matrix form as [24].

$$r^\times = \begin{bmatrix} 0 & -r_z & r_y \\ r_z & 0 & -r_x \\ -r_y & r_x & 0 \end{bmatrix}$$

2.3 Dynamics of Space Robot Manipulator System

The equation of motion of a free flying space robot manipulator system as shown in Fig. 1 is formulated using Euler-Lagrangian equation of motion. The potential energy is set to zero, as it is mentioned on assumption 1. Hence, the Lagrangian (L) is equals to the system kinetic energy (T).

$$L = T = \frac{1}{2} \left(\sum_{i=0}^{n} \left(\omega_i^T I_i \omega_i + m_i v_i^T v_i \right) \right) \tag{7}$$

where,

$$v_i = v_b + \omega_b^\times (r_i - r_0) + \sum_{j=1}^{i} \left(k_j^\times (r_i - p_j) \dot{\theta}_j \right) \tag{8}$$

$$\omega_i = \omega_b + \sum_{j=1}^{i} \left(k_j \dot{\theta}_j \right) \tag{9}$$

Substituting Eq. (8) and (9), into Eq. (7) yields;

$$L = \frac{1}{2} \left(\omega_b^T I_b \omega_b + m_b v_b^T v_b \right) + \frac{1}{2} \left(\omega_b + \sum_{j=1}^{i} k_j \dot{\theta}_j \right)^T I_i \left(\omega_b + \sum_{j=1}^{i} k_j \dot{\theta}_j \right)$$

$$+ \frac{1}{2} m_i \left(v_b + \omega_b \times (r_i - r_0) + \sum_{j=1}^{i} (k_j \times (r_i - p_j)) \dot{\theta}_j \right)^T \tag{10}$$

$$\left(v_b + \omega_b \times (r_i - r_0) + \sum_{j=1}^{i} (k_j \times (r_i - p_j)) \dot{\theta}_j \right)$$

By proper re-arrangement and simplification of Eq. (10) we will get;

$$L = T = \frac{1}{2} \begin{bmatrix} v_b^T & \omega_b^T & \dot{\theta}_i^T \end{bmatrix} \begin{bmatrix} H_v & H_{v\omega} & H_{vm} \\ H_{\omega v} & H_\omega & H_{\omega m} \\ H_{mv} & H_{m\omega} & H_m \end{bmatrix} \begin{bmatrix} v_b \\ \omega_b \\ \dot{\theta}_i \end{bmatrix} \tag{11}$$

where,

$$H_v = m_{tot}(I_{3,3})$$

$$H_{v\omega} = -\sum_{i=1}^{n} m_i r_{0i}^\times$$

$$H_{vm} = \sum_{i=1}^{n} (m_i J_{Ti})$$

$$H_\omega = I_0 + \sum_{i=1}^{n} (I_i - m_i r_{0i}^\times r_{0i}^\times)$$

$$H_{\omega m} = \sum_{i=1}^{n} (I_i J_{Ri} + m_i r_{0i} J_{Ti})$$

$$H_{\omega v} = H_{v\omega}^T$$

$$H_{\omega m} = H_{m\omega}^T$$

$$H_{mv} = H_{vm}^T$$

$$H_{m\omega} = H_{\omega m}^T$$

$$H_m = \sum_{i=1}^{n} J_{Ri}^T I_i J_{Ri} + m_i J_{Ti}^T J_{Ti}$$

$$J_{Ti} = \begin{bmatrix} k_1{}^\times (r_i - p_1) \dots k_i{}^\times (r_i - p_i) & 0_{3,n-i} \end{bmatrix}, \forall (1 \leq i \leq n)$$

$$J_{Ri} = \begin{bmatrix} k_1{}^\times \dots & k_i{}^\times & 0_{3,n-i} \end{bmatrix}, \forall (1 \leq i \leq n)$$

The ith link moment of inertia matrix with respect to inertial frame of reference is given by;

$$I_i = R_{Li}{}^{CM} I_i R_{Li}^T \tag{12}$$

where, $^{CM}I_i$ is moment of inertia at center of mass. Now select the manipulator joint angle vector, q and the spacecraft linear and angular position vector, $\dot{x}_b = [v_b^T, \omega_b^T]$ as a generalized coordinate, Eq. (11) can be re-written as;

$$L = T = \frac{1}{2} \begin{bmatrix} \dot{x}_b^T & \dot{q}_i^T \end{bmatrix} \begin{bmatrix} H_b & H_{bm} \\ H_{bm}^T & H_m \end{bmatrix} \begin{bmatrix} \dot{x}_b \\ \dot{q}_i \end{bmatrix} \tag{13}$$

$$H_b = \begin{bmatrix} H_v & H_{v\omega} \\ H_{\omega v}^T & H_\omega \end{bmatrix}$$

$$H_{bm} = \begin{bmatrix} H_{v\omega}^T \\ H_{\omega m}^T \end{bmatrix}$$

Where, $H_b \in \mathbb{R}^{6\times6}, H_m \in \mathbb{R}^{n\times n}$, are the inertia matrix of the spacecraft and the manipulator respectively, $H_{bm} \in \mathbb{R}^{6\times n}$, is the dynamic coupling inertia matrix, which represent the contribution of the manipulator to the base spacecraft and vice-versa: J_{Ti}, J_{Ri}, are the linear velocity and the angular velocity Jacobian matrix for the ith link center of mass respectively. Simplifying Eq. (13) will give as;

$$L = \frac{1}{2}\dot{x}_b^T H_b \dot{x}_b + \frac{1}{2}\dot{x}_b^T H_{bm}\dot{x}_b + \frac{1}{2}\dot{q}^T H_m \dot{q} + \frac{1}{2}\dot{q}^T H_{bm}^T \dot{q} \tag{14}$$

The Euler-Lagrangian equation of motion is given by;

$$\frac{d}{dt}(\frac{\partial L}{\partial \dot{x}_b}) - (\frac{\partial L}{\partial x_b}) = \begin{bmatrix} f_b \\ \tau_b \end{bmatrix} \tag{15}$$

$$\frac{d}{dt}(\frac{\partial L}{\partial \dot{q}}) - (\frac{\partial L}{\partial q}) = \tau_m \tag{16}$$

Now apply Euler-Lagrangian equation of motion onto Eq. (14) will result the dynamic equation of motion of a free flying space robot manipulator system given in Eq. (17)

$$\begin{bmatrix} H_b & H_{bm} \\ H_{mb} & H_m \end{bmatrix} \begin{bmatrix} \ddot{x}_b \\ \ddot{q} \end{bmatrix} + \begin{bmatrix} \dot{H}_b & \dot{H}_{bm} \\ \dot{H}_{mb} & \dot{H}_m \end{bmatrix} \begin{bmatrix} \dot{x}_b \\ \dot{q} \end{bmatrix} - \begin{bmatrix} C_b \\ C_m \end{bmatrix} = \begin{bmatrix} f_b \\ \tau_b \\ \tau_m \end{bmatrix} \tag{17}$$

With,

$$C_b = \frac{1}{2}\frac{\partial}{\partial x_b}(\dot{x}_b^T H_b \dot{x}_b + \dot{x}_b^T H_{bm}\ddot{x}_b + \dot{q}^T H_m \dot{q} + \dot{q}^T H_{bm}^T \dot{q})$$

$$C_m = \frac{1}{2}\frac{\partial}{\partial q}(\dot{x}_b^T H_b \dot{x}_b + \dot{x}_b^T H_{bm}\ddot{x}_b + \dot{q}^T H_m \dot{q} + \dot{q}^T H_{bm}^T \dot{q})$$

Equation (17) can be expressed in a canonical form as;

$$H\ddot{\varphi} + C\dot{\varphi} = \tau \tag{18}$$

With,

$$C = \dot{H} - \begin{bmatrix} C_b \\ C_m \end{bmatrix}$$

Where, $\ddot{\varphi} = [\ddot{x}_b, \ddot{q}]^T$ is a generalized acceleration vector and $\tau = [f_b, \tau_b, \tau_m]^T$ is a generalized force and torque vector term.

3   Controller Design

In this section, a free flying space robot manipulator control system based on a fuzzy sliding mode controller is designed in such away that, the manipulator can track a desired trajectory in joint space, while the position and attitude of the base spacecraft are stabilized in the direction that guarantee communication with ground station. If we take into account disturbances and uncertainty, the dynamic model given in Eq. (18) is modified to;

$$H\ddot{\varphi} + C\dot{\varphi} + \tau_d + \Delta u = \tau \qquad (19)$$

$\tau_d = [\tau_{d1}, \tau_{d2} \dots \tau_{dn}]^T$ represents the bounded disturbance force or torque vector applied on base spacecraft($|\tau_d| \le d_i$) and manipulator,Δu denotes the bounded uncertainty dynamics. i.e.,$|\Delta u| \le u_i$.

Lemma 1. *The inertial matrix H and the Coriolis force C in Eq. (19) are bounded with $\lambda I_{n,n} \le H \le \lambda_{max} I_{n,n}$ and $||C|| \le ||c\dot{\varphi}||$, where λ, λ_{max} and c are positive constants [25].*

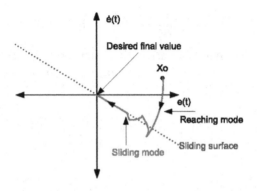

Fig. 2. Sliding surface manifold [27]

3.1   Sliding Mode Controller Design

SMC is one of the variable structure control system (VSCS) approaches. VSCS is a variable high-frequency switching controller [26]. Designing of a slide mode controller (SMC) majorly consists of two procedure;

1. Select sliding surface vector as;

$$s = \dot{e} + K_p e + K_i \int_0^t e(\tau)d\tau \qquad (20)$$

Where, K_p, K_i are a diagonal and positive definite matrix (sliding surface gains), $e, \dot{e} \in \mathbb{R}(6 + n \times 1)$ are an error and rate of error vectors respectively, and they are given as follow;

$$e = \varphi_{di} - \varphi_i \qquad (21)$$

$$\dot{e} = \dot{\varphi}_{di} - \dot{\varphi}_i \tag{22}$$

Where, $\varphi_{di}, \dot{\varphi}_{di}$ are desired linear, angular position and velocity of base and the manipulator respectively. $\varphi_i = [x, y, z, \phi, \theta, \psi, \theta_1, \theta_2, \ldots, \theta_n]$ are the actual base position and attitude and manipulator joint angles. Differentiating Eq. (20) with respect to time yields;

$$\dot{s} = \ddot{e} + K_p \dot{e} + K_i e \tag{23}$$

Solving for φ_i in Eq. (20), then substitute it with Eq. (22) into Eq. (23) and set $\dot{s} = 0$ in Eq. (23), the equivalent control law can be obtained as;

$$\tau_{eqv} = H(\ddot{\varphi}_{di} + K_p \dot{e} + K_i e) + C\dot{\varphi} + \tau_d + \Delta_u \tag{24}$$

Where, τ_{eqv} is the equivalent control law, considering disturbance and uncertainty that makes the system trajectory moving on the sliding surface as indicated in Fig. 2.

2. Select an appropriate reaching control law. In this work, a constant plus proportional reaching control law is adopted as expressed in Eq. (25).

$$\dot{s} = -K_s s - Gsgn(s) \tag{25}$$

Where, G, K_s are a diagonal and positive definite matrix switching gains and reaching control law gains respectively. $sgn(s)$ denote a signum function.

$$sgn(s) = \begin{cases} -1, & if \ s < 0 \\ 1, & if \ s > 0 \end{cases}$$

Finally, the total control law can be obtained as;

$$\tau_{tot} = \tau_{eqv} + H(Gsgns + K_s s) + \tau_d + \Delta_u \tag{26}$$

3.2   Fuzzy Logic Controller Design

In this subsection, a fuzzy logic controller based on zero order Sugeno inference mechanism is designed. The control law expressed in Eq. (26) suffers with chattering problem due, to a high frequency switching control law represent by $Gsgn(s)$.

In this paper, a singleton MF is used for the output variable since, it improves the efficiency of the defuzzification process or simplifies the computation required by the Mamdani method, that finds the centroid of a two dimensional function instead of integrating across the two dimensional function to find the centroid, it is better to use the weighted average of a few data points.

$$- Gsgn(s) = -G\Delta u \tag{27}$$

Now a fuzzy logic controller can be design to smooth the switching controller expressed in Eq. (27) with a FLC output Δu.

1. **Fuzzify Inputs:** Firstly, the sliding surfaces, s and the fuzzy logic controller outputs, Δu has taken as input and output variables of fuzzy inference system respectively. Then, a triangular and a singleton membership functions are assigned for fuzzy input and output variables respectively, as shown in Fig. (3) with a seven linguistic variables; Negative Large (NL), Negative Medium (NM), Negative Small (NS), Zero (Z), Positive Large (PL), Positive Medium (PM) and Positive Small (PS). The range of input and output variables has taken from $[-5\ 5]$. Additionally, the width of both input and output fuzzy MF $\beta = \frac{5}{3}$, as shown in Fig. 3a & 3b.

Remark 1. The input MF range is selected based on the range sliding surfaces from SMC simulation result and the output MF range is based on the range of signum function (since, FLC is chosen to substitute the switching control law of SMC and to produce a continuous control signal as shown in Fig. 3c.

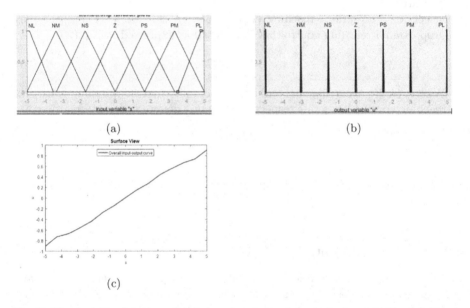

(a)

(b)

(c)

Fig. 3. Membership function of input and output of fuzzy logic controller:(a) Input membership function, (b) Output membership function (c) Overall input output curve

2. **Fuzzy rule construction:** The fuzzy rules are designed based on the methodology that the high frequency switching control law have direct and positive relationship with sliding manifold.
   if **s** is NL, then Δu is NL
   if **s** is NM, then Δu is NM
   if **s** is NS, then Δu is NS
   if **s** is Z, then Δu is Z

if **s** is PS, then Δu is PS
if **s** is PM, then Δu is PM
if **s** is PL, then Δu is PL

3. Computing the output of fuzzy logic controller as;

$$- G\Delta u = -G \sum_{i=-5}^{5} \frac{\mu_i(s)i}{\mu_i(s)} \tag{28}$$

where, $\mu_i(s)$ is the strength of the ith rule and i is the associated single membership function of Δu

3.3   Trajectory Planning

In this study, a fifth order polynomial path planning technique is adopted to plan the desired trajectory of the space robot base attitude and manipulator joint position.

$$\dot{\varphi}_i(t) = a_1 + 2a_2 t + 3a_3 t^2 + 4a_4 t^3 + 5a_5 t^4 \tag{29}$$

The velocity and acceleration can be obtained by taking the first & second derivative of Eq. (29) respectively.

$$\dot{\varphi}_i(t) = a_1 + 2a_2 t + 3a_3 t^2 + 4a_4 t^3 + 5a_5 t^4 \tag{30}$$

$$\ddot{\varphi}_i(t) = 2a_2 + 6a_3 t + 12a_4 t^2 + 20a_5 t^3 \tag{31}$$

where $t_0 \leq t \leq t_f$ and a_i, for $i = 0, 1 \ldots 5$ are coefficients of the polynomial and determined by solving the following polynomial equation. the a_i parameter can be found as;

$$\begin{bmatrix} a_0 \\ a_1 \\ a_2 \\ a_3 \\ a_4 \\ a_5 \end{bmatrix} = \begin{bmatrix} 1 & t_0 & t_0^2 & t_0^3 & t_0^4 & t_0^5 \\ 0 & 1 & 2t_0 & 3t_0^2 & 4t_0^3 & 5t_0^4 \\ 0 & 0 & 2 & 6t_0 & 12t_0{}^2 & 20t_0{}^3 \\ 1 & t_{tf} & t_{tf}^2 & t_{tf}^3 & t_{tf}^4 & t_{tf}^5 \\ 0 & 1 & 2t_{tf} & 3t_{tf}^2 & 4t_{tf}^3 & 5t_{tf}^4 \\ 0 & 0 & 2 & 6t_{tf} & 12t_{tf}^2 & 20t_{tf}^3 \end{bmatrix}^{-1} \begin{bmatrix} \varphi_i(0) \\ 0 \\ 0 \\ \varphi_i(tf) \\ 0 \\ 0 \end{bmatrix}$$

4   Stability Analysis

In this section the stability of the designed controller is proved using Lyapunov function technique (Fig. 4). Select a lyapunov function candidate as;

$$V = \frac{1}{2} s^T s \tag{32}$$

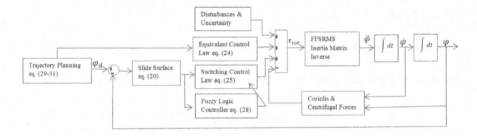

Fig. 4. Overall control system structure

Differentiating Eq. (32) with respect to time and substituting Eq. (19) and (26) into it gives;

$$
\begin{aligned}
\dot{V} &= s^T \dot{s} \\
&= s^T (\ddot{\varphi}_{di} - \ddot{\varphi}_i + K_p \dot{e} + K_i e) \\
&= s^T (\ddot{\varphi}_{di} - H^{-1}(\tau_{tot} - C\dot{\varphi} - \tau_d - \Delta_u) + K_p \dot{e} + K_i e) \\
&= s^T (\ddot{\varphi}_{di} - H^{-1}(\tau_{eqv} + G\Delta_f + K_s s - C\dot{\varphi} - \tau_d - \Delta_u) + K_p \dot{e} + K_i e) \\
&= s^T H^{-1} (-G\Delta_f - K_s s + \tau_d + \Delta_u) \\
&= s^T (H^{-1}(\tau_d + \Delta_u) - |s^T|H^{-1}(G) - s^T H^{-1} K_s s
\end{aligned}
\tag{33}
$$

By choosing appropriate values G and K_s which satisfy $\dot{V} \le 0$ and the matrix H^{-1} is positive definite, Eq. (33) is always less than zero. Therefore the stability is proven.

5   Simulation Results and Discussion

In this section, simulation of three link free flying space robot as shown in Fig. (5) is conducted using MATLAB. The initial and terminal time for trajectory plan-ing is set as $t_0 = 0$ and $t_f = 20$ second respectively. The inertia and DH param-eters of a FFSRMS are specified in Table 2 and Table 3 respectively.

Table 2. Kinematic ad dynamic parameters of the FFSRMS [29]

	$m(kg)$	$l(m)$	$a(m)$	$b(m)$	$I_{xx}(kgm^2)$	$I_{yy}(kgm^2)$	$I_{zz}(kgm^2)$
Base	100	–	–	0.5	30	30	30
Link1	8	0.5	0.25	0.25	0.2	0.0064	0.2
Link2	10	1	0.5	0.5	0.008	0.8	0.8
Link3	10	1	0.5	0.5	0.008	0.8	0.8

Table 3. DH parameters of the FFSRMS

	$\alpha(rad)$	$l(m)$	$d(m)$	$\theta(rad)$
Link1	0.5π	0.5	0.5	θ_1
Link2	0	1	0	θ_2
Link3	0	1	0	θ_3

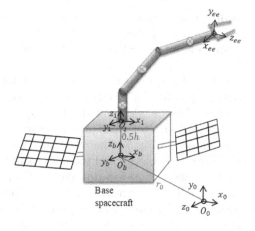

Fig. 5. Free flying space robot manipulator system

The initial configuration of space robot are given as;

$$[\phi_0, \theta_0, \psi_0]^T = [\frac{5\pi}{180}, \frac{20\pi}{180}, \frac{10\pi}{180}]^T (rad)$$

$$[x, y, z]^T = [0.11, 0.2, -0.3]^T (m)$$

$$[\theta_{10}, \theta_{20}, \theta_{30}]^T = [\frac{10\pi}{180}, \frac{10\pi}{180}, \frac{10\pi}{180}]^T (rad)$$

The final values of a FFSRMS are set to be;

$$[\phi_f, \theta_f, \psi_f] = [0, 0, 0](rad)$$

$$[x_f, y_f, z_f] = [0, 0, 0](m)$$

$$[\theta_{1f}, \theta_{2f}, \theta_{3f}] = [\frac{40\pi}{180}, \frac{60\pi}{180}, \frac{-30\pi}{180}](rad)$$

The change in mass of fuel due to thruster firing and unknown payload carried by the manipulator end-effector are considered as parametric uncertainty, as a result the mass of space robot in turn the mass and inertia parameters are also changed. For simulation the mass and inertia of link three and base spacecraft Table 2 are changed to $m_b = 90\,kg, m_3 = 11\,kg, I_0xx = I_0yy = I_0zz =$

$27\,\text{kgm}^2, I_{3\text{xx}} = 0.0088\,\text{kgm}^2, I_{3\text{yy}} = 0.88\,\text{kgm}^2, I_{3\text{zz}} = 0.88\,\text{kgm}^2$. The disturbance force or torques are chosen as [30].

$$\tau_{dm} = [0.1sin(1.2t), 0.108sin(0.1t), -0.1sin(0.6t)]$$

$$\tau_{db} = [0.4sin(0.3t), 0.3cos(0.1t), 0.2sin(0.3t)]$$

$$f_b = [0.1sin(0.3t), 0.01cos(0.1t), 0.02sin(0.3t)]$$

Where, $\tau_{db}, f_b, \tau_{dm}$ are disturbance force & torque on the base spacecraft positions, attitudes and manipulator joints respectively. The parameter of controller are selected by trail and error method which fulfill our desired objectives.
$K_p = 3I_{9,9}, K_i = 1.5I_{9,9}, K_s = 2I_{9,9}, G = 0.1I_{9,9}$.

Figure 6 shows the desired base spacecraft position, orientation and manipulator joint angle trajectories planned using fifth order polynomial path planning
technique. The manipulator joint angles and the base spacecraft positions &
orientations as well as the end effector position trajectory tracking capability
with an initial errors are shown in Fig. 7. The result indicates the manipulator
joint angles track their desired trajectories and the base spacecraft positions and
attitudes are re-oriented at their target location (zero). Figure 7d depicts that
the end effector successfully follow the prescribed path in Cartesian space. The
simulation results in Fig. 7b & 7c are used to demonstrate the situation in which
the base spacecraft must change its attitude or position to re-orient the earth
pointing antenna so as not to loss ground communication or to keep the solar
panel aligned with the sun for maximum power.

Due to an initial deviation of the system states, the manipulator joint torques
as well as, the base spacecraft thrust forces & attitude torques are high initially,
as shown in Fig. 8a, 8b, 8c respectively. However, the magnitude of this manipulator and base spacecraft torques and forces are bounded within $\pm0.5\,\text{Nm}$,
$\pm2\,\text{Nm}$ and $\pm1\,\text{N}$ as shown in Fig. 8 respectively at steady state. Practically the
base spacecraft attitudes are controlled by either reaction or moment wheel and
they have limited torque up to $\pm5\,\text{Nm}$ & $\pm1\,\text{Nm}$ respectively. Thus, the result
confirms this reality.

Figure 9 shows the sliding surface are asymptotically converges to the origin
and staying there over time. Therefore, all the states of the system (errors in
this case) are attracted to it and stay along with the slide surface.

Figure 10 shows the trajectory tracking errors of the manipulator joint angles
and the base spacecraft position & orientation angles as well as the thrust forces
and torques of the system due to the addition of external disturbances.

Figure 11 shows the control torques and thrust forces of the system with
FSMC & SMC. This results indicate that SMC has chattering problem. However,
this problem is solved by employing a fuzzy logic controller along with SMC, as
shown in Fig. 11. Table 4 shows the integral of time multiplied by absolute error
(ITAE) of the system with nominal model, disturbance and parametric variation
of both controller. The proposed controller has a relatively smaller ITAE than
the sliding mode controller, as indicated in Table 4. Thus, the results show that
FSMC has better trajectory tracking performance than SMC.

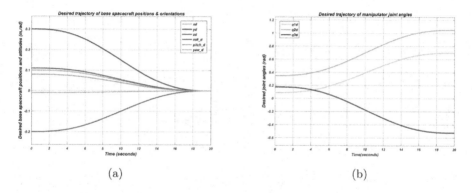

Fig. 6. The desired trajectory of base position and attitude as well as the manipulator joint angle designed using fifth order polynomial path planning technique:(a) Desired trajectory of spacecraft position and attitude (b) Desired joint angle trajectory

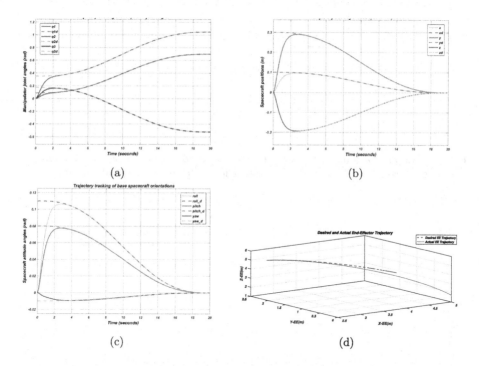

Fig. 7. Manipulator joint angles and Base spacecraft position & orientation and end effector trajectory tracking:(a) Manipulator joint angles trajectory tracking (b) Base spacecraft position trajectory tracking (c) Base spacecraft orientation trajectory tracking (d) end effector trajectory tracking

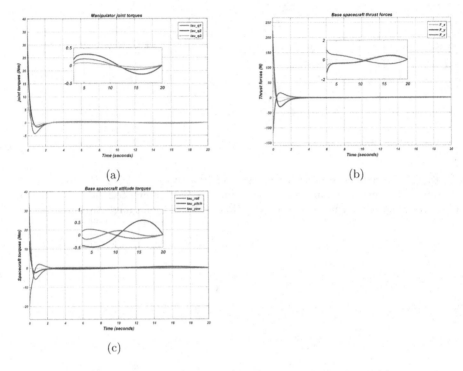

Fig. 8. Manipulator joint angle and base spacecraft torques & forces:(a) Manipulator joint angle torques (b) base spacecraft thrust forces (c) base spacecraft torques

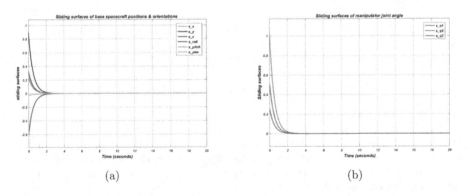

Fig. 9. Sliding surface of the base spacecraft positions and orientations and the manipulator joint angles:(a) Sliding surface of spacecraft positions and orientations (b) Sliding surface of the manipulator joint angles

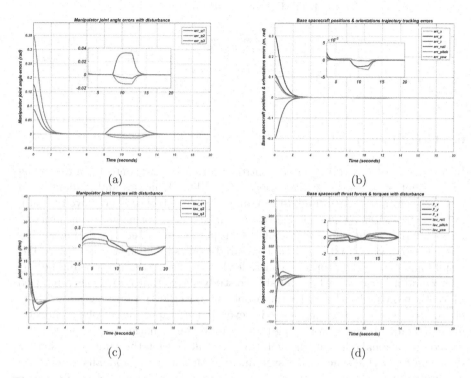

Fig. 10. Manipulator joint angle and base spacecraft position & attitude angle errors and the corresponding torques & forces with disturbance (a)Manipulator joint angle errors (b) base spacecraft position errors (c) Manipulator joint torques (d) base spacecraft forces & torques

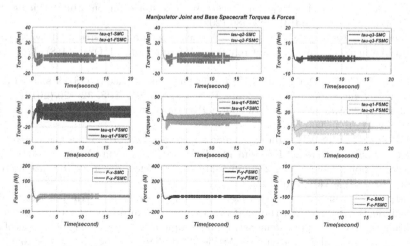

Fig. 11. Manipulator joint and base spacecraft torques & forces with FSMC & SMC

Table 4. ITAE for a FFSRMS with FSMC

	$x(m)$	$y(m)$	$z(m)$	$\phi(rad)$	$\theta(rad)$	$\psi(rad)$	$\theta_1(rad)$	$\theta_2(rad)$	$\theta_3(rad)$
FSMC-nominal	5.052×10^{-2}	1.516×10^{-1}	1.010×10^{-1}	5.752×10^{-2}	4.174×10^{-2}	2.822×10^{-2}	4.409×10^{-2}	1.763×10^{-1}	8.817×10^{-2}
SMC-nominal	6.933×10^{-2}	2.284×10^{-1}	1.482×10^{-1}	7.825×10^{-2}	5.425×10^{-2}	1.586×10^{-2}	5.951×10^{-2}	2.680×10^{-1}	1.229×10^{-1}
FSMC-disturbance	5.568×10^{-2}	1.553×10^{-1}	1.080×10^{-1}	1.005×10^{-1}	1.055×10^{-1}	2.224×10^{-2}	1.745×10^{-1}	5.247×10^{-1}	9.796×10^{-1}
SMC-disturbance	6.935×10^{-2}	2.284×10^{-1}	1.482×10^{-1}	8.314×10^{-2}	8.443×10^{-2}	1.570×10^{-2}	1.820×10^{-1}	7.333×10^{-1}	1.453×10^{0}
FSMC-parametric change	5.607×10^{-2}	1.676×10^{-1}	1.109×10^{-1}	6.019×10^{-2}	4.459×10^{-2}	2.659×10^{-2}	9.721×10^{-2}	2.232×10^{-1}	1.650×10^{-2}
SMC-parametric change	6.627×10^{-2}	2.164×10^{-1}	1.402×10^{-1}	7.520×10^{-2}	5.138×10^{-2}	1.525×10^{-2}	5.607×10^{-2}	2.464×10^{-1}	1.337×10^{-1}

6   Conclusion

In this paper a fuzzy sliding mode controller for trajectory control of a free flying space robot manipulator system was studied. The dynamics and kinematics of a the system has been formulated. A fuzzy logic controller is employed to substitutes the switching control algorithm, which causes for a chattering problem in SMC. To see the effectiveness of the proposed controller, simulation is conducted using MATLAB and it is compared with sliding mode controller. The result shows that, the proposed controller has better trajectory tracking performance and it is robust. The comparative study shows that FSMC has a reduced ITAE than SMC. Furthermore, the proposed controller(FSMC) has successfully eliminates the chattering problem. However, in this paper the coriolis /centerifugal matrix are computed following Euler-Lagrangian equation of motion which is not computational efficient. In the future, we will develop the same controller for dual arm free flying space robot system, which has a great capability to perform a complex tasks than a single arm.

References

1. Ding, X.L., Wang, Y.C., Wang, Y.B., Kun, X.: A review of structures, verification, and calibration technologies of space robotic systems for on-orbit servicing. Sci. Chin. Technol. Sci. **64**(3), 462–480 (2021)
2. Guang, Z., Heming, Z., Liang, B.: Attitude dynamics of spacecraft with time-varying inertia during on-orbit refueling. J. Guid. Control. Dyn. **41**(8), 1744–1754 (2018)
3. Li, W.J., Cheng, D.Y., Liu, X.G., et al.: On-orbit service (OOS) of spacecraft: a review of engineering developments. Prog. Aerosp. Sci. **108**, 32–120 (2019)
4. Mark, C.P., Kamath, S.: Review of active space debris removal methods. Space Policy **47**, 194–206 (2019)
5. Yoshida, K.: ETS-VII flight experiments for space robot dynamics and control. In: Rus, D., Singh, S. (eds.) Experimental Robotics VII. Lecture Notes in Control and Information Sciences, vol. 271, pp. 209–218. Springer, Berlin (2001). https://doi. org/10.1007/3-540-45118-8_22
6. https://www.nasa.gov/mission_pages/station/research/news/rrm_practice.html
7. Ogilvie, A., Allport, J., Hannah, M., Lymer, J.: Autonomous satellite servicing using the orbital express demonstration manipulator system. In: Proceedings of the 9th International Symposium on Artificial Intelligence, Robotics and Automation in Space (i-SAIRAS 2008), Los Angeles (2008)

8. Flores-Abad, A., Ma, O., Pham, K., Ulrich, S.: A review of space robotics technologies for on-orbit servicing. Prog. Aerosp. Sci. **68**, 1–26 (2014)
9. Mohamed, A., Saaj, C., Seddaoui, A., Nair, M.: Linear controllers for free-flying and controlled-floating space robots: a new perspective. Aeronaut. Aerosp. Open Access J. **4**(3), 97–114 (2020)
10. Gu, Y.-L., Yangsheng, X.: A normal form augmentation approach to adaptive control of space robot systems. Dyn. Control **5**(3), 275–294 (1995)
11. Parlaktuna, O., Ozkan, M.: Adaptive control of free-floating space robots in Cartesian coordinates. Adv. Rob. **18**(9), 943–959 (2004)
12. Xu, Y., Shum, H.-Y., Kanade, T., Lee, J.-J.: Parameterization and adaptive control of space robot systems. IEEE Transa. Aerosp. Electr. Syst. **30**(2), 435–451 (1994)
13. Abiko, S., Hirzinger, G.: Adaptive control for a torque controlled free-floating space robot with kinematic and dynamic model uncertainty. I:n 2009 IEEE/RSJ International Conference on Intelligent Robots and Systems, pp. 2359–2364. IEEE (2009)
14. Xie, L., Yu, X., Chen, L.: Robust fuzzy sliding mode control and vibration suppression of free-floating flexible-link and flexible-joints space manipulator with external interference and uncertain parameter. Robotica **40**(4), 1–23 (2021)
15. Liu, Y., Yan, W., Zhang, T., Yu, C., Tu, H.: Trajectory tracking for a dual-arm free-floating space robot with a class of general non-singular predefined-time terminal sliding mode. IEEE Trans. Syst. Man Cybern.: Syst. 52(5), 3273-3286 (2021)
16. Zhang, X., Liu, J., Gao, Q., Zhaojie, J.: Adaptive robust decoupling control of multi-arm space robots using time-delay estimation technique. Nonlinear Dyn. **100**(3), 2449–2467 (2020)
17. Jia, S., Shan, J.: Continuous integral sliding mode control for space manipulator with actuator uncertainties. Aerosp. Sci. Technol. **106**, 106192 (2020)
18. Wilde, M., Kwok Choon, S., Grompone, A., Romano, M.: Equations of motion of free-floating spacecraft-manipulator systems: an engineer's tutorial. Front. Rob. AI **5**, 41 (2018)
19. Markley, F.L., Crassidis, J.L.: Fundamentals of Spacecraft Attitude Determination and Control, vol. 1286. Springer, Cham (2014)
20. Longman, R.W., Lindberg, R.E., Zedd, M.F.: Satellite mounted robot manipulators - new kinematics and reaction moment compensation. Int. J. Rob. Res. **6**(3), 87–103 (1987)
21. Vafa, Z., Dubowsky, S.: On the dynamics of manipulators in space using the virtual manipulator approach. In: 1987 IEEE International Conference on Robotics and Automation, pp. 579–585 (1987)
22. Schaub, H., Junkins, J.L.: Analytical Mechanics of Aerospace Systems, 4th edn. AIAA Education Series, USA (2002)
23. Siciliano, B., Sciavicco, L., Villani, L., Oriolo, G.: Robotics. Springer-Verlag, London (2010)
24. Hughes, P.C.: Spacecraft Attitude Dynamics. Dover Publications, Mineola, NY (2004)
25. Spong, M.W., Hutchinson, S., Vidyasagar, M.: Robot Modeling and Control. John Wiley and Sons, New York (2006)
26. Edwards, C., Spurgeon, S.: Sliding Mode Control: Theory and Applications. CRC Press, Boca Raton (1998)
27. Holkar, K.S., Waghmare, L.M.: Sliding mode control with predictive PID sliding surface for improved performance. Int. J. Comput. Appl. **78**, 1–5 (2013)
28. Congqing, W., Pengfei, W., Xin, Z., Xiwu, P.: Composite sliding mode control for a free-floating space rigid-flexible coupling manipulator system. Int. J. Adv. Rob. Syst. **10**(2), 124 (2013)

29. Wang, X., Shi, L., Katupitiya, J.: Coordinated control of a dual-arm space robot to approach and synchronise with the motion of a spinning target in 3D space. Acta Astronaut. **176**, 99–110 (2020)
30. Zhang, Z., Li, X., Wang, X., Zhou, X., An, J., Li, Y.: TDE-based adaptive integral sliding mode control of space manipulator for space-debris active removal. Aerospace **9**(2), 125 (2022)

Prediction of Teff Yield Using a Machine Learning Approach

Adugna Necho Mulatu[(✉)] [iD] and Eneyachew Tamir

Computer Engineering Program, Faculty of Electrical and Computer Engineering, Bahir Dar
Institute of Technology, Bahir Dar University, Bahir Dar, Ethiopia
adugna.necho@bdu.edu.et, eneyachewt@gmail.com

Abstract. Teff is one of the main ingredients in everyday food for most Ethiopians, and its production mainly depends on natural conditions of the climate, unpredictable changes in the climate, and other growth factors. Teff production is extremely variable on different occasions and creates complex scenarios for prediction of yield. Traditional methods of prediction are incomplete and require field data collection, which is costly, with the result being poor prediction accuracy. Remotely sensed satellite image data has proven to be a reliable and real-time source of data for crop yield prediction; however, these data are enormous in size and difficult to interpret. Recently, machine-learning methods have been in use for processing satellite data, providing more accurate crop prediction results. However, these approaches are used in croplands covering vast areas or regions, requiring huge amounts of cropland mask data, which is not available in most developing countries, and may not provide accurate household level yield prediction. In this article, we proposed a machine learning based Teff Yield Prediction System for smaller cropland areas using publicly available multispectral satellite images, that represent spectral reflectance information related to the crop growth status collected from different satellites (Landsat-8, Sentinel-2). For this, we have prepared our own satellite image dataset for training. A Convolutional Neural Network was developed and trained to be fit for a regression task. A training loss of 3.3783 and a validation loss of 1.6212 were obtained; in other words, the model prediction accuracy was 98.38%. This shows that our model's performance is very promising.

Keywords: Teff yield · Multispectral satellite images · Machine learning · Convolutional neural network

1 Introduction

Crop farming in Ethiopia comprises large variations in both the variety of crops and growing areas. According to a study (Alemayehu 2012), small sized crop lands constitute 96% of farm land and yield a significant portion of total production for the major crops, including teff, maize, wheat, barley, and sorghum. Among these crops, teff is the most staple food in the country, and the same study shows that teff accounts for nearly 20%

B. H. Woldegiorgis et al. (Eds.): ICAST 2022, LNICST 455, pp. 159–176, 2023.
https://doi.org/10.1007/978-3-031-28725-1_10

of the harvested land. As one of the most important elements in everyday food for most Ethiopians (Lee 2018; Nandeshwar et al. 2020), the country is considered the largest teff producer in the world (Firdisa 2016; Tamirat and Tilahun 2020). In most places, it is sown by hand with the seeds left exposed (Sate and Tafese 2016). Teff covers the largest farmland (28% of crop area) in the country (FAO 2015; Firdisa 2016; Nascimento et al. 2018; Wato 2019), but its yearly production is very low as compared to other crops (Lakew and Berhanu 2019; Tesfahun 2018).

In terms of teff production area, the largest producing regions are Amhara (around 85%) and Oromia (around 87%) (Lee 2018). These regions also constitute the largest populations that consume this crop. Teff is a warm-season annual cereal that provides significant and unique nutritional qualities for health-conscious consumers. Nonetheless, it is not yet fully exploited and its contribution to food security requires further study. Teff is a low-risk crop that can withstand a variety of biotic and abiotic challenges. It can also continue to be a reliable source of food to meet the world's rapidly rising needs. Ethiopia is currently the world's largest producer of teff and it is only in this country that it is used as a staple crop, although other continents like the US and European countries are starting to promote its production and utilization (Lee 2018). Although there is a growing interest in the global market, in comparison to other important cereals, teff productivity is quite low (14.8 q/h) (CSA (Central Statistical Authority) 2016), having a number of factors to consider, including reduced fertility of soil, improper management of fertilizers and weeds, and irregular distribution of rainfall (Fenta 2018; Wakjira 2018; Tamirat and Tilahun 2020).

Teff is, and will continue to be, the most important and highly demanded crop due to its specific benefits and increasing international attention. As a result, increased production of this remarkable crop must be prioritized. Climate and other environmental changes have an impact on teff productivity, which is largely determined by natural climate conditions, which have a significant impact on teff yield.

Crop yield estimates before actual production is necessary in regions that depend on rain-fed agriculture with climatic uncertainties for taking various policy decisions. The results of yield estimates improve the timely availability of information for food security, allowing authorities to take necessary preparations to prevent famine, particularly during natural disaster years. Traditionally, in Ethiopia, the annual estimation of crop land area and production has been conducted by ECSA (Ethiopia Central Statistical Agency) at the national level (Fikre 2015).

In addition, estimation of crop yield in general is critical for food producers, policymakers, importers/exporters, seed producers, growers, and farmers in Ethiopia to improve national food security. However, due to several complicated aspects, the task is highly difficult and varies according to growing region and time of production. Identifying and addressing the difficulties is an overriding issue for understanding the stochastic nature of crop yield and devising ways to address them. Several crop yield prediction models have been developed to enhance the accuracy of yield estimations. However, getting accuracy is not easy due to the complexity and variability of natural settings, since many factors influence crop production and hence crop output.

Due to the limitations of traditional methods, modern methods such as artificial intelligence and machine learning (ML) have been introduced for efficient crop yield prediction. Considerable research efforts have been made in applying multivariate regression, decision trees, association rule mining, and artificial neural networks for crop prediction. Many key characteristics make ML methods to be potentially applied in yield prediction problems. These methods can be used to tackle complex and nonlinear real-world classification and regression problems. Not only does machine learning provide a powerful and flexible framework for data-driven decision making, but it also allows for the integration of expert knowledge into the system (Anna 2018).

In this research, we proposed a yield prediction model for teff crop using multispectral satellite images for training the model, downloaded from different sources in different years for the selected sample sites based on a ML technique, specifically a type of deep learning algorithm called the Convolutional Neural Network (CNN) model. The selected model is trained using a series of multiple-band satellite images cropped from the downloaded satellite images for specific plots of land for the main teff growing season, which is the duration from the sewing month to before harvest month (June 25–November 30 is considered) in the selected prominent Teff growing sites in the region. For ground truth data, average teff yield data was collected for the different plots for 10 years from 2010 to 2020. Primary data was collected on site by interviewing farmers. Data from CSA and other sources was also used to verify the correctness of the data provided by the farmers.

The main contributions of the paper include:

1. A dataset consisting of multiple bands of multispectral images showing temporal variation in biomass of Teff growth for different years, from different satellite sources
2. The use of freely available satellite images for crop prediction in developing countries (which may not afford commercial satellite data), which opens a window of further research for other applications
3. The use of modern deep learning technique for Teff yield prediction purpose
4. A basic CNN model was adjusted at the output layer (replacing sigmoid activation with ReLU) to be used for a regression problem (mostly, CNNs are effectively used for classification problems)

The rest of the paper is structured as follows. In Sect. 2, different related papers are reviewed. In Sect. 3, we discussed the problem setting and model of the proposed system. In Sect. 4, step-by-step procedures for developing and implementing the model are presented. Implementation, results, and analysis are presented in Sects. 5 and 6, respectively. Section 7 provides a conclusion and future work.

2 Related Works

In many countries, traditional crop production forecasting approaches have been used to collect data from field trips and reports. However, a number of problems and challenges are associated with these methods. The subjectiveness and insufficient ground truth data of these methods expose them to the introduction of large errors. Sampling and data collection is too costly and time consuming, and may not represent the real picture on the ground. Due to this, other methods like crop model-based monitoring systems and remote sensing-based forecasting methods (that are based on NDVI) have been drawing the attention of many researchers. Crop monitoring and yield forecasting with remote sensing in Ethiopia was largely done at the regional/national level, covering broad areas with low-resolution imagery (Rojas 2006; Greatrex 2012).

Crop yield estimation and crop growth monitoring using geospatial and remote sensing technology are used for clustered crop areas to save money and time but require accurate data, technology, and expertise. It assists growers, government agencies, and crop insurance firms in planning and contributing to the national goal of food security with the following limitations to this type of research: (1) for several years, some of the areas have lacked local level ground truth data; (2) the approach necessitates additional regression analyses based on selected machine learning algorithms; (3) future analyses based on a variety of variables such as climate, agronomic factors, crop factors (e.g., harvest index), and others, in addition to ground truth; and (4) their hypothesis was to explore the yield estimation under different scenarios based on the capability of GIS technology in conjunction with machine learning algorithms in clustered areas, however, this will require additional testing for fragmented cropping systems (Hailu et al. 2022).

Ethiopian crop agriculture grows a variety of crops in many regions of the country. According to Alemayehu's research, Ethiopia's agricultural economy depends primarily on five primary cereals such as teff, wheat, maize, sorghum, and barley, which cover 75% of total cultivated land and 29% of agricultural GDP in 2005/06 (14% of the country's overall GDP) (Alemayehu 2010). In Ethiopia, Abiy introduced a GIS and RS based crop yield prediction model for the maize crop in the south Tigray Zone derived from time series data of SPOT VEGETATION, actual and potential evapotranspiration, and rainfall estimate satellite data for the years 2003–2012. Through correlation analyses, he used the input data to validate the grain yield from CSA and processed the data to forecast maize yield and map it. He got a rainfall estimate and an average NDVI that associate to maize output with 85% and 80% of the variation, respectively, by validating the generated spectro-agro-meteorological yield model by comparing it to the estimated zone level yields from CSA (r2 = 0.88, RMSE = 1.405 qha1, and a 21% coefficient of variation) (Abiy 2014).

Aklilu et al. in (Fikre 2015) filled the gap of Abiy in (Abiy 2014) in maize yield prediction in the south Tigray Zone by developing a model to forecast wheat yield for the year 2014 for the east Arsi zone, Ethiopia using remote sensing and GIS using time series data of SPOT VEGETATION, actual and potential evapotranspiration, rainfall estimate, and satellite data for the years 2004–2013. He employed CSA ground truth data to evaluate the indices' strength, and used correlation analysis to find relationships between crop yield, spectral indices, and agrometeorological variables for wheat crops over the long wet season (Meher). He discovered indices with a strong relationship to wheat yield which are highly correlated with the average Normalized Difference Vegetation Index (NDVIa) and rainfall, with 96% and 89% correlations, respectively.

A machine learning approach has been introduced to solve the limitations of conventional, remote sensing, and geospatial methods in crop yield. One of the approaches is deep learning. The researchers (You et al. 2017) introduced a deep learning framework for crop yield prediction models based on histograms and a Deep Gaussian Process framework to avoid spatially correlated errors and inspire other applications in remote sensing and computational sustainability using publicly available remote sensing data to improve existing techniques using hand-crafted features and a novel dimensionality reduction technique to train a CNN or LSTM. They tested the method on county-level soybean yield prediction in the United States and found that it beat other methods.

Other researchers (Sun et al. 2019) introduced a deep CNN-LSTM model using weather data, MODIS Land Surface Temperature (LST) data, MODIS Surface Reflectance (SR) data, and historical soybean yield data to train the model for both end-of-season and in-season soybean yield prediction in the CONUS at the county level. All of the training data was combined and turned into histogram-based tensors for deep learning using the Google Earth Engine (GEE). The proposed model outperformed the pure CNN or LSTM model in both end-of-season and in-season scenarios, with an average RMSE of 329.53 from 2011 to 2015 and an R2 of 0.78 for the 5 years combined.

A deep learning framework based on CNNs and RNNs was proposed in (Khaki et al. 2020) for crop yield prediction utilizing environmental data and management strategies. Utilizing historical data, the CNN-RNN model was used to forecast corn and soybean yields over the whole Corn Belt (containing 13 states) in the United States using random forest (RF), deep fully connected neural networks (DFNN), and LASSO. It had a root-mean-square-error (RMSE) of 9% and 8% of their respective average yields, respectively, significantly surpassing all other methods examined.

The authors in (Debalke and Abebe 2022) introduced a maize yield forecast model using time series data from the Moderate Resolution Imaging Spectroradiometer NDVI, actual evapotranspiration, and potential evapotranspiration, and Climate Hazards Group Infrared Precipitation, with the indicators' correctness checked against official grain yield data from CSA in Kafa Zone, Ethiopia. The average NDVI and the Climatic Hazards Group Infrared Precipitation using station data reveal significant connections with maize productivity, with correlations of 84% and 89%, respectively. The spectro-agro meteorological yield model (r2 = 0.89, RMSE = 1.54 qha1, and 16.7% coefficient of variation) satisfactorily matched the CSA's predicted Zone level yields.

Another prediction system based on machine learning techniques such as decision tree, multivariate logistic regression, and k-nearest neighbor model was developed in (Cedric et al. 2022) to predict by combining climatic data, meteorological data, agricultural yields, and chemical data with hyper-parameter tuning methodology for the yield crops (rice, maize, cassava, seed cotton, yams, and bananas) at the country-level in West African countries throughout the year. The decision tree model performed well with a coefficient of determination (R2) of 95.3%, while the K-Nearest Neighbor model and logistic regression performed well with R2 = 93.15% and R2 = 89.78%, respectively. As a result, the decision tree and K-Nearest Neighbor models' prediction outputs are associated with the predicted data, demonstrating the efficacy of the model.

Ayalew et al. in (Ayalew et al. 2022) proposed a hybrid CNN-DNN model and compared with other machine learning algorithms such as the XGBoost machine learning (ML), CNN-DNN, CNN-XGBoost, CNN-RNN, and CNN-LSTM how performs against various performance criteria. They tested their findings using a publicly available soybean dataset with 395 characteristics, including meteorological and soil conditions, and 25,345 samples. The suggested model outperforms other models, with an RMSE of 0.266, MSE of 0.071, and MAE of 0.199. The model's predictions fit with an R2 of 0.87. The XGBoost model came in second place, taking less time to run than the other DL-based models.

In contrast to other approaches that use cropland masks that span broad areas, we generated and used an individual plot of land multispectral satellite image based dataset in our research. This enables farmers to anticipate teff yields based on individual households, which aids in agricultural field management. For prediction, our approach employs satellite biomass photos, which can be simply downloaded without the need for any additional software or physical field visits. CNN is most commonly employed for image classification, but it was also used for regression in our system, similar to the approach followed in (Adrian 2021), by modifying the fully connected output layer, activation function, and loss functions used in classification problems.

3 Problem Setting

The traditional technique of crop yield prediction in Ethiopia is practiced by collecting data based on field visits and reports which are subjective, costly, time consuming and prone to errors due to incomplete ground observation that leads to poor crop yield assessment and delay in reporting appropriate actions to be taken. This article proposed Teff Yield Prediction Model using Multispectral Satellite Imaging and Machine learning (ML) Approach with the following specific objectives: 1) Preparation of dataset that consists of multitude of multi-layer data from different sources 2) Devising data fusion framework for combining input data from different sources for model training 3) Exploring appropriate ways of combining ML and statistical methods for improved accuracy 4) Design of the proposed model and 5) Implementation and testing of the model.

For training the model, we downloaded multiple bands of multispectral images of the selected areas from different sources, in different years for the selected sites. The multispectral bands used are related to temporal variation in biomass. The machine learning model is trained using a series of multiple-band satellite images cropped from the downloaded satellite images for specific plots of land for the main Teff growing season, which is the duration from the sewing month to before harvest month (June 25–November 30 is considered) in the selected prominent Teff growing sites in the region. For ground truth data, average teff yield data was collected for the different plots for 10 years from 2010 to 2020, with primary data collected on site, by interviewing farmers. Data from CSA and other sources was also used to verify the correctness of the data provided by the farmers.

One of the biggest challenges in data collection was obtaining an exact source of data for the specific plot, since not enough documented information was available from the farmers, agricultural extension workers, or even from woreda agricultural offices. Therefore, we had to rely on the interviews with farmers and agricultural extension workers for ground truth data preparation. Another challenge was that a specific plot of land does not grow Teff continuously (as farmers use crop rotation, or leaving the land fallow to reduce soil stress). This creates a set of satellite image data outliers since different crops were grown and harvested during the rotation season, and since the biomass images of Teff and other crops/non-crops.

For each of the five selected regions, shown in Fig. 1, we have collected both satellite image and ground truth data for ten plots in each region. The minimum plot size considered was 1ha since we used free satellite data, which has poor resolution. Since the land management in the region is very fragmented, finding a plot with an area of 1ha or more was challenging. Therefore, we had to combine contiguous lands to get a larger area as used in (Mengesha et al. 2018). Totally, image data was collected for 500 satellite data from two satellites (Landsat L1, L2 and Sentinel A, B) with ground truth collected (five regions × 10 plots per region × 10 years for each plot). In each of the 500 satellite data sets, multispectral images the area of the selected plots depends on the size of the available contiguous land at each site.

Satellite data: One of the sources from which data is obtained is the Landsat Program. NASA and the US Geological Survey jointly operate the Landsat Program, which consists of a series of Earth-observing satellite missions. It is the world's largest collection of moderate-resolution remote sensing data that is continuously acquired. Landsat 8 and Landsat 9 are the currently orbiting, active satellites. The data has been obtained uninterruptedly starting from 1972 capturing different parts of the earth, and being a valuable source of information for research in various fields (U.S. Geological Survey 2020).

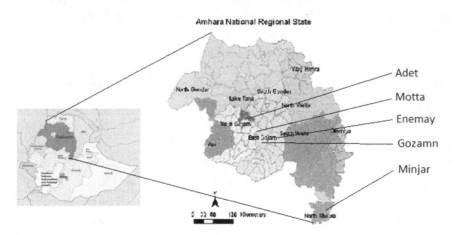

Fig. 1. Map of the study area.

For this article, we have used the Level-2 and Level-1 data from Landsat Collection 2 multispectral data. In addition, we have also used similar data downloaded from the Copernicus Sentinel-2 mission, which comprises a constellation of two polar-orbiting satellites that aim at monitoring variability in land surface conditions. This mission has a wide swath (290 km) and a high revisit time of 10 days at the equator with 1 satellite and 5 days with 2 satellites. We have downloaded data from both satellites (SA and SB) with a revisit time of nearly 5 days at the selected sites.

4 Materials and Methods

4.1 Data Sources

As previously stated, the dataset was created using multispectral temporal satellite images. As indicated in Table 1, the downloaded data includes a variety of remote sensing data, including multispectral images obtained at various bands. The whole zipped folder was downloaded for all times of visit within the given years (2005–2020). Each of the Landsat satellite's sensors was intended to collect data for a number of different frequencies in the electromagnetic spectrum. Landsat 8 measures a number of frequency bands across the electromagnetic spectrum (each range is called a band). There are numerous bands on Landsat 8. The full list of bands can be found in Table 1.

The images from the Landsat 8 Operational Land Imager (OLI) and Thermal Infrared Sensor (TIRS) have nine spectral bands, each with a spatial resolution of 30 m. Coastal and aerosol research benefit from the new band 1 (ultra-blue). The new band 9 is beneficial for detecting cirrus clouds. Band 8 (panchromatic) has a resolution of 15 m. Thermal bands 10 and 11 are collected at 100 m and are effective for delivering more accurate surface temperatures. The picture spans 170 km north to south and 183 km east to west (106 mi by 114 mi). Bands 1–9 are used to collect image data. Sentinel-2A Multispectral Instrument (MSI) data has spectral bands that are quite similar to Landsat 8 and 9 (with the exception of the thermal bands of Thermal Sentinel). The spectral bands in the Sentinel-2A Multispectral Instrument (MSI) data are fairly comparable to those in Landsat 8 and 9 (except for the thermal bands of the Thermal Infrared Sensor (TIRS)). Sample footprints acquired from Landsat and Sentinel sources are shown in Fig. 1.

Table 1. Landsat 8 list of bands

Bands	Wavelength (μm)	Resolution
B1-Coastal aerosol	0.433–0.453	30 30 m
B2-Blue	0.450–0.515	30 m
B3-Green	0.525–0.600	30 m
B4-Red	0.630–0.680	30 m
B5-Near Infrared (NIR)	0.845–0.885	30 m
B6-SWIR 1	1.560–1.660	30 m
B7-SWIR 2	2.100–2.300	30 m
B9-Cirrus	1.360–1.390	30 m
B10-Thermal Infrared (TIRS) 1	10.6–11.2	100 m
B11-Thermal Infrared (TIRS) 2	11.5–12.5	100 m

4.2 Data Collection

The coordinates used for downloading satellite data were acquired using a hand-held Garmin 70 GPS. In order to get a good resolution image and to compensate for the moderate resolution of the publicly available satellite images, we had to search for larger contiguous plots of land. The minimum contiguous/single plot assumed was 1 Ha (100 m × 100 m). However, this gave us very poor resolution of downloaded and cropped images (e.g. only 5 × 5 pixels with a 20-m resolution satellite). The satellite data was downloaded from the U.S. Geological Survey (USGS) Earth Explorer (EE). This online earth explorer portal has a number of features including searching and browsing data online, viewing images, downloading zipped data and exporting metadata. Through enhanced user interface. Downloading was done manually by entering the coordinates and other attribute values in the USGS portal. An example attribute setting is shown in the table below, for the Motta site, plot 1. A total of more than 500 GB of 10-year data was downloaded for the 5 sites, each with 10 plots, for the different months in the harvest season of each year. Ground truth yield data was collected on site by interviewing landowners and agricultural extension workers (Table 2).

Table 2. USGS portal attribute settings/ filter criteria for downloading the plot with four corner coordinates (11.0709, 37.8955, 11.0694, 37.8972) which represent East Gojjam – Motta – Plot1

Attribute	Setting
Search Criteria	
Cloud cover	0–40%
Geocoding method	Address/Space
Shape	Predefined area
Coordinates (opposite corners – Lat, Long)	11.0709, 37.8955 11.0694, 37.8972
Dataset	
Satellites	Landsat Collection-2 Level-2, Landsat 8-9 OLI/TIRS C2 L2
	Landsat Collection 2 Level-1, Landsat 8-9 OLI/TIRS C2 L1
	Sentinel-2

4.3 Dataset Preparation

Most research works on crop yield prediction focus on a wide area of cropland, or for a certain region, using cropland masks already available, and some are based on multispectral images taken in real time using unmanned aerial vehicles. However, both of these methods do not apply to teff since no prior data is available. On the other hand, predictions made for large areas give only a general picture of production information over a region and may not accurately indicate individual farmer yield information. Hence, we had to prepare our own dataset that represents specific plots of land, particularly for the teff crop. For this purpose, we used the data from multispectral bands of satellite images downloaded from the two satellites. The satellites used for remote sensing crop growth images have sophisticated sensing instruments that allow them to capture detailed information about the biomass on the earth's surface.

The Landsat-8 satellite uses Operational Land Imager (OLI) and the Thermal Infrared Sensor (TIRS) sensors for acquisition of seasonal coverage of the global landmass at a spatial resolution of 30 m (visible, NIR, SWIR) [landsat.gsfc.nasa.gov/satellites/landsat-8/]; 100 m (thermal); and 15 m (panchromatic) of different bands at different wavelengths of the electromagnetic spectrum. The downloaded data contains a large amount of information, but we only used the 9 bands: 0–9 of multispectral images. As mentioned above, the downloaded footprint covers a very large area (relative to a single plot), and the plots of land to be studied had to be cropped out of this footprint for each band, and separately stored. For cropping and analysis of the downloaded geospatial images, the open-source QGIS 3.24.2 desktop application was used. Every image was cropped manually, and a total of more than 45,000 images were cropped from the 5 sites. These images were saved in a hierarchical directory structure for clarity and ease of access. Figure 2 shows the results of sample cropped images. (The procedures for downloading and cropping and saving image data are described in Appendix-A).

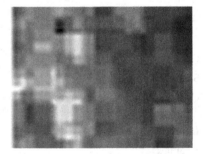

Fig. 2. Cropped images from Landsat L2 for the Gozamn area, site 1, captured on 2016-11-14, representing bands B1, B5 and B9

4.4   ML Models for Prediction

As discussed in the literature, machine learning models have proved to be much more efficient than traditional statistical models in prediction tasks. The models provide various supervised and unsupervised techniques to predict future outcomes from historical data by employing predictive models of regression. In regression, relationships between two or more variables (dependent/target and independent/predictor) are estimated, by discovering key patterns in data sets, and using simple linear regressions or complex and deep neural networks in a supervised manner.

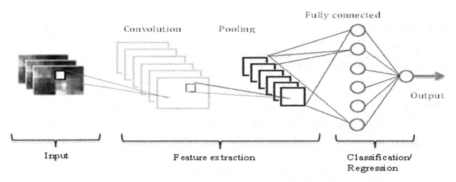

Fig. 3. The basic architecture of the CNN based teff yield prediction model

Deep learning, a branch of machine learning, is currently the most widely used model in various applications due to its various features, including its adjustable model parameters, powerful features learning ability, and end-to-end learning capability. Deep learning relies on model training based on an objective function that is related to the purpose of the model. A number of different types of deep learning models have been in use, but the Convolutional Neural Networks (CNNs) are becoming more and more efficiently employed in crop prediction tasks. CNNs can handle large amounts of complex labeled input, including image data, and computation.

The basic architecture of CNN consists of input, feature extraction, and classification/regression components (Fig. 3). Input data is represented as a multidimensional array and every portion of an input image is extracted by assigning input neurons, which are called receptive fields. The feature extraction layer consists of convolution, which produces feature maps, and a pooling layer, which reduces features. The classification layer consists of a fully connected network and an output layer. The convolution layer creates feature maps out of input data. Feature maps are filtered by applying a convolution operation using a convolution filter, which is a weight matrix that extracts certain features from the input image. The pooling layer decreases the size of the feature map by reducing the spatial size of images, which reduces computational cost. The Fully Connected (FC) layer consists of the weights and biases along with the neurons and is used to connect the neurons between two different layers.

The different steps of training algorithm:

1. Prepare dataset:
       Preprocess images (cloud cover removal, image resizing to 32x32 pixels).
2. Import necessary modules from Keras and other libraries
3. Load dataset:
       Divide dataset into training and testing sets;
       Load variables: train_input, train_target, test_input, test_target;
4. Process data:
       Normalize pixel intensity to [0,1]
       Reshape array into appropriate dimensions (instances, pixels(x,y), channels)
5. Create sequential model:
       Model = sequential()
       Init number of layers;
       For each layer,
              Create layers: CONVOLUTION, ReLU, Batch_N, MAX_ POOL
              X=Convolve_2D(train_input)
              X=Activation(ReLU)(train_input)
              X=Batch_Normalization(train_input)
              X=Max_Pooling2D(pool_size)
       Create dense layer for regression with linear activation
       Create single output layer
       Construct model: Model(inputs, x)
6. Compile model:
       Loss function = Mean_Sq_Err, Mean_Abs_Err
       Model_Optimizer = Adams
       Metrics = accuracy
7. Train model
       Model.fit(train_input, test_input, batch_size, num_ epochs)
8. Produce statistics:
       MSE,avg_prediction_result

5   Implementation

For implementing the prediction system, we have created a deep learning model with input, hidden, and output layers. The model was created using Keras and Tensorflow tools. The algorithm was implemented using Python and the Jupyter Notebook environment. Due to resource limitations on our local machines, a Google Colab notebook was used.

Most of the CNN models that exist in literature using images as input are for classification purposes. However, with some adjustments to the basic CNN architecture, CNNs can be used for regression problems similar to yield prediction. In our experiment, we used RMSE as the loss function, and replaced the sigmoid activation with ReLU to enable regression on a CNN. In addition, the output FC layer used for classification has to be changed to a single node, since only a single value of prediction is expected. The sigmoid function for the output layer is also replaced with linear regression. For optimization, we used the Adam optimizer with a learning rate of 0.001, which is fast and efficient, and has a higher convergence rate compared to other adaptive models.

Model definition: The implemented CNN model consists of the following layers: The input layer: consists of a sequence of preprocessed multispectral spatio-temporal images that represent the teff plant growth at different growth status. Since the cropped images are of variable size, the images are resized to 32×32 pixels. So the input consists of 32×32 pixels, single channel image data. The convolution layer: extracts valuable features from the input image. Multiple layers of convolutions were tested, using 3×3 filters. Pooling layer: Max pooling operation with 2×2 windows was used. Model hyper

parameters: In CNN (and other ML models), there are a number of hyper parameters that are required to configure the model. The research in (Aszemi and Dominic 2019) summarizes two classes of hyper parameters listed below:

a) Hyper parameters that determine the network architecture:

- *Kernel Size* – convolution filter size.
- *Stride* – number of steps the kernel moves over the input image.
- *Padding* – adding 0s at the borders of the image to account for filtering the edge of the image.
- *Hidden layers—layers* the sequence of convolution layers between input and output layers.
- *Activation functions* – allow the model to learn nonlinear prediction boundaries.

b) Hyper parameters that determine the network training:

- *Learning rate* – determines how weights are updated.
- *Momentum* – determines how the previous weight update influences the current weight update.
- *Number of epochs* – number of iterations of the entire training.
- *Batch size* – the number of inputs to be input before the updating weight.

These parameters, which are external to the model, greatly influence model performance. However, it is very difficult to have optimal values for a specific model. So the parameters are usually set by rules of thumb or heuristics, or based on previous research recommendations. The hyper parameters used and their settings are listed in Table 3.

Table 3. Hyper parameter settings for training the CNN model

Hyper parameter	Value
Kernel size	5 × 5
Stride	2
Padding	Yes, 0
Number of hidden layers (COV- > ReLU- > POOLING)	12
Activation functions	ReLU for conv. Layers, linear for output
Learning rate	0.01
Batch size	30
Number of epochs	30, 100, 200, 300

6 Results and Analysis

The proposed model training was done by varying the number of epochs while keeping other parameters fixed. Table 4 shows training loss and validation loss for the different epochs used for training. The results show that clearly, for lower epochs, larger values of loss were obtained. The very promising thing is that both training and validation losses were rapidly reduced as the epoch increased. This is also shown in Fig. 4a–d. However, at 300 epochs, unstable validation loss is observed as shown in Fig. 4d, so the best result obtained was at 200 epochs, with a training loss of 3.3783 and a validation loss of 1.6212.

Table 4. MSE for the different epochs tested during training

No. of epochs	30	100	200	300
Training loss (%)	55.15	13.42	3.38	1.24
Validation loss (%)	8.05	2.37	1.62	1.23

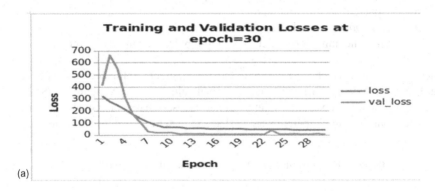

(a)

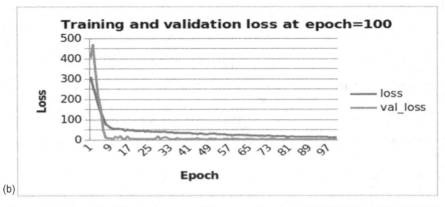

(b)

Fig. 4. Training and validation losses at different epochs of: (a) 30, (b) 100, (c) 200, (d) 300

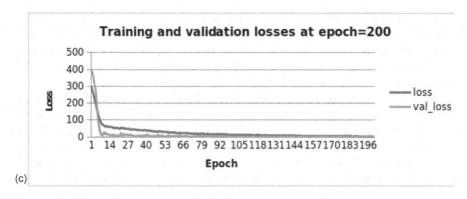

(c)

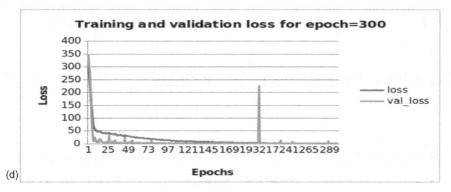

(d)

Fig. 4. (*continued*)

7   Conclusion and Future Work

In this research, we have implemented a machine learning based teff yield prediction system using multispectral satellite images collected from different satellites (Landsat-8, Sentinel-2). For this, we have prepared our own satellite image dataset for training. A CNN network was trained, with some modifications of the model to fit for regression. A training loss of 3.3783 and a validation loss of 1.6212 were obtained; in other words, the model prediction accuracy was 98.38. This shows that our model's performance is very promising. The results can be further improved and used for implementing household level yield prediction systems, such as mobile apps, so that farmers can use it for cropland management.

This work is only a start in teff yield prediction using a machine learning approach and can be extended or improved in a number of ways. A machine learning model is as accurate as the quality of training data. We have used only representative data and fewer samples from sample sites. The accuracy was obtained using satellite image data only. This accuracy can be further improved in a number of aspects. First, a large amount of dataset is required for training the CNN, so preparing such a dataset that incorporates many teff-growing regions in the country is a huge task that requires expertise and budget. On the other hand, one can extend this work for better results by incorporating environmental and soil data with the image data for training, using data fusion techniques.

Implementing and testing machine learning-based regression algorithms other than CNN could also be another future work. The trained model can also be used as a good starting point for prediction of teff yield in other regions of the country, by applying techniques like transfer learning. Apart from that, the dataset can be enriched by using more accurate remote sensing images of specific croplands taken using unmanned aerial vehicles. Due to time and budget constraints, we were not able to develop a mobile app for household level prediction purposes. In the future, we plan to develop such an app that integrates with the system, and provides yield prediction at any stage of the growth of teff to the farmers, by acquiring the most recent available satellite image data.

Acknowledgments. This research was fully funded by Bahir Dar Institute of Technology, BiT, Bahir Dar University, and we would like to acknowledge BiT for providing the funding.

Appendix-A: Training Progress for 30 Epochs

```
Epoch 1/30
120/120 [==============================] -5s 16ms/step-loss: 305.3471-val_loss: 382.2482
Epoch 2/30
120/120 [==============================] -2s 14ms/step-loss: 256.0030-val_loss: 381.8294
Epoch 3/30
120/120 [==============================] -2s 13ms/step-loss: 216.2533-val_loss: 327.8867
Epoch 4/30
120/120 [==============================] -2s 13ms/step-loss: 174.1100-val_loss: 233.1403
Epoch 5/30
120/120 [==============================] -2s 14ms/step-loss: 136.8061-val_loss: 134.1799
Epoch 6/30
120/120 [==============================] -2s 14ms/step-loss: 107.5622-val_loss: 55.5419
Epoch 7/30
120/120 [==============================] -2s 13ms/step-loss: 87.1361-val_loss: 22.9324
Epoch 8/30
120/120 [==============================] -2s 14ms/step-loss: 74.8587-val_loss: 12.3104
Epoch 9/30
120/120 [==============================] -2s 14ms/step-loss: 62.5622-val_loss: 10.0009
Epoch 10/30
120/120 [==============================] -2s 14ms/step-loss: 64.1516-al_loss: 10.5025
Epoch 11/30
120/120 [==============================] -2s 13ms/step-loss: 60.1122-val_loss: 6.8759
Epoch 12/30
120/120 [==============================] -2s 13ms/step-loss: 59.9180-val_loss: 9.5608
Epoch 13/30
120/120[==============================] -2s 14ms/step-loss: 55.1472-val_loss: 8.0470
```

Appendix-B: Sample Ground Truth Data from CSA

Places	Year	2013/14	2014/15	2015/16	2016/17	2017/18	2018/19	2019/20	2020/21
	Ethiopia	14.65	15.75	15.6	16.64	17.48	17.56	18.5	18.82
	Amhara	14.95	15.83	16.07	16.99	17.92	18	18.94	19.3
	North Shewa	15.13	16.88	16.51	17.8			18.95	21.06
	p1	14	15.2	16.4	17.2			19.3	20.1
	p2	15	16.2	16.3	17.1			19.2	20.9
CSA	p10	15.5	16.1	16	17.3			18.7	21.66
Minjar	East Gojam	17.14	17.39	18.53	19.23			20.76	22.65
	p1	17.3	17.4	18.3	19.1			20.8	22.5
.	p2	17.2	16.1	18.8	18.31			20.3	22.1
	p10	17.1	17.3	17.7	20.1			21.3	22.42
	p1	18.1	18.2	18.3	20.3			22.1	23.2
Mota	p2	17.2	17.4	18.1	19.4			19.8	23.1
	p10	17.6	17.9	18.7	18.9			20.3	22.7
	p1	17.14	17.39	18.53	19.23			20.76	22.65
	p2	17.1	17.4	18.6	19.3			20.8	22.7
	p10	18.1	17.6	19.22				20.91	22.5
	West Gojam	13.58	15.77	16.1	17.22			20.57	20.4
	p1	13.8	14.6	16.2	17.1			19.88	20.5
	p2	14.1	15.5	16.51	17.3			19.45	20.4
Enebisie	p10	13.6	15.77	17.2	18.44			19.33	20

References

Taffesse, A.S., Dorosh, P., Gemessa, S.A.: Crop production in Ethiopia: regional patterns and trends. In: Food and Agriculture in Ethiopia: Progress and Policy Challenges, pp. 53–83 (2012)

Chilingaryan, A., Sukkarieh, S., Whelan, B.: Machine learning approaches for crop yield prediction and nitrogen status estimation in precision agriculture: a review. Comp. Electr. Agric. **151**, 61–69 (2018)

Rojas, O.: Operational maize yield model development and validation based on Remote sensing and agrometeorological data in Kenya. In: Proceedings of Remote Sensing Support to Crop yield Forecast and Area Estimates Workshop (2006)

Greatrex, H.: The Application of Seasonal Rainfall Forecasts and Satellite Rainfall Estimates to Seasonal Crop Yield Forecasting for Africa. PhD Dissertation, University of Reading, Reading, UK (2012)

Wgderes, A.: Maize Yield Forecasting in South Tigray, M.Sc. Thesis Addis Ababa University. Addis Ababa, Ethiopia (2014)

Fikre, A.: Wheat Yield Forecast Using Remote Sensing and GIS In East Arsi Zone, Ethiopia (2015)

Lee, H.J.: Teff, a rising global crop: current status of teff production and value chain. Open Agric. J. **12**(1), 185–193 (2018). https://doi.org/10.2174/1874331501812010185

Debalke, D.B., Abebe, J.T.: Maize yield forecast using GIS and remote sensing in Kaffa Zone, South West Ethiopia. Environ. Syst. Res. **11**(1), 1–16 (2022). https://doi.org/10.1186/s40068-022-00249-5

Cedric, L.S., et al.: Crops yield prediction based on machine learning models: case of West African countries. Smart Agric. Technol. 100049 (2022)

Sun, J., Di, L., Sun, Z., Shen, Y., Lai, Z.: County-level soybean yield prediction using deep CNN-LSTM model. Sensors **19**(20), 4363 (2019). https://doi.org/10.3390/s19204363

You, J., Li, X., Low, M., Lobell, D., Ermon, S.: Deep gaussian process for crop yield prediction based on remote sensing data. In: Thirty-First AAAI conference on Artificial Intelligence (2017)

Hailu, S., Getachew, T., Habtamu, S., Leulseged, T.: Crop yield estimation of teff (Eragrostis tef Zuccagni) using geospatial technology and machine learning algorithm in the central highlands of Ethiopia. Sustain. Agric. Res. 11(1), 1–34 (2022)

Sewnet, H., Tesfaye, G., Shiferaw, H., Tamene, L.: Phenology based Time Series LAI as a Proxy for Teff Crop Yield Estimation: A Case in Major Teff (Eragrostis) Growing Zones of Ethiopia, p. 30 (2021)

Tesfaye, G., Sewnet, H., Shiferaw, H, Desta, L., Abera, W., Gudeta, K.: Tef yield estimation based on satellite-derived light use efficiency model. A case study in the top tef growing zones of Ethiopia, p. 38 (2021)

Pantazi, X.E., Moshou, D., Alexandridis, T., Whetton, R.L., Mouazen, A.M.: Wheat yield prediction using machine learning and advanced sensing techniques. Comput. Electron. Agric. 121, 57–65 (2016). https://doi.org/10.1016/j.compag.2015.11.018

Battude, M.: Estimating maize biomass and yield over large areas using high spatial and temporal resolution Sentinel-2 like remote sensing data. Remote Sens. Environ. 184, 668–681 (2016). https://doi.org/10.1016/j.rse.2016.07.030

Fenta, A.: Effect of Teff variety and rates of nitrogen fertilizer application on growth and yield components under Jimma condition. ARPN J. Agri. Biol. Sci. 13(3), 42–43 (2018)

Nandeshwar, B.C., et al.: Teff (Eragrostis tef [Zucc] Trotter): an emerging global demanding crop. Agric. Observer 1(1), 28–33 (2020)

Tadesse, D., Alem, T., Wossen, T.: Evaluation of improved varieties of Teff in West Belesa, Northwest Ethiopia. Rev. Plant Stud. 3, 1–6 (2016). https://doi.org/10.18488/journal.69/2016. 3.1/69.1.1.6

Tamirat, W., Tilahun, N.: The response of Teff [Eragrostis teff (Zucc.) trotter] to nitrogen fertilizer and development of Barley (Hordeum vulgare L.) at Bore district, Southern Oromia, Ethiopia. American J. Life Sci. 2(5), 260–266 (2020). https://doi.org/10.11648/j.ajls.20140205.12

Tesfahun, W.: Tef yield response to NPS fertilizer and methods of sowing in East Shewa, Ethiopia. J. Agri. Sci. Sri Lanka 13(2), 162–173 (2018). https://doi.org/10.4038/jas.v13i2.8340

Wato, T.: Effects of nitrogen fertilizer rate and inter-row spacing on yield and yield components of teff [Eragrostis teff (Zucc.) Trotter] in Limo district, Southern Ethiopia. Int. J. Plant Soil Sci. 31(3), 1–12 (2019). https://doi.org/10.9734/IJPSS/2019/v31i330211

Khaki, S., Wang, L., Archontoulis, S.V.: A CNN-RNN framework for crop yield prediction. Front. Plant Sci. 10, 1750 (2020). https://doi.org/10.3389/fpls.2019.01750

Wato, T., Moral, M.T. (Reviewing ed.): Tef [Eragrostis tef (Zucc)] grain yield response to nitrogen fertilizer rates in East Badewacho district, Hadiya Zone, Southern Ethiopia. Cogent Food Agric. 7, 1 (2021). https://doi.org/10.1080/23311932.2021.1909203

Cochrane, L., Bekele, Y.W.: Average crop yield (2001–2017) in Ethiopia: trends at national, regional and zonal levels. Data Brief 16, 1025 (2018)

Alexandros, O., Cagatay, C., Ayalew, K.: Hybrid deep learning-based models for crop yield prediction. Appl. Artif. Intell. (2022). https://doi.org/10.1080/08839514.2022.2031823

Meshesha, D.T., Abeje, M.: Developing crop yield forecasting models for four major Ethiopian agricultural commodities. Remote Sens. Appl. Soc. Environ. support52938517302434. https://doi.org/10.1016/j.rsase.2018.05.001 (2018)

Rosebrock, A.: https://pyimagesearch.com/2019/01/28/keras-regression-and-cnns/. Accessed 7 May 2022(2021)

U.S. Geological Survey: https://pubs.er.usgs.gov/publication/fs20153081. Accessed 7 May 2022. Revised (April 8, 2020)

Aszemi, N.M., Dominic, P.D.D.: Hyperparameter optimization in convolutional neural networks using genetic algorithms. Int. J. Adv. Comput. Sci. Appl 10(6), 269–278 (2019)

The US Geological Survey: www.usgs.org

Amharic Character Recognition Using Deep Convolutional Neural Network

Achamie Aynalem[(✉)]

Bahir Dar Institute of Technology, Bahir Dar, Ethiopia
aynacham@gmail.com

Abstract. Amharic is the working language in the Federal Democratic Republic of Ethiopia. The Amharic alphabet has a large number of symbols and there is a close resemblance among shapes of the different symbols available in the language which challenged the task of machine-based optical character recognition systems in the language. The absence of a standardized labeled dataset for the Amharic language created additional barriers for different researchers. Our aim in this paper is to design a deep convolutional neural network based architecture that could extract features and classify Amharic characters with significant confidence of accuracy that could be utilized for real-world applications. A total of 90,000 characters are prepared for training the proposed architecture and an additional of 25,000 characters are reserved for testing purpose. Due to the occurrence of a large number of symbols and a close resemblance in the shapes of the different characters available in the language, a relatively complex convolutional Neural Network is utilized to capture those features and categorize them into the correct characters. Dropout layers are utilized to avoid overfitting. The character recognition system proposed in this paper achieved an accuracy of 99.27% on the testing dataset which is a significant improvement for the Amharic language. The implementation was done using Tensorflow on Keras neural network layers and Opencv in python to pre-process image data which enables us to make the system readily available for software developers as an API.

Keywords: Amharic recognition · Character recognition · Deep convolutional neural networks · Deep learning application

1 Introduction

Amharic which is one of widely spoken Semitic language is the working language in the Federal Democratic republic of Ethiopia. Alphabets of the Amharic language which are also called Ethiopic alphabet is an indigenous writing alphabet used for writing different languages in Ethiopia including Amharic, Geez, Tigrinya, Agaw and several other languages in Ethiopia, Eritrea and northern parts of Sudan. The Amharic alphabets are believed to have been derived from the ancient Geez language which is today confined to only for church services in the Ethiopian and Eritrean Orthodox and Catholic churches.

B. H. Woldegiorgis et al. (Eds.): ICAST 2022, LNICST 455, pp. 177–195, 2023.
https://doi.org/10.1007/978-3-031-28725-1_11

Different historical, political, socio-cultural and academic documents are found written in different languages using the Ethiopic alphabet.

Many of the ancient scientific, historical and sacred books of Christianity and Judaism are also found on different parts of the world written in the Geez language using the Ethiopic alphabet [1].

In this decade, researchers have made a major breakthrough in creating deep learning based models for character recognition mainly on languages such as English, Chinese and some other Latino based languages. However, there is no much research on languages which use the Ethiopic alphabet like Amharic and Tigrinya [2].

Amharic alphabet has large number of symbols and there is a close resemblance among shapes of the different symbols available in the language which made machine based character recognition systems of the language challenging. The absence of a standardised labelled dataset for Amharic language created additional barriers for different researchers. In this research paper, we have proposed a model for Amharic character recognition system using Deep Convolutional Neural Network. The model has multiple convolutional layers that could capture the variations in the features of closely resembling characters in the language.

1.1 Features in the Amharic Alphabet

The Amharic has 34 base characters. Unlike English where different sounds of a consonant are written with different combinations of the five vowels, in Amharic language, the different sounds of a base character are represented with different symbols. There are seven basic vowels in Amharic which are shown in the following figure (Fig. 1).

Fig. 1. The seven vowels in Amharic

The language has an additional 8th vowel ኧwhich is not used quite often, but available in conservative writers of the language and in other languages like Geez and Tigrinya.

Each of the 34 base characters produce a total of 34 * 8 = 272 symbols which represent the different sounds generated by combination of those base characters with the eight vowels. A partial view of the Amharic character set is shown in Fig. 2 below.

Taking the Ethiopic base character U(hä)as an example, we can generate the eight symbols derived from this base character. The 8 symbols generated with combination of this base character with the 8 vowels are U(hä), U-(hu), ሂ(hee), ሃ(ha), ሄ(hae), U(heh), U(ho)and ኋ(hua). In addition to these alphabetic symbols, there are more than 10 punctuation marks and 20 numeric symbols resulting more than 300 symbols in the language.

In addition to the large number of distinct symbols in the language, most of the Ethiopic alphabets have a higher degree of morphological structure similarities making the recognition process much difficult to recognize the characters as expected. Sometimes, even for our own eyes, it might be difficult to differentiate accurately among

								k	ህ	ኩ	ኪ	ነ	ኬ	ክ	ኮ
ሀ	ሁ	ሂ	ሃ	ሄ	ህ	ሆ		[k]	kä	ku	ki	ka	ke	ka	ko
ha	hu	hi	ha	he	ha	ho									
ለ	ሉ	ሊ	ላ	ሌ	ል	ሎ		k̠	ኸ	ኹ	ኺ	ኻ	ኼ	ኽ	ኾ
lä	lu	li	la	le	la	lo		[h]	k̠ä	k̠u	k̠i	k̠a	k̠e	k̠a	k̠o
ሐ	ሑ	ሒ	ሓ	ሔ	ሕ	ሖ		w	ወ	ዉ	ዊ	ዋ	ዌ	ው	ዎ
ḥa	ḥu	ḥi	ḥa	ḥe	ḥa	ḥo		[w]	wä	wu	wi	wa	we	wa	wo
መ	ሙ	ሚ	ማ	ሜ	ም	ሞ		·ʿ	አ	ኡ	ኢ	ኣ	ኤ	እ	ኦ
mä	mu	mi	ma	me	ma	mo		[ʔ]	·a	·u	·i	·a	·e	·a	·o
ሠ	ሡ	ሢ	ሣ	ሤ	ሥ	ሦ		z	ዘ	ዙ	ዚ	ዛ	ዜ	ዝ	ዞ
śä	śu	śi	śa	śe	śa	śo		[z]	zä	zu	zi	za	ze	za	zo

Fig. 2. A partial view of the Amharic alphabet

slightly deformed characters in the language. We can take an example of " ለ" and " ሐ", " ይ" and " ጸ", " ገ" and " ጎ", or " ከ" and " ኸ" which looks alike. The different symbols for vowel combinations of a base character is usually derived by adding small strokes at the right of the base or introducing a small structural variation on the base character.

For example consider the base character, ለ(le)whose eight derivations are ለ(le), ሉ(lu), ሊ(li), ላ(a), ሌ(lĕ), ል(lĭ), ሎ(lo)and ሏ(lua).

As we can see, there is a close resemblance in the morphological structure among most of the eight derivative symbols of the base character. Image processing algorithms should be carefully designed during recognition of Amharic characters in order not to remove or erode these little extensions or strokes available in the alphabet.

1.2 Offline Character Recognition System

The process of character recognition involves the conversion of images containing handwritten or printed texts into machine-readable text format.

Character recognition complexity depends on the distinct shapes, strokes and the number of characters available in the language.

In general, there are two types of text recognition approaches: off-line and on-line systems. Off-line text recognition system involves converting already existing scanned image or image captured through digital camera into text; whereas on-line text recognition system involves converting the different strokes and lines emanating from a real time stream of data captured through different transducers such as electromagnetic or pressure sensitive touch pens of tablet into sequences of characters.

In offline character recognition system the document is first generated, digitized, stored in computer hard disk and then it is processed latter. It is not real a time process.

Off-line handwritten character recognition refers to the process of recognizing characters in a document that have already been scanned or captured through a digital camera which might be from a sheet of paper or a label plate and are then stored in digital format image.

The character recognition process involves extracting different features from an image containing characters; and approximating some groups of extracted features to the nearest resembling character based on a previously "learned knowledge" of feature-to-character map.

1.3 The Deep Convolutional Neural Network

In the recent years, deep neural networks brought key breakthrough in different application areas like computer vision, image recognition, natural language processing and in speech recognition. In the past decade, deep networks have enabled machines to recognize images, speech and even play games at accuracy even impossible for humans. Likewise, a number of Artificial Neural Network Based methods have been used for character recognition for different languages.

Morphological/Rank/Linear Neural Network (MRL-NN) [3], where the different combinations of inputs in every node is formed by hybrid linear and nonlinear (of the morphological/rank type) operations, was studied for handwritten digit recognition. A hybrid Multilayer Perceptron Support Vector Machine (MLP-SVM) model was used for recognition of English numerals [4] and Chinese character [5] recognition. Support Vector Machine (SVM) with Radial Basis Function (RBF) network was also used for character recognition of English language [6].

In the last decade, Convolutional Neural Networks (CNN) is found efficient for handwritten character recognition due to its capability of capturing spatial information [7].

CNN exploits local spatial correlation by introducing local connectivity constraints between neurons of adjacent layers, which made CNN very well-suited for classification problems. Convolutional Neural Networks also provides some degree of translational invariance which made them further suitable for extracting and classifying patterns from an image. A CNN based model has shown a significant improvement on the accuracy of English character recognition. A CNN based model was tested on UNIPEN [6] English character dataset and found recognition rates of 93.7% and 90.2% for lowercase and uppercase characters, respectively [8].

A Deep Convolutional Neural Network (DCNN) contains multiple neural network layers. There are particularly three different types of layers used in convolutional neural network model; these are the convolutional layers, pooling layers and fully connected layers as shown in Fig. 3 below.

The Convolutional Layer: This layer is used to extract local features from matrix that come from an input image using different filters. This mimics receptive neurons in biological nervous system of living organisms. The convolutional layer will enable us to extract different features from an input image by utilizing several filters parallely.

Pooling Layers, (aka down sampling). Performs dimensionality reduction, which involves reducing the number of parameters from the input by applying an aggregation function to the values generated at the convolutional layer. Two types of pooling filters or matrices are available; max pooling and average pooling filters.

Fully Connected Layer: The last stage in a convolutional neural network is commonly made of one or more fully connected layers. The fully connected layer performs classification of features that are extracted from the input image at using the previous layers of the network. The convolutional and pooling layers usually utilise the ReLu activation functions for extraction of features from the input image, whereas, fully connected layers usually implement a softmax activation function which always produce either 0 or 1 for classification of features extracted at the previous stages.

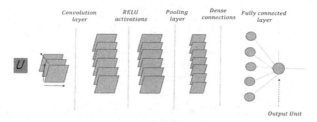

Fig. 3. Components of a convolutional Neural Network

In this paper, we have developed a DCNN based deep learning model for recognition of Amharic text.

Deep learning algorithms depends heavily on data. Different machine learning algorithms use different dataset and the dataset is highly specific to the particular model. Dataset is one of the most crucial aspects that makes algorithm training possible and it highly affect the performance of a machine learning algorithm. To achieve high level of accuracy, large amount of dataset is needed for training and henceforth powerful computing power is needed to train deep neural networks. Data preparation is one of the most difficult steps in most machine learning project.

There is no sufficient public accessible dataset for Amharic character recognition. In the absence of such dataset, we have prepared a dataset of Amharic characters collected from different news and magazines published in Amharic language.

During the dataset preparation, first we have collected the images that contain the desired characters using a scanner. After the image is accessed, different image processing algorithms were used to improve quality of the image; then line and character segmentation algorithms were used to separate segments of images that contain characters. These segments of pixels that represent characters are finally stored as a matrix of those fragments within sub folders with labels that represent the classes of each fragment.

The DCNN is used to classify those segments to the correct characters and produce a binary encoded output for each class of characters by comparing them to an already trained state of the network.

Finally those binary coded outputs of the model are post-processed to produce human understandable outputs such as decoding binary coded outputs to human readable symbols, representing character spacing, word spacing and line breaking. However, the post classification process is not implemented in this research paper in order not to scatter quite broad idea in a single paper.

The basic phases of data preparation and classification stages are shown in the following Fig. 4.

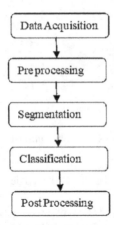

Fig. 4. Block diagram of the character recognition system

2 Data Acquisition

In this research we have collected more than 120 pages of Amharic document from different fiction books, Addis Zemen magazine, Ethiopian church documents from www. eotmk website and from Ethiopian Code of Civil rights. Those documents were also synthetically reformatted with different font families available in Amharic in order to generate images of characters with different shapes.

3 Image Pre-processing

Most of the time the image captured through a camera or scanner may not have the required quality for extracting features that represent characters. Therefore, first the acquired image should pass through a series of image pre-processing algorithms in order to enhance the quality of images required for preparing a dataset.

3.1 Grey Scale Conversion

Image captured through camera or scanner has always three channels which represent the Red, Green and Blue channels of the image. Since colour does not have significant information content except adding some aesthetic value to the document, the RGB scale input image should be converted to a gray scale image which will subsequently reduce the processing time and complexity of the neural network used for the recognition process.

3.2 Noise Removal

Image acquired through scanner or digital cameras is usually associated with some noise due to factors like malfunctioned devices, misalignment of paper, poor light intensity during image capturing or scanning,

Different filters such as median filter and Gaussian filter with different kernels are tested for removing noise which could be approximately modelled with salt and paper noise. However, due to a common feature of thin horizontal extensions available in Amharic alphabet, those filters have the tendency to average out those extensions to the background pixel or degrade them (Fig. 5).

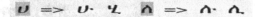

Fig. 5. Thin horizontal extension in Amharic alphabet

An adaptive Gaussian filter with an asymmetric kernel resulted in relatively better performance for removing such salt and paper noise with reduced degrading effect on the lateral extensions. In order to minimize this erosion of small lateral extensions, we have used asymmetric kernel of (5, 3) for the Gaussian filter. Relatively small value is deliberately used in the x-axis which enabled us to preserve those small horizontal strokes commonly available in the language.

3.3 Binary Image Conversion

Once the images are converted to grayscale and the irrelevant noises are removed from the images, the image is converted to binary scale image where the image is converted into binary images so that the image is composed of pixels having a colour value of either black (pixel value of 0) or white (pixel value of 255). An adaptive threshold filter is used to classify pixels into white or dark value adaptively depending on the local variance using a kernel size of 5. The [5] kernel is used to select local matrix from the region along the image surface and a threshold filter is applied on this local matrix to classify pixels to either 0 (text pixel) or 1 (background pixel) depending on a local threshold value. The kernel or window is then slided horizontally and vertically with the step size until the whole pixel available within the image are classified to 0 or 1. This could effectively classify pixels into dark and white values from an image with variable spacial background intensity since the threshold value is calculated locally depending on the local pixel variance. The input image as shown in Fig. 6/A has background with variable intensity along the vertical axis. This variable background is effectively removed using the adaptive threshold as shown in Fig. 6/B below.

184   A. Aynalem

Fig. 6. A) Input Image with variable background. B) Binary scale conversion using adaptive threshold followed by background inversion.

During this adaptive threshold binarization process, there are noise introduced into the image due to misclassification of some background pixels as text pixels which result in small black pixels randomly distributed across the image. Therefore, the resulting binary image is further passed through Gaussian noise in order to remove such salt and paper noise introduced during the binarization process. A global, inverted threshold algorithm is then used to correct smoothening or blurring effect introduced by the Gaussian filter and to invert the colour scale such that the background is represented with black (pixel values of 0) and the foreground or characters with white (pixel values of 255).

3.4 Skew Correction

Sometimes due to misalignment of paper during the scanning process, or while capturing through camera, lines of the text are rotated randomly with small angle. This creates sever problem in subsequent stages during segmentation of lines of characters from the image. A skew detection and correction algorithm is utilized. The algorithm first finds all the pixel coordinates that are part of the foreground image. Then the rotation angle of the rectangle is calculated based on the collected pixel coordinates. Then by using the calculated rotation angle, we will find the rotation matrix with respect to the centre coordinates of the image which will be used to rotate the input binary image as shown in Fig. 7.

Fig. 7. Skew corrected image

3.5 Segmentation

Segmentation is the process of breaking the whole image into subparts to process them further. Segmentation is a critical step in the handwritten recognition process and it highly affects the accuracy of the recognition process. There are three levels of segmentation in character recognition.

Line Segmentation
This involves cropping each line of text out of the image and store them in a sorted manner. Thus, using those lines, words and characters can be extracted from each line with the original order. Horizontal histogram projection is used to segment the individual lines of text from the image.

During the horizontal projection

- Rows that represented a text pixel in a line have high number of foreground pixels, which corresponds to higher peak values in the histogram.
- Rows that represent the gaps in-between the lines have high number of background pixels, which correspond to lower peaks in the histogram.

Therefore, rows that correspond to lower peaks in the histogram can be selected as the segmenting lines to separate the lines.

Figure 8 shows vertically expanded view of the histogram of the first 9 lines. As shown in the figure, the picks of the histogram represent lines that represent text since text are represented with white pixels (255) whereas, pixels that represent gaps between lines will have minimum point in the histogram since there will not be text along those lines. Partitioning the image regions horizontally using a threshold pixel value could effectively segment the image into lines as shown in Fig. 9.

Fig. 8. Histogram of lines in horizontal projection

Fig. 9. Segmented lines

Word and Character Segmentation

Once the lines are segmented successfully word and character segmentation is carried out to crop out segments of each character from an image which contain a line of characters. Vertical histogram projection technique resulted in poor result during segmenting characters from within a line due to too narrow space in between characters within a word or due to overlapping pixel of neighbouring characters of a word. The problem is much worse in texts that contain oblique font styles. Consider Amharic text fragments ቡታ፣ መንgwhere the first character has a forward stroke, whereas the second character has backward structural extension which make the gap between the neighbouring characters zero. The problem becomes worse when the text is in italic font style resulting an overlapped regions between characters. This made isolating characters using vertical projection technique inefficient. For the character segmentation, we have used contour detection segmentation method which involve the following steps;

1. Accept the list of lines which are returned from the line segmentation stage.
2. Find all contours/characters on the line and return a list of coordinates for each contour.

3. Crop the characters from the threshold text line images using the returned coordinates.
4. Iterate through each line.

Fig. 10. Segmented characters

This could segment characters of a line into separate chunks as shown in Fig. 10.

4   Dataset Preparation

Due to existence of relatively large number of symbols and resemblance in the morphological shapes among the different characters available in the language, we have designed relatively complex convolutional Neural Network with significantly larger number of trainable parameters which would be able to map features available in the language. Such neural network models with large number of parameters require large and diversified dataset in order to be able to capture those large number of diverse features and categorize them into the correct character; and avoid over fitting of the model which is a cause of degraded performance of a model on a real-world data.

However, there is limited public accessible datasets for character recognition in Amharic language. There is one dataset prepared by Birhanu Hailu Belay which contains the first 231 synthetically generated text-line images with Power Geez and Visual Geez fonts with each image 32 * 32 size. The dataset does not contain all characters available in the language; and numeric symbols and punctuation marks are not included, either.

Therefore, we have prepared an additional dataset that contain Amharic characters, numerals and punctuation marks. We have collected 105 pages of Amharic text collected from Amharic fiction books, Addis Zemen Gazetta and Ethiopian Code of civil Rights and from Ethiopian Orthodox church documents. Those documents are reformatted using different Amharic font families and segmented into 32 * 32 separate character segments to generate a dataset which contain most of the characters, numeric symbols and punctuation marks in Amharic language. This dataset then merged with the dataset which is found from the aforementioned source and then shuffled and augmented to a representative dataset.

5   Data Augmentation

Data segmentation is a set of techniques to artificially increase the amount and quality of data by generating new slightly modified data points from the existing data. Data augmentation techniques enable machine learning models to be more robust by creating variations that the model may came across in the real world. We have increased size and "quality" of the dataset using 5° rotation, adding some Gaussian noise and rescaling with a ±5% factor. The resulting dataset is shuffled to produce a randomly distributed dataset.

This dataset is finally separated into groups of their respective symbols, with images of different characters placed in different folders where folders named in an extended alphabetical order. Labels for each class of the dataset will be subsequently generated from folder labels using Tensorflow.

A total of 90,000 characters are prepared for training the proposed architecture and an additional 25,000 characters are reserved for testing purpose.

6 DCNN Architecture for Recognition

Naturally, the number of trainable **parameters available in a neural network model is proportional** to the **complexity** of the task the model has to perform; requiring a **proportional amount of examples or labelled dataset**, to get good performance. In this research paper, the proposed model for recognition of Amharic characters is Deep Convolutional Network. A CNN primarily composed of two components: a feature extractor module which is composed of convolutional and pooling layers with a ReLu activation function followed by a trainable DNN classifier.

The convolutional neural network contains the following layers:

Input layer which accepts batches of image matrices of 32 * 32 size.
Convolution layer will take images from an input layer or from preceding layers and convolve its input with a specified number of filters to create feature maps. The number of feature maps extracted is equal to the specified number of filters used at a convolutional layer. Convolutional layers are generally used with the ReLu activation function.

As stated previously, the Amharic character set contains relatively large number of symbols with each symbol containing highly resembling morphological features. In order to be able to map such features, we have selected three stages of feature extraction with each stage extracting different depths of features. Each stage is consists of two convolutional layers separated with batch normalization layers followed by a maxpooling layer and dropout layer.
Pooling layer will down sample the 2D activation maps along the height and width dimensions. Pooling operators aggregate the values of a sub-matrix generated by a convolutional filter into one single output value that represent the "meaning" associated with that special region.
Dropout layer will regularize weights in the fully connected layers of a convolutional neural network and feed forward networks to avoid overfitting. Too much cooperation between neurons makes the neurons dependent on each other and they fail to learn distinct features which usually result in doing classification well on a training dataset but produce incorrect prediction if tested on somewhat different dataset, which is an overfitting problem.

In order to solve this overfitting problem in Deep Neural Networks, a specified proportion of neural network units is randomly dropped at training time for each training sample in a mini batch. This makes the remaining neurons learn important features all by themselves without relying on cooperation from other neurons. In this model, we have used four dropout layers each in between the three stages and one in between the dense connected layer and the output layer.

We have used padding of 2 for both dimensions (width and height) at each convolutional layer; and maxpooling layers with a pool size of (2,2) at the end of each convolutional stage.

Fully connected layers are dense feed forward neural layers that receive different sets of weights from the preceding layers. Each neuron in this layer will be connected either to all the neurons in the previous layer or to all the neurons in the next layer. There are two fully connected layers; one hidden layer and the output layer. For classification, the output neurons receive inputs from the final hidden fully connected layers and generate the appropriate class for the particular input pattern.

We have considered 300 Amharic characters, punctuation marks and numeric digits; therefore the output layer should have 300 neurons which produce one-hot encoded output for each character in the Amharic language (Fig. 11).

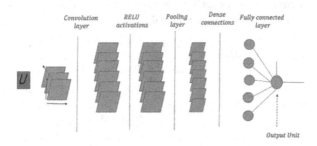

Fig. 11. A general structure of the convolutional Neural Network Model

7 Training CNN Model

The process of training a neural network is simply the process of updating the weights of the neural networks. Unlike, the Multi-Layer Perceptions (MLP), in CNN neurons share the same weights among them and also, they are followed with pooling layers. In which the sharing of weights among the neurons and pooling layers will help to decrease the overall weights of the neural network and computational power. Commonly an activation layer particularly sigmoid and ReLU activation function is placed between the convolutional and pooling layers. Once the image passes through these processes, the features from the image will be extracted and then unzipped into a 1D vector which will then be used by the successive dense layers of the network for classifying the input imag. Consider I is a two-dimensional image vector and K is filtering window which has a size of w*h, then the convolution process can be given as:

$$(1+K)_{ij} = \sum_{m=0}^{w-1} \sum_{n=0}^{h-1} km, n * Ii + m, j + n + b \qquad (1)$$

The weights in the neural network architecture are calculated using this convolution operation which is why this architecture is named Convolutional Network.

Backpropagation:
It is one of the most commonly used training algorithms in neural network. The process of training a neural network requires a greater computational power so, we need to make our design and training process as efficient as possible. The process of training CNNs can be also called convolution operations which utilize the mathematical correlation operation. Training a neural model involves updating the weight and bias parameters available in the network. In the case of backpropagation, the weights are not the only things to be updated also the deltas should be updated.

The updates to the weights can be computed using the expression;

$$\frac{\partial E}{\partial w_{m',n'}^{l}}, \tag{2}$$

The gradient components of each weight can be calculated using a chain role which is expressed using the following mathematical expression.

$$\frac{\partial E}{\partial w_{m',n'}^{l}} = \sum_{i=0}^{H-k_1} \sum_{j=0}^{H-k_2} \delta_{i,j}^{l} o_{i+m',j+n'}^{l-1} \tag{3}$$

$$\frac{\partial E}{\partial w_{m',n'}^{l}} = rot_{180°}\{\delta_{i,j}^{l}\} * o_{m',n'}^{l-1} \tag{4}$$

The summation represents all the sum of the gradients from $\delta_{i,j}^{l}$ coming from the output layer l in which the double summation is for the shared weights of the filter kernel. The chaining role used in the above equation is used for the optimization of the backpropagation training process. As can be understood from this discussion, in the backpropagation training method, we use a chain role to optimize the training process of the CNN architecture. Stochastic Gradient Descent (SGD) algorithm utilize the Backpropagation in order to calculate the gradient or derivative of the loss function.

Then, after the inputs to each neuron is calculated using the forward propagation, the output of the network is used in backward fashion to update the weights based on the chosen loss function.

The total error of prediction of the different classes is calculated using the mean squared error equation or using cross entropy from the network output y for a true output t as shown in the following equations.

$$E = \frac{1}{2} \sum_{p} (t - y)^2 \tag{5}$$

$$E = -\sum_{p} y \log(t) \tag{6}$$

Training the DCNN was handled interactively using Tensorflow.

Table 1. Different parameters in the proposed model

NO_CLASS = 300

Layer (type)	Output Shape	No. of Parameters
Stage I Conv.		
conv2d_0 (Conv2D)	(None, 28, 28, 20)	200
normalizer_0 (Batch Normalization)	(None, 28, 28, 20)	80
conv2d_1 (Conv2D)	(None, 28, 28, 20)	3620
normalizer_1 (Batch Normalization)	(None, 28, 28, 20)	80
max_pooling_0 (MaxPooling2D)	(None, 14, 14, 20)	0
dropout (Dropout)	(None, 14, 14, 20)	0
Stage II Conv.		
conv2d_2 (Conv2D)	(None, 14, 14, 64)	11584
normalizer_2 Batch Normalization)	(None, 14, 14, 64)	256
conv2d_3 (Conv2D)	(None, 14, 14, 64)	36928
normalizer_3(Batch Normalization)	(None, 14, 14, 64)	256
max_pooling_1 (MaxPooling 2D)	(None, 7, 7, 64)	0
dropout_1 (Dropout)	(None, 7, 7, 64)	0
Stage III Conv.		
conv2d_4 (Conv2D)	(None, 7, 7, 128)	73856
normalizer_4 (Batch Normaliz'n)	(None, 7, 7, 128)	512
conv2d_5 (Conv2D)	(None, 7, 7 128)	147584
normalizer_5 (Bath Normaliz'n)	(None, 7, 7, 128)	512
max_pooling_2 (MaxPooling 2D)	(None, 3, 3, 128)	0
dropout_2 (Dropout)	(None, 3, 3, 128)	0
Dense Layer		
flatten (Flatten)	(None, 1152)	0
dense (Dense)	(None, 500)	576500
dropout_3 (Dropout)	(None, 500)	0
output (Dense)	(None, 300)	150300

Total parameters: 1,002, 268

Output of the network parameters generated during the training process is shown in the following Table 1.

The input to the model is batches of images with their respective labels which contained dataset of the Amharic characters and one-hot encoded labels of each character. Each of the characters are reshaped to a size of 32 × 32 pixels. This relatively large pixel size is chosen in order not to lose small features differences between the different characters available in the language. The images are fed in batches as four-dimensional tensors where the first dimension is specific to the image index in the batch, second and third dimensions are specific to the height and width of the image. The model contains three successive feature extraction stages. Each of those stages contains convolutional layers which could be able to extract different levels of features successively; and maximum pooling layer.

8 Experimental Result and Discussion

The experiments were conducted using Tensorflow with keraz neural layers in Python. The input of the network is a dataset of 32 * 32 sized images. Images for each symbol are stored in separate folders with each folder labelled in ascending order of the Amharic characters in the Unicode encoding order. Then, during the data reading stage, those images are read and one-hot encoded labels for each image are generated from folder names. Therefore, image fragments which are found within the same folder represent instances of the same character and are all associated with the same output label. This process of fetching data and associating them to output labels is handled in a separate python routine that generates training and testing datasets from input images and their labels is handled in a separate python routine.

The resulting dataset is passed to the DCNN model and trained using the 'Adam' optimizer algorithm, which took about one hour to train the model for 20 epochs on a local machine. The model took significantly longer time for the training process due to the large number of training parameters available in the model and the slow processor speed of the local machine. The output of the training process on the Pycharm terminal is displayed in the following table.

Table 2. Training the proposed model for 20 epochs

Epoch 1/20
271s 567ms/step - loss: 0.3892 - accuracy: 0.8875 - val_loss: 5.7527 - val_accuracy: 0.1126

Epoch 16/20
- 233s 534ms/step - loss: 0.0261 - accuracy: 0.9920 - val_loss: 0.0465 - val_accuracy: 0.9872
Epoch 17/20
- 231s 530ms/step - loss: 0.0263 - accuracy: 0.9921 - val_loss: 0.1024 - val_accuracy: 0.9760
Epoch 18/20
- 233s 534ms/step - loss: 0.0285 - accuracy: 0.9915 - val_loss: 0.0364 - val_accuracy: 0.9911
Epoch 19/20
- 231s 530ms/step - loss: 0.0253 - accuracy: 0.9923 - val_loss: 0.0309 - val_accuracy: 0.9925
Epoch 20/20
- 245s 570ms/step - loss: 0.0253 - accuracy: 0.9932 - val_loss: 0.0286 - val_accuracy: 0.9925

Do you want to save state? y/n: y
- 15s 111ms/step - loss: 0.0297 - accuracy: 0.9927
==
Test loss: 0.0292 - Test accuracy: 0.9927
==

The model was run tuned with hyper parameters shown in Table 2 above. The choice of the hyper parameters is empirical with a focus on the learning behavior of the model.

As shown in the following figures, the model rapidly converges around 8 epochs and gained only a little improvement in the accuracy after the 8th epoch. Therefore, around 25 min are required to train the model with a reasonable accuracy of prediction, which is achievable at the 8th epoch. The loss curve shows a drop from 5.75 to 0.0286 on the validation dataset and to 0.0292 on the test dataset; whereas the accuracy jumped to 99.32% on the training dataset and to 99.27% on the test dataset, as shown in Table 2 above.

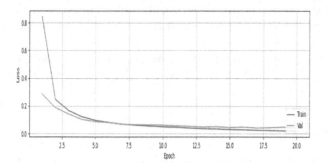

Fig. 12. Loss of the DNN model for 20 epochs

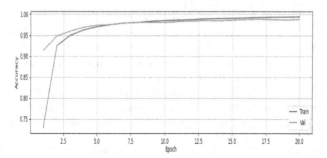

Fig. 13. Accuracy of the DNN model for 20 epochs

As we can see from the graphs at Fig. 12 and 13, it could achieve a prediction accuracy of 99.27% on the test dataset. This model misclassified around 7 characters out of a text containing 1000 characters, which is very good accuracy of prediction in comparison to previous models which have been proposed for Amharic recognition [10].

9 Conclusion and Future Work

The main objective of this research is to design a CNN based character recognition architecture for the Amharic language. The research work included image pre-processing techniques that could be utilized for the implementation of the system for real world applications. We have prepared a dataset merged with an existing database to generate a representative dataset in further research on the recognition of the Amharic language. The dataset contains 90,000 training images and 25,000 testing images consisting of symbols of most of the Amharic characters, punctuation marks and numeric digits.

As we can see from Table 2 shown above, we could achieve an accuracy of 99.27% on the test dataset. This model misclassified less than 7 characters from a text containing 1000 characters. In contrast to previous works by different individuals where most of the works on recognition of Amharic or Geez language which have been done using MATLAB, in this Amharic text recognition system, image pre-processing and recognition algorithms are all implemented using Tensorflow 2.x with tf.keras network models using the python language. This would enable us to create an API for android, desktop-based or web application software that needs to integrate recognition of Ethiopic languages.

We have compiled a representative dataset and will try to make it available by putting it on public accessible repositories for further research on Ethiopic languages recognition.

References

1. Yemisrach, B.: Ethiopian historical antiques (1999). Retrieved 10 May 2022
2. Chirag, I.P., Ripal, P., Palak, P.: Handwritten character recognition neural network. Int. J. Sci. Eng. Res. **2** (2011)
3. Int. J. Eng. Comput. Sci. **4**(5), 11729–11732 (2015)
4. Pessoa, L.F.C., Maragos, P.: Neural networks with hybrid morphological/rank/linear nodes: a unifying framework with applications to handwritten character recognition. Pattern Recogn. **33**(6), 945–960 (2000)
5. Bellili, A., Gilloux, M., Gallinari, P.: An MLP-SVM combination architecture for offline handwritten digit recognition. IJDAR **5**, 244–252 (2003)
6. Dong, J., Krzyżak, A., Suen, C.Y.: An improved handwritten Chinese character recognition system using support vector machine. Pattern Recogn. Lett. **26**(12), 1849–1856 (2005)
7. Theodoridis, S., Koutroumbas, K.: Pattern recognition and neural networks. In: Paliouras, G., Karkaletsis, V., Spyropoulos, C.D. (eds.) ACAI 1999. LNCS (LNAI), vol. 2049, pp. 169–195. Springer, Heidelberg (2001). https://doi.org/10.1007/3-540-44673-7_8
8. Guyon, L., Schomaker, R., Plamondon, R., Liberman, M., Janet, S.: Unipen project of on-line data exchange and recognizer benchmarks. In: Proeedings. of 12th International. Conference on Pattern Recognition (ICPR), vol. 2, pp. 29–33. IEEE (1994)
9. Yuan, A., Bai, G., Jiao, L., Liu, Y.: Offline handwritten English character recognition based on convolutional neural network. In: 10th IAPR International Workshop on Document Analysis Systems (DAS), pp. 125–129 (2012). https://doi.org/10.1109/DAS.2012.61
10. Siranesh, G., Menore, T.: Ancient Ethiopic manuscript recognition using deep learning artifcial neural network (Unpublished master's thesis). Addis Ababa University (2016)
11. Ahangar, R.G., Ahangar, M.F.: Handwritten farsi character recognition using artificial neural network. Int. J. Comput. Sci. Inform. Security **4** (2009)
12. Christos, S.: Dimitrios, S.: Neural Networks. https://www.doc.ic.ac.uk/~nd/surprise_96/journal/vol4/cs11/report.htmlpril (2018)
13. Ahmad, I., Fink, G.: Class-based contextual modeling for handwritten Arabic text recognition. In: 2016 15th international conference on frontiers in handwriting recognition (ICFHR) (2016)

14. Shafi, M.: Optical character recognition of printed Persian/Arabic documents. https://scholar.uwindsor.ca/etd/5179 (2014). Retrieved 16 Dec 2018
15. Kavallieratou, E., Sgarbas, K., Fakotakis, N., Kokkinakis, G.: Handwritten word recognition based on structural characteristics and lexical support. In: 7th International Conference on Document Analysis and Recognition (2003)
16. Nasien, D., Haron, H., Yuhaniz, S.S.: Support vector machine (SVM) for english handwritten character recognition. In: 2010 Second International Conference on Computer Engineering and Applications, Bali, Indonesia, pp. 249–252 (2010)

Assistive Smart Cane Technology for Visually Impaired Peoples: A Review

Getnet Ayele Kebede(✉) 🆔 and Yosef Kassa Shiferaw

Faculty of Mechanical and Industrial Engineering, Department of Automotive and Electro Mechanical Engineering, Bahir Dar University, BIT, Bahir Dar, Ethiopia
beget13@gmail.com

Abstract. Smart cane technology is an assistive technology that allows visually impaired people to walk more freely and independently. This study reviewed various researchers' works on cane development methodologies. This review's goal is to determine the full cane configuration of hardware parts, software architecture, and cane structure. We discussed object detection methods, object identification methods, flame detection methods, water detection methods, and location tracking methods. Recently, many researchers have focused on the key development of a smart cane using a computer vision system with Python and Yolo V5 deep learning algorithms to identify objects or obstacles in cane users' paths. The hardware part is used to connect sensors to the Raspberry Pi module, which is mostly used as a controller. The ergonomics of cane structure are cane tip and handle shape, which is the key future of cane design. Finally, this study concludes that the most effective methods and materials for making and improving smart cane are described.

Keywords: Smart cane · Cane ergonomics · Deep learning · Location tracking · Feed-back system · Visually impaired

1 Introduction

According to a WHO study, at least 2.2 billion people worldwide have a near-or distant visual impairment [1]. Over 39 million people were completely blind, including 19 million children (below 13 years of age). Around 80% of people with vision impairment or blindness live in low- and middle-income countries, where expensive assistive technologies are out of reach. Those who are visually impaired face numerous challenges in their daily lives, both socially and economically, as well as psychologically [2]. Because of this, they have difficulty living independently, especially if they cannot move freely. To solve this problem, different assistive technologies were implemented. Smart cane is one of those technologies that is rapidly growing [3]. A smart cane is an assistive electronic technology that allows visually impaired people to walk independently without encountering any obstacles. Smart canes are used to simplify the day-to-day activities of visually impaired people [4]. Smart cane can be developed in different electronic devices, mechanical components, and manufacturing processes for different tasks [5].

Published by Springer Nature Switzerland AG 2023. All Rights Reserved
B. H. Woldegiorgis et al. (Eds.): ICAST 2022, LNICST 455, pp. 196–208, 2023.
https://doi.org/10.1007/978-3-031-28725-1_12

Previously visually impaired people utilize a white cane or a guide dog as a temporary aid. When a guided dog is utilized, the person can go to the places where the dog has been trained to go. It's considerably more difficult to care for a dog when you're blind. In many ways, technological developments will benefit the blind. In the realm of electronics, a lot of work is being done to produce navigation support for visually impaired people [6]. An Electronic Travel Aid (ETA) gadget has recently been invented that captures ambient data and communicates it to the user, allowing for autonomous movement [7]. ETAs, in general, use one or more sensors to identify impediments in the user's way and offer feedback to the participant via audio or hepatic devices. The primary benefit of an ETA over a conventional white cane is its ability to identify obstructions above the waistline, enabling more independent and secure walking [8].

Obstacle detection and object recognition in the user's micro-environment have been intensively investigated, and a variety of technologies have been employed to warn the user of the presence of an obstacle in real time. Vision-based, sensor-based, and crowd-assisted technologies are the three basic categories in which these technologies can be classed. The most effective technology option in this category is computer vision. It is utilized in urban contexts to detect and identify obstacles and signs (such as crosswalks, traffic lights, and road signage). Yolo object detection and deep learning algorithms are the most commonly used and effective methods in object detection and identification systems for visually impaired systems.

2 Review of Smart Cane Hardware Part

2.1 Environmental Modeling and Configuration

Microcontroller
Various studies employ a variety of object detection mechanisms via a variety of sensors and controllers. Before the cane is activated, a specialized camera module with a PIR sensor is used in the design to find moving objects or people. A photo is taken and cross-referenced with the database when the PIR sensor picks up a moving object. Additionally, there is an ultrasonic sensor that is utilized to identify objects [9]. The configuration diagram is shown in Fig. 1.

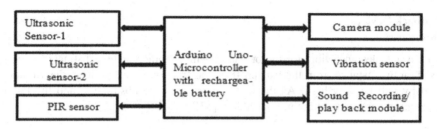

Fig. 1. Configuration of cane circuit with Arduino Uno controller [9].

Recently, a smart cane has been developed with a microprocessor for processing data gotten from sensors and cameras to detect objects [10]. A mini-and single-board computer called a Raspberry Pi can be used to learn programming and carry out practical tasks. It is an excellent piece of equipment for electronic work [11]. The Raspberry Pi is a controller that outperforms the Arduino controller. It can also be used to detect obstructions, lights, potholes, and traffic lights, among other things. Because it makes use of the camera module, this controller is preferred. The object is detected by this camera module, which then delivers the information to the controller [11]. This study, the Raspberry Pi processor is configured with many sensors and a camera as shown in the Fig. 2.

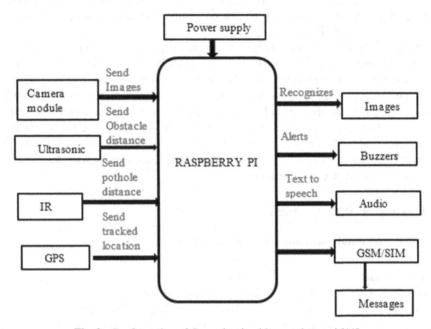

Fig. 2. Configuration of Cane circuit with a raspberry pi [11].

The Raspberry Pi 4 features a faster processor and can connect to the raspberry pi camera, GPS module, and GSM modules [12].

Feedback Devices
Vibration Motor

In the system design, a vibration motor is used, which vibrates at three distinct intensities based on the distance from the barrier. If the obstruction is really close, the vibration intensity will be very high. As the distance between the obstruction and the motor increases, the intensity of the motor reduces [13].

Buzzer

A buzzer is a sound-signaling device. It may be a mechanical, electromechanical, or piezoelectric device. By applying DC voltage, it produces a steady single tone. This type can be employed where enormous sound volumes are required by using a well-built resonant system [14]. Many of the most common varieties are grouped by type, sound level, frequency, rated voltage, size, and packaging type at father electronics [13] (Fig. 3).

Fig. 3. Buzzer [13]

Audio device
The Raspberry pi receives information from the input devices. A text file is given to the Raspberry Pi after the detection. The output information in text format is then delivered as input to Google Text to Speech. Then this turns the text data into an audio signal, and the participant receives it via Bluetooth earphones [11].

Input Devices
Ultrasonic sensor
An ultrasonic sensor will detect an obstruction in front of the smart cane and will function in conjunction with a buzzer to provide an audio indication to the participant when an impediment is identified [15]. When a fall is detected, the buzzer will also sound [16]. The non-contact measuring range of the HC-SR04 ultrasonic sensor is 2 cm to 400 cm, with a variable precision of 3 mm. The module includes ultrasonic transmitters and receivers and a control circuit. A 5V and 15 mA power supply is required. The signal from the IO sensor is greatly increased for at least 10 s [6] (Fig. 4).

Fig. 4. Ultrasonic sensor [6]

Water sensor
Water Detector Whether a blind person is wearing shoes or not, he/she will not notice if there is any water on the floor, and the moist surface can cause slipping [17]. To prevent slipping, a water sensor can be placed at the end of the cane. When water sensors detect the presence of water, a signal is sent to the processor, which causes the buzzer to sound,

Fig. 5. Water level sensor [18].

alerting the blind person. As a result, the blind individual will be more cautious [18].
(Fig. 5)

Flame sensor

A flame sensor, also known as a heat sensor, detects heat radiation from a great distance
[15]. Because blind people can't always perceive heat, we equipped the stick with a heat
sensor that detects the presence of heat, such as a fire or something burning, and then
alerts the blind person by making the buzzer peep with a different noise, allowing the
blind to avoid danger sooner [18] (Fig. 6).

Fig. 6. Flame Sensor [15]

2.2 Cane Ergonomics

To design an ergonomically good smart cane, different conditions must be considered.
Most of them are dependent on each other. The interdependence and relations of those
conditions are described in Table 1. The table describes the interdependence of one
property over the other out of 9 values, and then the total dependence and independence
are calculated.

Table 1. Relationship Matrix of Independence [19]

Feature	Steadiness	Safety	weight	Battery life	price	Transport ability	Simplicity	Ergonomics	Total dependency
Steadiness		9	9	0	3	3	1	1	26
Safety	3		0	9	0	0	9	0	21

(continued)

Table 1. (*continued*)

Feature	Steadiness	Safety	weight	Battery life	price	Transport ability	Simplicity	Ergonomics	Total dependency
Weight	3	0		3	0	3	3	3	15
Battery life	0	9	3		3	3	1	0	19
Price	1	9	0	9		0	9	3	31
Transportability	3	1	3	1	0		1	3	12
Simplicity	0	9	3	1	9	3		1	26
Ergonomics	3	3	3	0	1	9	1		20
Total independency	13	40	21	23	16	21	25	11	170

Handle Shape

The shape model of the smart cane handle part may vary for different constructions. The most common smart cane handle shapes are shown in Figs. 7, 8 and 9. Since most sensors and controllers are packed in the handle, design of the handle is the most critical issue [20].

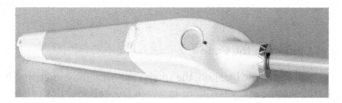

Fig. 7. Wewak cane

Fig. 8. Ultra-cane

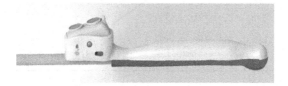

Fig. 9. BAWA Cane

Tip of the Cane

The most common tip of smart canes are marshmallow and sphere type tip as shown in Figs. 10 and 11.

Fig. 10. Marshmallow type cane tip

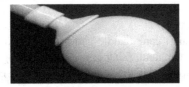

Fig. 11. Sphere type cane tip

2.3 Position of Camera in the Cane

As the person goes towards the location, the smart cane with the camera module may be tilted. Figure 12 depicts the many positions in which the walking stick can be used when the individual is walking. As a result, when the user walks normally, the stick makes an angle of 300–400 degrees, ensuring coverage of 1500–1600 FOV [9]. The camera should record all of the things in the area, even if it is angled. This is beneficial because if the individual drops the cane, the angle will be abrupt. It's being examined as a means of delivering an alarm to families about the situation [9].

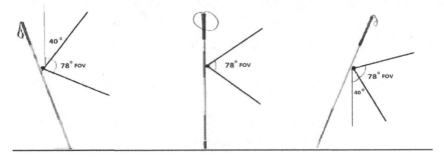

Fig. 12. Different positions of the smart cane during movement [9].

2.4 Power Source and Charging System

The Raspberry Pi is an excellent choice for a portable device, and the Pi Sugar battery provides a solderless alternative for the battery [14]. The Pi Sugar battery module is connected to the backside of the Pi Zero via little spring pins. The battery is the same size as the Pi Zero board. Because direct contact between the battery and the board is undesirable, a magnet is utilized to attract the battery to the board and maintain the distance between the two [21] (Fig. 13).

Fig. 13. Pi sugar battery for raspberry pi [21].

3 Review of Software Design of Cane

3.1 Object Detection Algorisms

YOLO is built on a Convolutional Neural Network (CNN) that recognizes things [5]. Yolo RCNNs are early CNNs that accurately detect objects based on how YOLO works in two phases. It begins by identifying the region in the photos that needs to be categorized. Second, use a CNN to classify this region of interest. It calculates the probability and bounding box for each region in the image [17]. The specific class of image is identified based on the confidence in each of the bounding boxes and the correlating probabilities of these boxes [22]. The general steps for yolo object detection in deep learnings are.

- Segmentation: The image is segmented into individual character sub-images.
- Pre-processing: Skew Correction, Linearization, and Noise Removal are the three processes in pre-processing. The skew correction is carried out until the horizontal axis is aligned. For further processing, the noise created during capture is eliminated.
- Feature Extraction: The text is extracted by detecting the edges of the photos.
- Character recognition: Individual characters can be identified.
- Text-to-speech conversion: Image characters are converted into vocal output via a speaker [23].

The Pi-camera turns on when the push button is pressed; otherwise, the camera remains off. The text from a picture is captured by the Pi-camera and analyzed using OCR technology. An image of alphanumeric characters is recognized and transformed into speech [5] (Fig. 14).

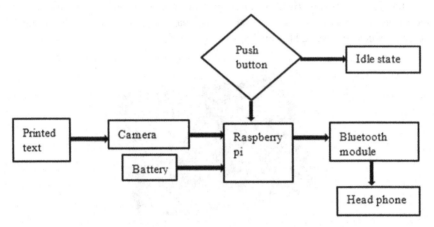

Fig. 14. General working principles of object detection [21].

The steps to Haar algorithmic feature extraction approaches for object recognition.

1. To train the classifier, a large number of positive and negative images, both with and without faces, are required.
2. Next, using a type of 'filter' known as Haar features, extract features from photos.
3. The idea is to look at one part of the image by applying this filter to it.
4. Then, for each window, the pixel light intensities of the white and black parts are added together.
5. The extracted feature's value is the total of the two summations subtracted [19] (Fig. 15).

Alphanumeric Detection

"Text recognition" is another name for optical character recognition. OCR's basic functions are to examine scanned text documents and then use the information for data processing [25]. The printed text images are scanned using an optical scanner or specific circuitry, and software analyzes the acquired image characters for further processing. OCR converts non-editable documents into editable ones, allowing the user to search for the content by characters, just like in a word document [26] (Fig. 16).

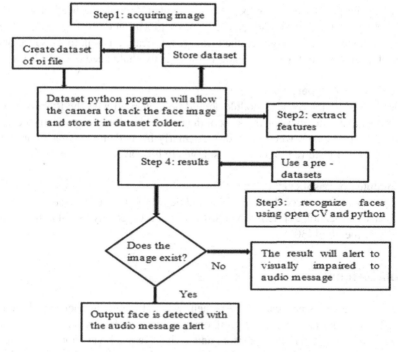

Fig. 15. Steps of smart system of cane [24].

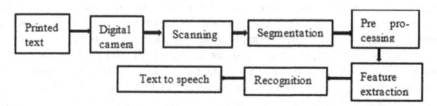

Fig. 16. Alpha numeric detection [9].

Noise Reduction

Noise reduction is a technique for avoiding unnecessary noise from the core audio signals received from Bluetooth headphones' microphone. Individuals who are partially deaf will benefit from this technique [25]. They find it difficult to communicate with the speaker in many places, such as marketplaces and roads, because they can't hear the speaker's sound or the person talking to the challenged person's speech adequately owing to background noise. Deep learning approaches can effectively filter out disturbances from the fundamental audio stream [27].

3.2 Navigation and Location Tracking

This navigation system module aids visually impaired people in navigating without assistance. The user must communicate their destination via voice message. With the use of GPS, the position of their location is provided through the headset as a voice note. GSM is also utilized to send the user's current location to their caretaker via text message [28]. During an emergency, the message notification is sent to the saved phone number of a caretaker to track the location of a visually challenged person [29].

GPS module for raspberry pi

The Global Positioning System module should be integrated with the Raspberry Pi to enable proper and exact location tracking of the user. This allows relatives to receive the address of the participant when needed, particularly in sudden emergency conditions. The GPS data is transferred to the processor, where it is saved in a database [20].

GSM module for Raspberry Pi

The Raspberry Pi is connected to the GSM module, whose major function is to communicate the user's location to relatives via SMS, which is especially useful in the event of a disaster for the blind [30].

4 Conclusions

These studies showed smart cane development with different approaches and methodologies. The primary problem is detecting obstacles and guiding in cane configurations with constraints. However, using sensors and giving feedback via haptic response is a solution for smart cane applications. The most recent object identification approaches are processed in computer vision systems and deep learning algorithms using high-pixel cameras and ultrasonic sensors. Due to its good accuracy in object identification, Yolo deep learning is the most widely used method. Various types of microprocessors are suitable for analyzing for smart cane applications, particularly the Raspberry Pi module, which is suitable for computer vision systems due to its high processing capacity, compact design, and compatibility with various modules. And the cane guiding system and location tracking system use a GPS module integrated with a GSM-sim module to send location information to their caretaker. Sensors are also utilized for detecting and measuring the obstacle objects with a feedback system of buzzers and vibration motors. The Smart Cane's other future is an audio feed-back system that uses Bluetooth earphones to notify the user of path conditions. In general, many limitations must be addressed in the development of smart cane technology, such as cane affordability, battery charging system and durability, object detection accuracy, and cane ergonomics. As a result, future development of smart cane technology will necessitate addressing those issues in order to improve the lives of visually impaired people.

References

1. WHO: Visual Impairment and Blindness. https://www.who.int/news-room/fact-sheets/detail/blindness-and-visual-impairment
2. Joshi, R.C., Yadav, S., Dutta, M.K., Travieso-Gonzalez, C.M.: Efficient multi-object detection and smart navigation using artificial intelligence for visually impaired people. Entropy **22**(9), 941 (2020)

3. El-Taher, F.E., Taha, A., Courtney, J., Mckeever, S.: A Systematic review of urban navigation systems for visually impaired people. Sensors **21**(9), 3103 (2021)
4. Mind, P., Palkar, G., Mahamuni, A., Sahare, S.: Smart stick for visually impaired. Int. J. Eng. Res. Technol. **10**(06), 196–198 (2021)
5. Kramomthong, P., Pintavirooj, C., Paing, M.P.: Smart cane for assisting visually impaired people and the blind. In: 2021 13th Biomedical Engineering International Conference (BMEiCON), pp. 1–5. Ayutthaya, Thailand (2021)
6. Sharmila, K., Boomika, D.: Smart walking cane using IOT. Int. J. Adv. Eng. Sci. Inform. Technol. **4**(4) (2021)
7. Beingolea, J.R., Zea-Vargas, M.A., Huallpa, R., Vilca, X., Bolivar, R., Rendulich, J.: Assistive devices: technology development for the visually impaired. Designs **5**(4), 75 (2021)
8. dos Santos, A.D.P., Medola, F.O., Cinelli, M.J., Garcia Ramirez, A.R., Sandnes, F.E.: Are electronic white canes better than traditional canes? A comparative study with blind and blindfolded participants. Univ. Access Inf. Soc. **20**(1), 93–103 (2020). https://doi.org/10.1007/s10209-020-00712-z
9. Lavanya Narayani, T., Sivapalanirajan, M., Keerthika, B., Ananthi, M., Arunarani, M.: Design of Smart Cane with integrated camera module for visually impaired people. In: 2021 International Conference on Artificial Intelligence and Smart Systems (ICAIS), pp. 999–1004. Coimbatore, India (2021)
10. Abu-Abdoun, D.I., Alsyouf, I., Mushtaha, E., Ibrahim, I., Al-Ali, M.: Developing and designing an innovative assistive product for visually impaired people: smart cane. In: 2022 Advances in Science and Engineering Technology International Conferences (ASET), pp. 1–6. Dubai, United Arab Emirates (2022)
11. Laki, S., Stoyanov, R., Kis, D., Soulé, R., Vörös, P., Zilberman, N.: P4Pi: P4 on Raspberry Pi for networking education. ACM SIGCOMM Comput. Commun. Rev. **51**(3), 17–21 (2021). https://doi.org/10.1145/3477482.3477486
12. Khan, M.A.: Motion based smart assistant for visually impaired people. Indian J. Sci. Technol. **13**(16), 1612–1618 (2020)
13. Al-Fahoum, A.S., Al-Hmoud, H.B., Al-Fraihat, A.A.: A smart infrared microcontroller-based blind guidance system. Act. Passive Electron. Compon. **2013**, 1–7 (2013)
14. Midi, N.S., Idris, N.A.M., Yusoff, S.H., Ripah, N.A.M.: Evaluation of energy harvesting for smart cane application. In: 2021 8th International Conference on Computer and Communication Engineering (ICCCE), pp. 110–114. Kuala Lumpur, Malaysia (2021)
15. Omoregbee, H.O., Olanipekun, M.U., Kalesanwo, A., Muraina, O.A.: Design and construction of a smart ultrasonic walking stick for the visually impaired. In: 2021 Southern African Universities Power Engineering Conference/Robotics and Mechatronics/Pattern Recognition Association of South Africa (SAUPEC/RobMech/PRASA), pp. 1–7. Potchefstroom, South Africa (2021)
16. Wang, T., Grobler, R., Monacelli, E.: EVAL Cane: an IoT based smart cane for the evaluation of walking gait and environment. In: 2020 IEEE International Symposium on Broadband Multimedia Systems and Broadcasting (BMSB). IEEE (2020)
17. Nazri, N.M.A., Fauzi, S.S.M., Adderley, R., Gining, J.M., Razak, T.R., Jamaluddin, M.N.F.: Smart cane for visually impaired with obstacle, water detection and GPS. Int. J. Comput. Dig. Syst. **10**(1), 1347–1354 (2021)
18. Alissa, H.A., Qawaqzeh, M., Abu-zaideh, S., Shamsudin, S.A.: The integrated technology design to guide the blind person to navigate safely. J. Theor. Appl. Inform. Technol. **98**(23), 3718–3728 (2020)
19. Frizziero, L., Liverani, A., Donnici, G., Papaleo, P., Leon-Cardenas, C.: Smart cane developed with DFSS, QFD, and SDE for the visually impaired. Inventions **6**(3), 58 (2021)
20. More, P.R., Raut, P.S., Waghmode, P.M.: Virtual eye for visually blind people. Int. J. **5**(12), 562–566 (2021)

21. Chaudary, B., Pohjolainen, S., Aziz, S., Arhippainen, L., Pulli, P.: Teleguidance-based remote navigation assistance for visually impaired and blind people-usability and user experience. Virtual Real **24**, 1–18 (2021)
22. Mavarkar, P.R., Mundargi, Z.K.: Real time smart bind stick using artificial intelligence. Indian J. Artif. Intell. Neural Network. **1**(1), 9–13 (2020)
23. Rahman, M.W., Tashfia, S.S., Islam, R., Hasan, M.M., Sultan, S.I., Mia, S., Rahman, M.M.: The architectural design of smart blind assistant using IoT with deep learning paradigm. Internet of Things **13**, 100344 (2021). https://doi.org/10.1016/j.iot.2020.100344
24. Al-Muqbali, F., Al-Tourshi, N., Al-Kiyumi, K., Hajmohideen, F.: Smart Technologies for Visually Impaired: Assisting and conquering infirmity of blind people using AI Technologies. In: 2020 12th Annual Undergraduate Research Conference on Applied Computing (URC), pp. 1–4. Dubai, United Arab Emirates (2020)
25. Tang, J., et al.: Design and optimization of an assistive cane with visual odometry for blind people to detect obstacles with hollow section. IEEE Sens. J. **21**(21), 24759–24770 (2021)
26. Ramisetti, C.: An ultrasonic sensor-based blind stick analysis with instant accident alert for blind people. In: 2022 International Conference on Computer Communication and Informatics (ICCCI), pp. 1–13. Coimbatore, India (2022)
27. Venugopalan, A.K., Varughese, S.K., Jacob, S. M., George, A., Joseph, J.: AI based audio recognition system for visually and audibly challenged. In: 2020 International Conference on Smart Electronics and Communication (ICOSEC), pp. 297–304. Trichy, India (2020)
28. Subbiah, S., Ramya, S., Parvathy Krishna, G., Nayagam, S.: Smart cane for visually impaired based on IOT. In: 2019 3rd International Conference on Computing and Communications Technologies (ICCCT), pp. 50–53. Chennai, India (2019)
29. Johari, R., Gaurav, N.K., Chaudhary, S., Pramanik, A.: START: smart stick based on TLC algorithm in IoT network for visually challenged persons. In: 2020 Fourth International Conference on I-SMAC (IoT in Social, Mobile, Analytics and Cloud) (I-SMAC), pp. 605–610. Palladam, India (2020)
30. Dhanalakshmi, K.S., Sai Reddy, G.K., Vineesh Reddy, M., Sahul, N.B., Sreenath, K., Pazhani, A.A.J.: Development of smart stick for visually challenged people. Ann. Rom. Soc. Cell Biol. **25**, 17078–17090 (2021)

A Headphone-Based Heart Rate and Heart Rate Variability Monitoring Unit

Gashaye Lewtie Hailu[(✉)]

Bahir Dar Institute of Technology, Bahir Dar University, Bahir Dar, Ethiopia
gashay2009@gmail.com

Abstract. The Heart Rate (HR) is detected using the Reflectance Photoplethysmography (PPG) Technique, which is a non-invasive approach for measuring changes in blood volume in tissue using an appropriate light source and detector. This study presents a new headphone-based Heart Rate and Heart Rate Variability (HRV) monitoring device. The system is portable and built into conventional headphones. Because the device is built into the headphone, the HR and HRV are recorded without causing any disruption or inconvenience to the person wearing it. The captured data is wirelessly transferred to and displayed on a cell phone through Bluetooth communication. In this work, an android interface was developed and used to show data obtained in the form of beats per minute (BPM) using the MIT App Inventor 2 Bluetooth connectivity application. The device is powered by 5 V by the Intex Wireless Roaming Headset, which contains two AAA batteries that provide 1.5 V each. Because the headphone only provides 3 V, we need a DC-DC Boost Converter Step Up Module 1–5 V to 5 V 500 mA to provide the appropriate supply voltage to the system. Tests on various persons utilizing the prototype system built demonstrated the scheme's usefulness. Data from eight friends were used in the assessment study. The accuracy of the heart rate measuring device was evaluated with -0.5 ± 2.13 BPM as a tolerance during the normal resting period using the standard device "WEAL Pulse Oximeter Fingertip" manufactured by WEAL.

Keywords: Photoplethysmography (PPG) · Pulse rate · Reflectance · Transmittance · Massachusetts Institute of Technology (MIT)) App inventor · Infrared sensor · Signal conditioning circuit · Filters

1 Introduction

Heart rate describes the number of heartbeats per minute [1–4]. However, the beat of a healthy heart is not regular [4]. Fluctuations in the time intervals of adjacent heartbeats are referred to as heart rate variability (HRV) [4]. Measuring heart rate variability (HRV) can help determine the autonomic nervous system's functioning and the ratio of sympathetic and parasympathetic activity [2]. To assist us in adjusting to environmental and psychological difficulties, HRV is an emergent trait of interdependent regulatory systems that function on various time scales [3]. The regulation of autonomic balance,

© ICST Institute for Computer Sciences, Social Informatics and Telecommunications Engineering 2023
Published by Springer Nature Switzerland AG 2023. All Rights Reserved
B. H. Woldegiorgis et al. (Eds.): ICAST 2022, LNICST 455, pp. 209–226, 2023.
https://doi.org/10.1007/978-3-031-28725-1_13

blood pressure (BP), gas exchange, gut, heart, and vascular tone the size of the blood vessels that control BP are all reflected in HRV. It may also represent the modulation of facial muscles [4].

Noninvasive wearable heart rate monitoring units have recently been used for real-time distant and uninterrupted medical monitoring, owing to a large number of benefits for both physicians and patients through medium- and long-term recording of health indicators during daily life activities [5]. The measurement of heart rate (HR) and heart rate variability (HRV) is an important aspect of any person's cardiovascular health [5–8]. The traditional way of measuring heart rate is to physically place fingers or thumbs over the wrist to detect the arterial pulse [8]. In this method, the felt pulse is counted for one minute, and the number of beats per minute (BPM) is roughly calculated or assessed [8]. However, this strategy is unsuitable for continual inspection [9, 10].

According to [5, 11], ECGs are often recorded using torso-mounted, disposable adhesive electrodes. ECG signals may be distorted during long-term acquisitions during the wearing subject's typical activities as a result of sweat, contact, or movement artifacts [4, 5, 12]. When electrodes are affixed to their skin, people may feel uneasy and exhibit very little compliance. Drying the electrode-applied conductive gel and allergy reactions may affect the time-dependent results. Standard electrode placement necessitates a skilled operator because improper electrode location could lead to a flawed recording.

Photoplethysmography (PPG) has been developed as an alternative method for measuring HRV through the surrogate monitoring of pulse rate variability to address some of these issues and to promote the dissemination of portable and easily wearable devices (PRV) [5]. Due to the high signal amplitude that may be achieved in comparison to other sites, PPG sensors are usually placed on the fingers because the high signal amplitude on the fingertip can be accomplished compared with other body surfaces [8, 9]. According to [10], this arrangement is not appropriate for continuous measuring in daily activities involving the use of the fingers.

The utilization of downsizing technologies, as well as the integration of a PPG sensor, microcontroller, Bluetooth, and DC-DC converter incorporated into headphones near the ear, can help to eliminate motion artifacts caused by torsion or contact. PPG sensors and devices are non-invasive since they have no electrical interaction with the human body [13]. They are also less expensive and require less maintenance than ECG equipment.

A method of measuring blood volume variations caused by the pulsing of blood that occurs with each heartbeat is called plethysmography. Small, wearable pulse rate sensors have been created using photoplethysmography (PPG) technology [14]. These gadgets, made of infrared light-emitting diodes (LEDs) and photodetectors, provide a straightforward, dependable, and affordable method of noninvasively measuring the pulse rate [11, 16]. PPG uses sturdy, low-maintenance optical sensors that are reasonably priced [11]. In this study, it is the perfect portable device because it uses very little power and can be supplied by a battery pack with a 5 V power supply.

From the PPG signal, a variety of clinically important characteristics can be extracted, including heart rate, respiratory rate, respiratory-induced intensity variations (RIIV), ventilatory volumes, and autonomic dysfunction [15].

The most basic PPG sensor comprises an infrared LED and a photodetector that is housed inside the same plastic housing in Fig. 1 [15]. The sensor is attached to the surface of the body using items such as headphones, earbuds, eyeglasses, etc. The sensor can be either a reflecting or a transmitting kind, as indicated in the figure below.

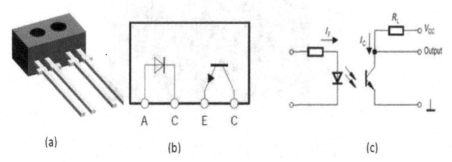

Fig. 1. Reflective optical sensor with transistor output (TCRT1000). (a) TCRT1000 IC; (b) TCRT1000 symbol; (c) TCRT1000 circuit.

In this study, the low-cost, non-intrusive Heart Rate (HR) Monitoring System with Headphones contains a Bluetooth interface that links to the microcontroller is designed, which shows the pulse rate on an Android mobile. The maximum and minimum beat rates per minute are displayed if the monitoring device is connected to the Android phone via an app written using App Inventor 2. Everyone nowadays has access to a home environment where they can monitor and study their heart rate.

The user is free to concentrate solely on exercising with his or her Android phone while running, driving, biking, or performing another activity without having to constantly check their heart rate. This improved form of exercise is shaping the future of driving, training, and exercise because the gadget is less expensive, portable, and safe than constantly monitoring yourself. Additionally, it reflects several physiological states such as biological workload, stress at work, and attention to tasks, as well as the autonomic nervous system's active state.

The number of heartbeats in total that take place over the course of a minute is referred to as heart rate. However, Heart rate variability (HRV) is the term used to describe the normal variation in heart rate that takes place throughout time as shown in Table 2. The variation in the peak-to-peak interval is used to calculate it. Calculate the heart rate for each beat as illustrated in (Fig. 2) below by using the LabView oscilloscope to measure the interval between peaks of a PPG signal. The oscilloscope's time per division is set to 200 ms. We can calculate the heart rate variability using the PPG graph in Fig. 2 below using time domain analysis.

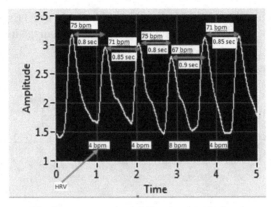

Fig. 2. Heart rate variability from PPG signal

Heart rate variability time-domain indices assess the amount of HRV detected over monitoring periods ranging from 1 min to 24 h [16]. The variation of beat-to-beat intervals, also known as R-R intervals, is referred to as HRV [7]. The average heart rate variability is expressed using the standard deviation of all RR intervals [7]:

$$HRV = \sqrt{\frac{\sum_{i=1}^{N}\left(RR_i - \overline{RR}\right)^2}{N-1}}$$

2   Materials and Methods

The volunteer's eight healthy friends took part in the study, in which they were monitored using wearable devices while performing various activities. The proposed system components are integrated with the headphone, as shown in Fig. 3 below.

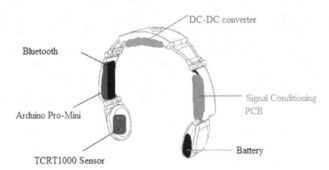

Fig. 3. The position of the system's components on the headphones is suggested.

The integrated components in Fig. 3 include the battery, sensor, signal conditioning circuit, Bluetooth, Arduino pro-mini, and DC-DC step-up converter.

2.1 General Block Diagram of the Proposed System

The general block design of the proposed system's hardware is depicted in Fig. 4 and uses reflecting photoplethysmography (PPG) to measure heart rate and heart rate variability through headphones. The PPG signal is made up of a large DC component as well as a pulsatile (AC) component. The AC component, which contains vital information such as heart rate, is much weaker than the DC component, and the PPG signal from the phototransistor is weak and noisy. To obtain a photoplethysmograph (PPG) signal, we need an amplifier and filter circuits to boost and clean the signal. The Arduino pro-mini board, which is used for signal processing, is the heart of the project and is responsible for all of the major digitalization of the PPG signal and beat rate calculation. The average heart rate in one minute is calculated using an Arduino pro-mini microcontroller, and the beat per minute is displayed on an Android phone via Bluetooth communication. If someone wants to use the LCD instead of the Android phone display, there is another display system in this device.

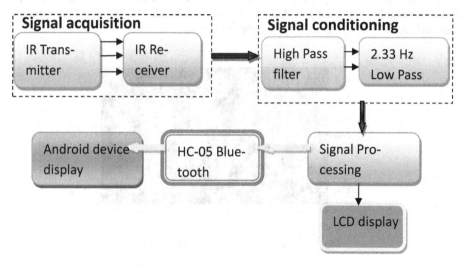

Fig. 4. The block diagram of the system

2.2 The Sensor Output Signal

The output voltage from the phototransistor without a filtering circuit consists of the pulsatile AC signal and slow varying high DC voltage. The output waveform of the sensor is shown below on different surfaces of the body.

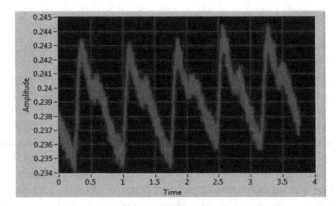

Fig. 5. Signal from a fingertip sensor output

A fingertip Comparing the PPG signal without a signal conditioning circuit, from the sensor to other body surfaces depicted in Fig. 5 above, the AC signal is considerably high and the DC voltage is low and slowly fluctuating.

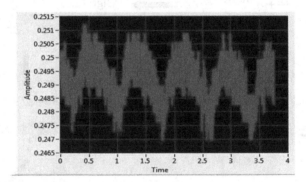

Fig. 6. Near-ear surface sensor output signal

The output signal from the sensor placed close to the ear is depicted in Fig. 6 above. In comparison to the sensor placed on the fingertip, the slow-varying DC voltage is larger and the AC signal is lower. We can get rid of the AC signal and slowly vary the DC voltage by using a signal conditioning circuit.

2.3 Signal Conditioning

The two-stage signal condition circuits are used to eliminate unwanted signals. High-pass filter and low-pass filter circuits are cascaded in the first-stage and second-stage signal conditioning circuits. The high pass signal conditioning circuit produces the necessary information, which will smother the significant DC component from the sensor. The HPF's cut-off frequency is set at 0.5 Hz. The high pass filter's cut-off frequency is determined by the resistor R1 and capacitor C1, as shown in Fig. 8. However, low pass signal conditioning will eliminate the significant DC component and enhance the weak pulsatile AC component, which conveys the necessary data. The cut-off frequency is determined by the resistor R2 and capacitor C2 as indicated in Fig. 8. The cutoff frequency in the circuit above was set to roughly 2.33 kHz, which corresponded to a maximum heart rate of 140 bpm. To achieve this, we used a resistor R2 of value 680 k and a capacitor C2 of value 100 nF.

As depicted in Fig. 7, the PPG output signal from the first stage signal conditioning circuit contains a low, slowly fluctuating DC voltage and a high AC signal.

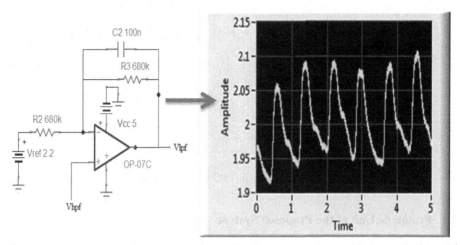

Fig. 7. PPG signal from the first stage of signal conditioning

The output signal from the first signal conditioning stage is not a pure sinusoidal waveform and is too faint to be used for beat calculation. As a result, HPF/LPF is employed for additional filtering and amplification in the second stage. The Second Stage high pass circuit, which is essentially a replica of the First Stage circuit, now receives the output from the active Low Pass Filter.

Similar HPF and varied gains from the first stage LPF circuits are also present in the second stage. A second Op-amp that is set up as a non-inverting buffer with G2 gain is now fed the two-step amplified and filtered signal. The necessary analog PPG signal is provided at the output of the second step. The PPG signal's amplitude, which is visible in the second stage's output, can be managed using the Gain G2.

Therefore, the two cascaded stages' combined voltage gain is $G = G1*G2 = 101*20.4 = 2063.28$. Beats per minute (BPM) = 60*f, which is the relationship between the frequency (f) of these pulses and heart rate (BPM).

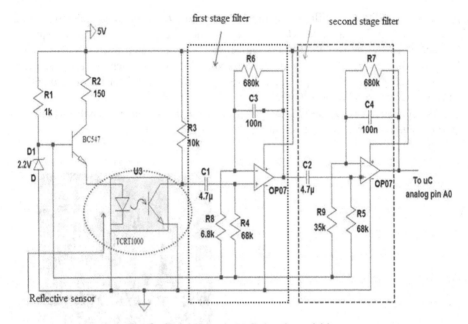

Fig. 8. Full schematic PPG signal acquisition

2.4 Prototype Unit of the Proposed System

Figure 9 shows the hardware circuit created from the circuit layout shown in Fig. 8 using a breadboard and the NI ELVIS II. The laptop uses a LabVIEW oscilloscope to display the waveform. Without taking into account the current and power used by the Bluetooth and Arduino pro tiny LEDs, the hardware circuit uses up to 160mA of current and roughly 1.2W of power. Due to the high power consumption of Bluetooth and Arduino Pro Mini devices, the batteries can only power the circuit for a few lengths of time.

Fig. 9. Prototype hardware setup on NI ELVIS board

2.5 Data Acquisition System

The workspace for the initial design and development of the hardware was NI ELVIS, and DAQ from LABVIEW was utilized for signal conditioning with a band-pass filter block and computer-based output signal display. Before being used by the signal conditioning circuit, the sensor's output signal is digitally filtered using the filter express VI from LABVIEW.

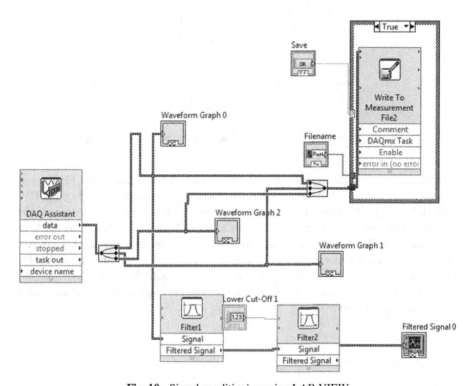

Fig. 10. Signal conditioning using LAB VIEW

PPG signal from the signal conditioning circuit is depicted in Fig. 11 using the LabVIEW waveform Graph 0 display depicted in Fig. 10.

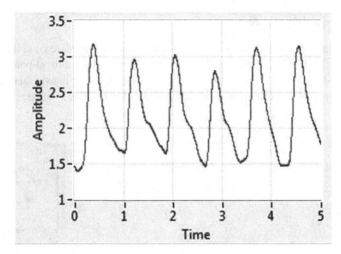

Fig. 11. PPG signal from the second stage signal conditioning circuit

In this study, to calculate heart rate and heart rate variability, first, the PPG signal has been converted into digital pulses that are close to transistor-transistor logic (TTL), as illustrated in Fig. 12. The condition of TTL outputs is often limited to narrower limits of 0.0 V to 0.4 V for a "low" and 2.4 V to VCC for a "high," offering at least 0.4 V of noise protection. The analog PPG signal is given to Arduino pro mini analog pin A0. For this, we identify peaks in the PPG waveform when the slope of the curve changes from positive to negative and the size of the signal exceeds 80% of the threshold. So when the ADC reaches more than the threshold value, the signal will count as a pulse unless ADC values drop to 0.

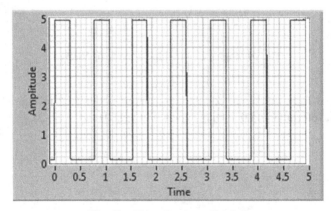

Fig. 12. TTL pulse from Arduino

2.6   Trial PPG Signal Waveform with Car

A brief test using a prototype arrangement on the car was done to determine the sensor's position. The primary goal of the test is to determine whether the PPG signal is stable and whether the heart rate varies while driving on the three body surfaces of the temple, nose, and near ear, as indicated in Fig. 13 below. As seen in Fig. 10, throughout this test, we recorded the data utilizing the LabVIEW block referred to write-to measurement file.

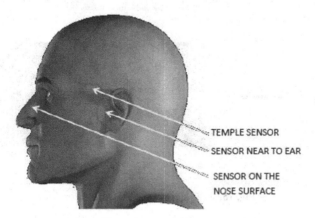

Fig. 13. Sensor position during car test

As shown in Fig. 14, the three sensors were attached to the driver's eyeglasses and signal conditioning circuit, power supply module, Bluetooth, etc. held by the back seat of the car using a breadboard, Elvis board, and laptop.

Fig. 14. Experimental test setup in the car

PPG signal from Temple Sensor
The PPG signal from the Temple sensor remained rather constant during the trial and was less impacted by the road's potholes, as seen in Fig. 15.

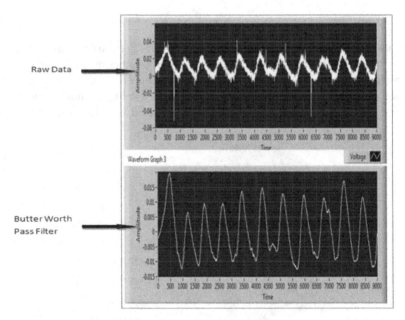

Fig. 15. PPG signal from Temple sensor

PPG signal from near Ear Sensor

According to Fig. 16, the PPG signal from the ear sensor was steady and not as significantly impacted by road imperfections. However, the sensor placed on the ear is not comfortable for continuous monitoring.

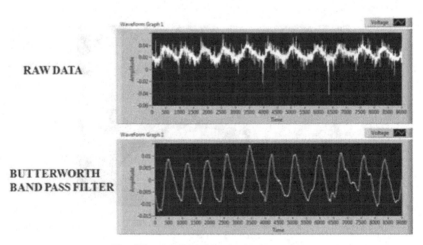

Fig. 16. PPG signal from near Ear sensor.

PPG signal from the surface of the Nose.
The PPG signal from the nose's surface is not very steady and is heavily influenced by road imperfections, as illustrated in Fig. 17.

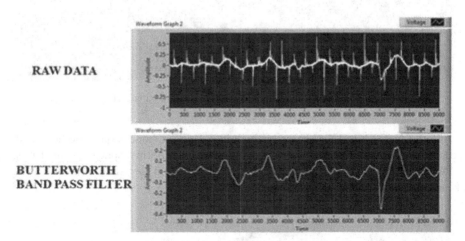

RAW DATA

BUTTERWORTH BAND PASS FILTER

Fig. 17. PPG signal from Nose surface sensor.

2.7 Powering of the Board

In this study, a Bluetooth and Arduino board are mounted on the headband of an Intex Wireless Roaming headset, which uses two AAA batteries, each of which provides 1.5 V for sensor integration on the ear pad and holding the signal conditioning board. The maximum output voltage from these headphones is 3 V, however, we require a 5 V source to operate them. The DC-DC Boost Converter Step-Up Module 1–5 V to 5 V 500 mA is used to obtain the necessary voltage from headphones to power the board.

2.8 Printed Circuit Board (PCB) Design and Fabrication

Components have been installed on a dual-layer PCB, as shown in Fig. 18 PCB layouts, on both the component and solder sides.

Capacitor and resistor placement is on the component side, while SMD Op Amp, SMD transistor, and SMD Zener diode soldering is done on the solder side, as illustrated in Fig. 18 from left to right. This PCB has seven holes for the TP3-TP6 connector for the IR TCRT1000 sensor, 5 V, GND, and PPG output (TP7). The 5V and GND are linked to the DC-DC Boost Converter Step Up Module 1–5 V to 5 V 500 mA, TP7 is attached to the analog port (A0) of the Arduino Pro Mini, and TP3–TP6 is either directly wired or connected via a jumper to the TCRT1000 sensor (Fig. 19).

Fig. 18. Dual layers PCB of signal conditioning part

Both Bluetooth and the sensor are built into the headphone earpad and headband, respectively. Figure 20 below shows the PPG signal waveform displayed on the laptop using LabVIEW software and the heart rate displayed on the Android mobile using Bluetooth and the MIT-app Inventor app.

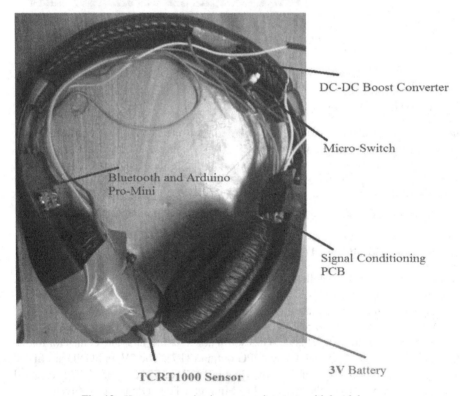

Fig. 19. Heart rate monitoring system integrate with headphones

The headphone, as shown in Fig. 20, is a component of the system; the HR and HRV are recorded without any disruption or inconvenience to the person wearing the device.

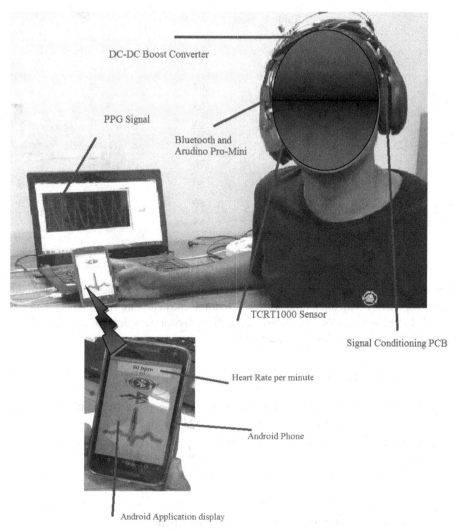

Fig. 20. Heart rate monitoring system and android display

The recorded data is wirelessly transmitted to a cellphone and displayed via Bluetooth communication.

3 Result and Discussion

3.1 Measuring the Performance of the System

Eight volunteers were used in this trial to test and validate the device's accuracy in measuring heart rate with a tolerance of ±3 beats during a typical time of rest using WEAL's standard device, the "WEAL Pulse Oximeter Fingertip." Table 1 below displays the heart rate readings obtained from a few friends to compare to their actual heart rates.

Table 1. Heart rate readings

Subject	Age	Male/Female	Actual heart rate (bpm)	Measured heart rate (bpm)	Difference (bpm)
A	33	M	60	62	-2
B	33	M	60	63	-3
C	24	M	92	93	-1
D	25	M	76	78	-2
E	28	M	78	76	+ 2
F	25	M	78	80	-2
G	24	M	82	80	+ 2
H	28	M	90	88	+ 2

3.2 Measurements of Heart Rate Variability (HRV)

In the second phase, the heart rate variability was obtained from my heart rate at different periods of the day with fixed time intervals. To know the heart rate variation the data was recorded at hour intervals of the day. The data showed a considerable change in heart rate activeness to sleep shown in Table 2 below.

Table 2. Heart rate variation of a subject at a fixed time interval

Time	7:00 AM	9:00 AM	11:00 AM	1:00 AM	3:00 PM	5:00 PM	7:00 PM	9:00 PM	11:00 AM	1:00 AM	3:00 AM	1:00 AM
Measured HR(bpm)	84	87	84	84	90	80	81	84	83	80	75	88

3.3 Analysis of the Differences Between Measurement

In this study, Bland and Altman used a graphical method to quantify the difference between actual and measured heart rates. For the analysis, the mean of the difference in measurement methods and the standard deviation was obtained to represent mean bias and the limits of agreement. Second, the data points can be restricted using + 2 SD to demonstrate a 95% confidence interval (CI; precisely defined: mean 1.96 SD) of distributed data [17]. The mean difference (mean bias) for this measurement was -0.5, with a standard deviation of 2.13. To understand the dispersion of variables, create a scatterplot with an X-axis (average) and a Y-axis (difference). The upper limit in this study can be calculated using mean + $1.96 \times SD$ ($-0.5 + 1.96 \times 2.13 = 3.69$) and the lower limit using mean $-1.96 \times SD$ ($-0.5-1.96 \times 2.13 = -4.69$). The following is an appropriate statement to use in the manuscript: The Bland-Altman plot revealed a mean bias $\pm$ SD between the first and second heart rate measurement as -0.5 ± 2.13 PBM, with the limits of agreement set at -4.69 and 3.69 (Fig. 21).

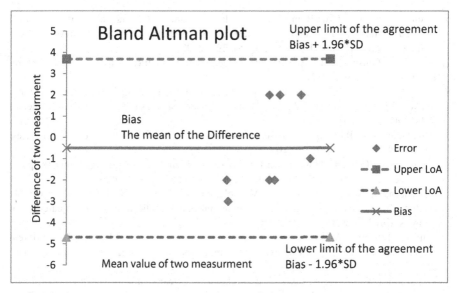

Fig. 21. Bland altman plot

4 Conclusion

This paper describes how a Reflectance PPG sensor-based heart rate monitoring system was put into practice utilizing a low-cost Pro Mini Arduino board, HC-05 Bluetooth, and other readily accessible components integrated into the headphone. Through HC-05 Bluetooth connectivity, data transmission was accomplished from the Arduino to the smartphone A minute's worth of heartbeats are counted, and the results are displayed on an Android device using the MIT App Inventor 2 application. This product's design aims to create a tool that can serve as a personal trainer, sleepy driving detector, and anger detector. Using his or her Android phone, the motorist can monitor their level of drowsiness at any moment. Hospitals can also use this system. Here, by the suggested system, a doctor is not required to be present when the heart rate is being monitored. The patient's heart rate can be sent to the doctor remotely.

References

1. Goroso, D.G., et al.: Remote monitoring of heart rate variability for obese children. Biomed. Signal Process. Control **66**, 102453 (2021)
2. Shi, K., et al.: Contactless analysis of heart rate variability during cold pressure test using radar interferometry and bidirectional LSTM networks. Sci. Rep. **11**(1), 1–13 (2021)
3. Ma, Z., et al.: A low-power heart rate sensor with an adaptive heartbeat locked loop. In: 2021 IEEE International Symposium on Circuits and Systems (ISCAS), pp. 1–5. IEEE (2021)
4. Hinde, K., White, G., Armstrong, N.: Wearable devices are suitable for monitoring twenty-four-hour heart rate variability in military populations. Sensors **21**(4), 1061 (2021)
5. Vescio, B., Salsone, M., Gambardella, A., Quattrone, A.: Comparison between electrocardiographic and earlobe pulse photoplethysmographic detection for evaluating heart rate variability in healthy subjects in short-and long-term recordings. Sensors **18**(3), 844 (2018)

6. Prieto-Avalos, G., Cruz-Ramos, N.A., Alor-Hernández, G., Sánche Cervantes, J.L., Rodríguez-Mazahua, L., Guarneros-Nolasco, L.R.: Wearable Devices for Physical Monitoring of Heart: a Review. Biosensors **12**(5), 292 (2022)
7. Shaffer, F., Ginsberg, J.P.: An overview of heart rate variability metrics and norm. Front. Public Health **5**, 258 (2017)
8. Chow, H.W., Yang, C.C.: Accuracy of optical heart rate sensing technology in wearable fitness trackers for young and older adults: Validation and comparison study. JMIR Mhealth Uhealth **8**(4), e14707 (2020)
9. Khamitkar, S.S., Rafi, M.: IoT-based system for heart rate monitoring. Int. J. Eng. Res. Technol. **9**(07), 1563–1571 (2020)
10. Khan, M.M., Tazin, T., Hossain, T.: Development of wireless monitoring system for pulse rate: a new approach. Multi. Dig. Publishing Inst. Proc. **67**(1), 13 (2020)
11. Castaneda, D., Esparza, A., Ghamari, M., Soltanpur, C., Nazeran, H.: A review on wearable photoplethysmography sensors and their potential future applications in health care. Int. J. Biosens. Bioelectron. **4**(4), 195 (2018)
12. Přibil, J., Přibilová, A., Frollo, I.: Comparative measurement of the PPG signal on different human body positions by sensors working in reflection and transmission modes. Eng. Proc. **2**(1), 69 (2020)
13. He, J., Choi, W., Yang, Y., Lu, J., Wu, X., Peng, K.: Detection of driver drowsiness using wearable devices: a feasibility study of the proximity sensor. Appl. Ergon. **65**, 473–480 (2017)
14. Ludwig, M., Hoffmann, K., Endler, S., Asteroth, A., Wiemeyer, J.: Measurement, prediction, and control of individual heart rate responses to exercise—Basics and options for wearable devices. Front. Physiol. **9**, 778 (2018)
15. Kao, Y.H., Chao, P.C.P., Wey, C.L.: Design and validation of a new PPG module to acquire high-quality physiological signals for high-accuracy biomedical sensing. IEEE J. Sel. Top. Quantum Electron. **25**(1), 1–10 (2018)
16. L'Her, E., N'Guyen, Q.-T., Pateau, V., Bodenes, L., Lellouche, F.: Photoplethysmographic determination of the respiratory rate in acutely ill patients: validation of a new algorithm and implementation into a biomedical device. Ann. Intensive Care **9**(1), 1–10 (2019)
17. Campos, L.A., Pereira, V.L., Jr., Muralikrishna, A., Albarwani, S., Brás, S., Gouveia, S.: Mathematical biomarkers for the autonomic regulation of the cardiovascular system. Front. Physiol. **4**, 279 (2013)
18. Doğan, N.Ö.: Bland-Altman analysis: A paradigm to understand correlation and agreement. Turk. J. Emerg. Med. **18**(4), 139–141 (2018)

Amharic Sentence-Level Word Sense Disambiguation Using Transfer Learning

Neima Mossa[1]([⊠]) and Million Meshesha[2]

[1] Faculty of Computing, Bahir Dar Institute of Technology, Bahir Dar University, Bahir Dar,
Ethiopia
`neimamussa32@gmail.com`
[2] School of Information Science, Addis Ababa University, Addis Ababa, Ethiopia

Abstract. Word sense disambiguation (WSD) plays an important role, in increasing the performance of NLP applications such as information extraction, information retrieval, and machine translation. The manual disambiguation process by humans is tedious, prone to errors, and expensive. Recent research in Amharic WSD used mostly handcrafted rules. Such works do not help to learn different representations of the target word from data automatically. Moreover, such a manual disambiguation approach looks at a limited length of surrounding words from the sentence. The main drawback of previous works is that the sense of the word will not be detected from the synset list unless the word is explicitly mentioned. Our study explores and designs the Amharic WSD model by employing transformer-based contextual embeddings, namely AmRoBERTa. As there is no standard sense-tagged Amharic text dataset for the Amharic WSD task, we first compiled 800 ambiguous words. Furthermore, we collect more than 33k sentences that contain those ambiguous words. The 33k sentences are used to finetune our transformer based AmRoBERTa model. We conduct two types of annotation for our WSD experiments. First, using linguistic experts, we annotate 10k sentences for 7 types of word relations (synonymy, hyponymy, hypernymy, meronomy, holonomy, toponymy, and homonymy). For the WSD disambiguation experiment, we first choose 10 target words and annotate a total of 1000 sentences with their correct sense using the WebAnno annotation tool. For the classification task, the CNN, Bi-LSTM, and BERT-based classification models achieve an accuracy of 90%, 88%, and 93% respectively. For the WSD task, we have employed two experiments. When we use the masking technique of the pre-trained contextual embedding to find the correct sense, it attains 70% accuracy. However, when we use the FLAIR document embedding framework to embed the target sentences and glosses separately and compute the similarities, our model was able to achieve 71% accuracy to correctly disambiguate target words.

Keywords: Word sense disambiguation · Transfer learning · Neural network · Pre-trained language model · Natural language preprocessing · Morphological analyzer · Amharic WSD

© ICST Institute for Computer Sciences, Social Informatics and Telecommunications Engineering 2023
Published by Springer Nature Switzerland AG 2023. All Rights Reserved
B. H. Woldegiorgis et al. (Eds.): ICAST 2022, LNICST 455, pp. 227–238, 2023.
https://doi.org/10.1007/978-3-031-28725-1_14

1 Introduction

Natural language processing (NLP) is a field of artificial intelligence that assists computers in understanding, interpreting, and manipulating human language. Natural language is now being used to exchange information among humans and has now reached the extent of being an evolution criterion for technology (Reta 2015). To properly access and understand the information on the internet, there is a need for people all over the world to be able to use their language. This requires the existence of NLP applications such as machine translation, information retrieval, information extraction, and others. These downstream NLP applications rely on tools such as word sense disambiguation for their reasonable performance.

Most of the words in natural languages are polysemic, which means that they have several meanings (Hassen 2015). Amharic is one of the languages that have many words with multiple meanings. It is like other Semitic languages with a morphologically complex structure (Senay 2021). The ability to recognize the meaning of a word from its context and solve the ambiguity is one of the most difficult problems in natural language processing (Alian et al. 2016). Ambiguity is defined as a word, term, notation, sign, or symbol interpreted in more than one way (Mindaye et al. 2010). Word Sense Disambiguation is a hard and challenging task in NLP, intending to determine the exact sense of an ambiguous word in a particular context (Huang et al. 2019). When WSD is used in conjunction with other NLP approaches, it improves the efficiency of identifying accurate keywords for use as features in classification, searching, and many more NLP application (Senay 2021).

Knowledge-based, corpus-based, and hybrid machine learning methods are the main categories of approaches for WSD tasks (Pal and Saha 2015). Knowledge-based WSD approaches are based on different knowledge sources such as machine-readable dictionaries (WordNet), thesauri, etc. LESK, semantic similarity, selection preference, and heuristic are the main algorithms for knowledge-based approaches. There are two sets of data for training and testing in supervised approaches. This approach to WSD systems employs machine learning techniques based on manually created sense-annotated data. The training set, which consists of examples related to the target word, could be used to learn a classifier. The supervised approach includes techniques such as Naïve Bays, decision lists, and K-nearest neighbor algorithms. Unsupervised WSD methods do not rely on external knowledge sources, machine-readable dictionaries, or sense-annotated data sets, rather, they use the information found in un-annotated corpora to differentiate the word meaning.

Recently, contextual embedding methods like BERT, ELMO, and GPT-2/3 learn sequence-level semantics by considering the sequence of all the words in the input sentence (Chawla et al. 2019). These methods are characterized by their high performance, and the ability to extract a lot of information from raw text. These recent language models, especially the BERT model is trained to predict the masked word(s) of the input sentence (El-razzaz et al. 2021). To weigh, the relationship between each word in the input sentence and the other words in the same sentence, BERT learns self-attention by giving a vector for each word. The vector represents the relationship of one word with other words in the input sentences and is used to generate word embedding. In this work, we have employed AmRoBERTa, a RoBERTa model trained for Amharic (Yimam et al. 2021).

2 Related Works

The research by Kassie (2009) tried to demonstrate WSD for Amharic language using semantic vector analysis. A total of 865 words were selected from the Ethiopian Amharic language legal statute documents. Instead of using sense-tagged words, the researcher evaluates WSD using pseudo-code words (artificial words). The developed algorithm outperformed the one used by Lucene, according to their comparison of the two. The achieved result is an average precision and recall of 58% and 82%, respectively. The author recommended developing resources such as Corpora, Thesaurus, and WordNet, that could be useful to advance the research in information retrieval, and word sense disambiguation.

Mekonnen (2010) conducted the Amharic WSD study using a corpus-based, supervised machine-learning approach. The author used the Naïve Bayes algorithm for Amharic WSD to classify a word to its correct sense using Weka 3.62 package in both the training and testing phases. A total of 1045 English sense examples for the five ambiguous words were gathered from the British National Corpus (BNC). The dictionary is used to translate the sense illustrations back into Amharic. For each sense of the ambiguous word, a total of 100 sentences were collected where the accuracy achieved ranged from 70% to 83.5% for all classifiers.

Assemu (2011) tried to develop corpus-based Amharic WSD through the use of unsupervised machine learning. A total of 1045 English sense examples for the five ambiguous words were gathered from the British National Corpus (BNC). Using the Amharic-English dictionary, the sense examples were converted to Amharic and prepared for experimentation. The result showed that the accuracy of unsupervised Amharic WSD is state-of-the-art result than the supervised machine learning approach, with an accuracy of 83.2% and 70.1%, respectively. For better Amharic WSD, the researcher recommended using linguistic tools like the Thesaurus, Lexicon from WordNet, machine-readable dictionaries, and machine translation tools.

Wassie (2014) utilized a semi-supervised learning strategy, and present a WSD prototype model for Amharic words. Unsupervised machine learning approach for clustering based on instance similarity and supervised machine learning approach after unlabeled data are applied. To cover all the senses of each target word available, annotated corpora are highly insufficient. The development of the Adaboost Bagging and ADtree algorithms perform at 84.90%, 81.25%, and 88.45%, respectively. The author concludes that Semi-supervised learning using bootstrapping algorithm performs better.

The research by Hassen (2015) developed an Amharic WSD knowledge-based approach based on WordNet to extract knowledge from word definitions and relationships between words and senses. They manually created the Amharic WordNet for this study and chose 2000 words, including ambiguous words. They carried out two tests to compare Amharic WordNet's impact with and without a morphological analyzer, and the results showed an accuracy of 57.5% and 80%, respectively. A two-word window on either side of the ambiguous word is sufficient for Amharic WSD, according to their research into the optimal window size. In this experiment, they have concluded that Amharic WordNet with a morphological analyzer can have better accuracy than without a morphological analyzer. They recommended automatic the development of Amharic WordNet and to apply a hybrid approach.

Tesema, Tesfaye and Kibebew (2016) applied supervised machine learning techniques to a corpus of Afaan Oromo language to automatically gather disambiguation information. This method is known as a corpus-based approach to disambiguation. To determine the prior probability and likelihood ratio of the sense in the provided context, they have utilized the Naïve Bayes approach. A total of 1240 Afaan Oromo sense examples were gathered for the chosen five ambiguous words, and the sense examples were manually tagged with their appropriate senses. The author used a corpus of Afaan Oromo sentences based on the five selected ambiguous words to acquire disambiguation information automatically. The system attains an accuracy of 79%, and it was discovered that the Afaan Oromo WSD can handle four words on either side of an ambiguous target word.

Siraj (2017) attempts to develop a system for word WSD that uses data from Word-Net and tagged example sentences to determine the sense of ambiguous Amharic words. Information from WordNet was extracted using the LESK algorithm and Python programming. The WordNet is made up of 17 ambiguous words from various classes, along with developed synonyms and glossary definitions. Based solely on the Jaccard Coefficient and Cosine Similarity, Amharic WSD's accuracy performance reached 84.52% percent and 85.96%, respectively. The average accuracy of the Jaccard Coefficient with Lesk scores is 89.83% which is a better result, compared to cosine similarity with LESK (86.69%). The researcher suggests for future work to use the Adaptive LESK algorithm and improve the performance of the WSD system.

Mulugeta (2019) attempts to develop an Amharic WSD system that uses Amharic WordNet hierarchy as a knowledge base. They use context to gloss overlap augmented semantic space approach. Most previous research on Amharic WSD focused on verb class; yet, Mulugeta (2019) tried to solve all open classes (verb, noun, adverb, and adjective) by developing WordNet. The WordNet contains about 250 synsets and does not include all relationships for single-sense words in the WordNet. The main challenge in this study was the unavailability of lexicon resources (WordNet), and the stemmer algorithm used in the preprocessing does not cover all exceptions and has limitations in returning the root word. Experimental result shows that context-to-gloss followed by augmented semantic space has achieved the highest recall of 87% and 79% for three target words at word and sentence level respectively. And the highest average accuracy of 80% and 75% at word-level and sentence level are achieved by this approach. Their recommendation is to develop a better stemmer or morphological analyzer and fully constructed WordNet containing relationships for non-ambiguous words.

Tadesse (2021) proposed a machine learning based WSD model for the Wolaita language. A total of 2797 sense instances were gathered to complete the investigation. Language specialists assessed the acquired data before creating five datasets for five ambiguous words, including "Doona," "Ayfiya," "Aadhdha," "Naaga," and "Ogiya." They used quantitative and experimental research to discover the ideal machine combination algorithms for learning and methods for extracting features. AdaBoost classifier utilizing BOW, TF-IDF, and Wor2Vec features as an extraction approach and the Support Vector Classifier, Bagging, Random Forest Classifier, and AdaBoost as classifier for the five datasets. In this study, precision and recall were used as the primary metrics for evaluation. Support Vector Classifier and Bagging classifiers with TF-IDF obtain an accuracy of 83.22% and 82.82%, respectively.

Recently, Senay (2021) has developed Amharic WSD by using a deep-learning approach. A total of 159 ambiguous words, 1214 synsets, and 2164 sentence datasets were used to create three distinct deep learning algorithms in three separate experiments. As a methodology, they used a design science research strategy. The author used different deep learning models for classification such as LSTM, CNN, and Bi-LSTM that are trained on the dataset using different hyperparameters. The results showed that LSTM, CNN, and Bi-LSTM obtained 94%, 95%, and 96% accuracy during the third experiment, respectively. But for disambiguation, they used handcrafted rules without applying any model. To increase the performance of the model, using lemmatization in the preprocessing, and using an attention mechanism are recommended.

Generally, Amharic word sense disambiguation was done by different researchers using different machine learning approaches. However, there is no easy and automatic Amharic word sense disambiguation, and there is no research that used the transfer learning algorithm for the disambiguation purpose. Generally, most of the literature tries to develop Amharic WSD but there is a gap in solving the problems of word sense. Most of them follow a manual approach for extracting word sense. Recent research used handcrafted rules or directly fetching the meaning of an ambiguous word from the synset list or in the WordNet but did not learn different representations from data automatically. The WSD developed by researchers requires manually labeled sense examples for every word sense. Previous researches also require defining features explicitly; but transfer learning algorithms aim to learn different representations from data automatically (Bouhriz et al. 2016); solve ambiguity problem based on sentence semantics. In this research, we attempt to employ transfer learning for Amharic WSD.

3 Amharic Language

Amharic is one of the northern Semitic languages in the part of the Afro-Asiatic families and it becomes a countless contribution in the area of literature in the 17th century up to the 19th century (Kebede at el. 1993). After Arabic, Amharic (አማርኛ) is the second most broadly spoken Semitic language (Gezmu et al., 2019). In addition, the language has a significant number of speakers in all regional states of the country (Salawu and Aseres 2015) and also in Canada, the USA, Eritrea, and Sweden (Mulugeta 2019).

3.1 Amharic Writing System

The Amharic language has its own alphabet, known as ፊደል/fidäl, which was inherited from the Geez. ፊደል/Fidäl is a syllabary writing system in which the consonants and vowels coexist within each graphic symbol. Unlike most Semitic scripts such as Arabic and Hebrew, Amharic fidäl is written from left to right. The writing system consists of 231 core characters, 33 consonants, each of which has 7 orders depending on the vowel with which it is combined, and some additional orders of ' ፊደል'/ fidäl are called dikala hoheyat/ ዲቃላ ሆሄያት(Getaneh 2020).

To separate each word and sentence in a formal Amharic writing system, the main punctuation marks are discussed as follow. The Ethiopic comma (፡) to separate words, Ethiopic full stop (።) to end the sentence, Ethiopic semicolon (፤) to separate Amharic words or phrases with similar concepts, the Ethiopic double dash (፥) to separate Amharic sentences with a similar concept and Ethiopic question mark (?) to end the question are the main unique Ethiopic punctuation marks. Nowadays, the Ethiopian modern writing system uses a single space rather than an Ethiopic comma (፡) to separate words.

3.2 Ambiguity in Amharic Language

Different scholars define ambiguity in a different way. According to Mindaye et al. (2010), ambiguity is described as the attribute of being ambiguous, where a word, term, notation, sign, symbol, phrase, sentence, or any other form used for communication is deemed ambiguous if it can be understood in more than one manner. Amare (2001) also define ambiguity as the quality of any thought, idea, statement, or claim whose meaning, intention, or interpretation cannot be determined decisively by a set of rules or processes.

Based on the study of Amare (2001), there are six types of ambiguities in Amharic language, namely Lexical Ambiguity, phonological ambiguity, structural ambiguity, referential ambiguity, semantic ambiguity, and orthographic ambiguity. These ambiguities are summarized below.

Lexical Ambiguity: Lexical ambiguity occurs when a lexical unit falls into separate part-of-speech categories with different senses, or when a lexical unit has more than one sense, all of which fall into the same part-of-speech category (Abate and Menzel 2007).
Phonological Ambiguity: The placement of pause within the word may lead to phonological ambiguity. When speakers use pauses and without pauses during speaking leads to ambiguity (multiple meanings) of a word (Kassie 2009, Mekonnen 2010).
Semantic Ambiguity: It determines the possible meanings of a sentence by focusing on the interactions among word-level meanings in the sentence. Polysemy, idiomatic and metaphorical word relations in a sentence are causes of semantic Ambiguity (Siraj 2017, Hassen 2015).
Syntactic Ambiguity: Structural ambiguity can give more than one meaning by the order of the word and holds more than one possible position or arrangement in the grammatical structure of the sentence.
Orthographic Ambiguity: Geminate and non-geminate sounds are causes of orthographic Ambiguity. This type of ambiguity can be solved using the context meaning of the sentence (Kassie 2009, Assemu 2011).
Referential Ambiguity: This ambiguity arises when a pronoun stands for more than one possible antecedent. a pronoun is understood by default even if it is not written grammatically.

4 Methodology

Algorithm: For this research we compared three models CNN, BiLSTM, and BERT to classify weather the word is ambiguous or not. Our experimental result showed that

BERT has better result than CNN and BILSTM because BERT used self-attention-based transformer architecture, which, in combination with a masked language modeling target, allows to train the model to see all left and right contexts of a target word at the same time (Chawla et al. 2019). After identifying whether the word is ambiguous or not the next task is assigning the meaning of ambiguous word. So, to disambiguate the ambiguous word we apply the AmRoBERTa model with the flair document embedding technique. It is a recent transfer learning approach that gives better performance in the available datasets (Yimam et al. 2021).

Dataset Collection and Preparation.   Since there are no labeled datasets available for Amharic word sense disambiguation, the main task for this thesis work is to prepare labeled datasets for WSD. We have collected 10k sentences and 800 ambiguous words from Amharic news, Amharic dictionary, Amharic Quran, Amharic Bible, Abissinica online dictionary and Amharic textbooks (from grade 7–12). A total of 33,297 sentences are used to finetune the AmRoBERTa model (transfer learning). The collected data passes through data preprocessing to prepare the data for experimentation. Data preprocessing is critical for improving the performance of the model. To make our data more suitable for the experiment, we use various data preprocessing techniques such as tokenization, stopword removal, special character removal, normalization, and morphological analysis.

Dataset annotation: In our study, we selected annotators to keep the nature and behaviors of Amharic language texts and to acquire quality and reliable data. We annotate both relationship of the sentence and the sense of the word in the sentence. For the dataset annotation, we have done two different annotations. The first annotation is to know whether the data set contains all the selected relationships of a word or not. Therefore, we selected three Amharic language and linguistic experts to annotate the data. The experts annotated the relationship between the sentences.

The second annotation is for disambiguation or to know the sense of the word. For this task, we have also used the WebAnno annotation tool to annotate the ambiguous word in the sentence. We selected two annotators and one curator from Amharic language native speakers. The main advantage of the WebAnno annotation tool is getting the value for inter annotation agreement (such as Fleiss kappa, and Cohen's kappa) is easy. We used Cohen's kappa as a measure of inter-annotator agreement.

5   Result and Discussion

5.1   Experimental Result of CNN Model

We have trained the CNN model with 2 dense layers with sigmoid activation functions and binary_crossentropy loss functions We also used 0.00001 for the learning rate, 64 batch-size, and a dropout rate of 0.2, which are optimal for our experiment (Fig. 1).

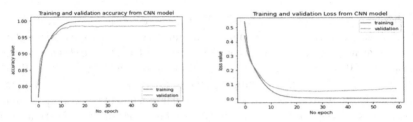

Fig. 1. Training and validation accuracy and loss graph for CNN model.

5.2 Experimental Result of BiLSTM Model

Experimental results of the Bi-LSTM model were analyzed and interpreted. We have trained the Bi-LSTM model with 2 dense layers with sigmoid activation functions and binary_crossentropy loss functions. We employed 64 neurons in the first dense, for a total of 128 neurons in both the forward and backward directions. We used, the maximum dropout rate of 0.2, the training epoch value of the model is 60, the learning rate that changes the weight of the training algorithm and we set the value of 0.00001. We set the batch-size to 64 (Fig. 2).

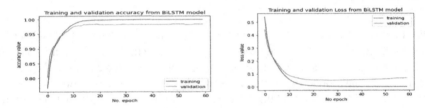

Fig. 2. Training and validation accuracy and loss graph from BiLSTM model

5.3 Experimental Result of BERT Model

We have used 60 epochs to train the model with a 0.00001 learning rate. To reduce overfitting, we set the dropout rate to 0.2. We have also used the Adam optimizer, RELU for the hidden layer, and Sigmond for the output layer is used as an activation function. To build the model we have used three dense layers, for the first dense we have used 64 neurons and a 0.2 dropout-rate. For the second dense layer we used 32 neurons. Lastly for the output layer we have used 2 neurons (Fig. 3).

For this research we select BERT for classification because BERT is better than both CNN and BiLSTM algorithms for semantic understanding.

5.4 Experimental Result of Disambiguation Model

We In our research, we have used the finetuned AmRoBERTa model with the FLAIR document embedding technique to disambiguate Amharic words in the given sentence.

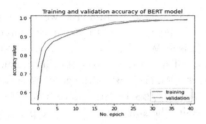

Fig. 3. Training and validation accuracy and loss graph from BERT model

AmRoBERTa fine-tuning: We fine-tuned the AmRoBERTa model using 33,297 sentences and 800 ambiguous words. When we train the model, we have used a maximum of eight contextual meanings for a single ambiguous word. Our experiment is conducted using an epoch of 200 and a batch_size of 64 using an NVIDIA GeForce RTX 1080/2080 Ti generations of GPU server, where each GPU has 12GB memory, with 32 CPU cores and 252 RAM to run our experiments. We have conducted our experiment with 100 and 150 epochs but the performance was not optimal. We set it to 200 epochs which is the optimal iteration for our data set. We have also experimented with batch-size of 32, 64, and 248. But we have selected batch-size 64 as the optimal batch size because when the batch-size is below 64 it takes more training time. When the batch-size is more than 64, there is faster training, but the performance is low.

AmRoBERTA with masking: AmRoBERTa model handles the context through masked language modeling by randomly masking the 15% of the sentence in each epoch of iteration. With a proper finetuning, our assumption is that, if we mask the ambiguous word, it should predict the correct word with the right sense. From the experiment, we take the following sentence predictions as an example.

Example: በአስተያየቱ ልክ ለለውጥ መትጋት ከምንም በላይ መሰረታዊ ነጥብ ነው፡፡From this sentence the ambiguous word ልክ (lik) is disambiguated as follow.

```
In [93]:   sentt='በአስተያየቱ <mask> ለለውጥ መትጋት ከምንም በላይ መሰረታዊ ነጥብ ነው'
           predictions = fill_mask(sent)
           #print(prediction)
           for i in range(5):
               print(predictions[i]['token_str'])

           ልክ
           መጠን
           ደረጃ
           ትክክል
           ሰርአት
```

Based on our experimental result, The sentence " በአስተያየቱ ልክ ለለውጥ መትጋት ከምንም በላይመሰረታዊ ነጥብ ነው፡፡" the model masks the ambiguous word ልክ(lk) then the top 4 meaning of the masked word are predicted.

Word Sense Disambiguation with Flair embedding technique: For this experiment, we have used the finetuned pre-trained contextual model to disambiguate the correct sense of the ambiguous words. We have used the fine-tuned **AmRoBERTa** model with the FLAIR document embedding technique. For the disambiguation task, we have followed a similar approach as Huang et al. (2019), where we have to prepare the target sentence and gloss sentence pairs. However, there is no WordNet for Amharic to employ for this task. Hence, we have selected 10 words that are previously annotated using the WebAnno annotation tool. These words are ዋና(Wana), መንገድ(Menged), ሳለ(Sale), አካል(Akal), ዋጋ(Waga), ገና(Gena), ቀና(Qena), ሀቅ(haq), ሃይል(Hayil), and ልክ(Lik). Then we constructed a gloss for 10 words, which contains the ambiguous word and possible senses with examples sentences. During disambiguation, we select a target sentence that contains ambiguous words where the sense is already annotated by the annotators. We use the FLAIR document embedding with the finetuned contextual pre-trained model to compute the similarity between the target sentence and the glosses. The sense which has a high similarity value with the target sentence would be the correct meaning of the ambiguous word. Based on the given sentence in the gloss, the model disambiguates the target word into its correct sense. Example 1 below shows a target sentence and glosses as examples.

Example 1: The sentence: " እያንዳንዱ ዋና ሃሳብ አንድ አንቀፅ ውስጥ ስፋፅል።" is disambiguate as follow.

Target sentence: እያንዳዱ ዋና ሐሳብ ራሱን በቻለ አንድ አንቀፅ ውስጥ ስፋፅል ።

ጥብጥ: የአብሰሙለን ዋና ሃሳብ አስረፊ ። ዋናዋን ነገር ብቻ ነጣና 0.5702
ዐይነተኛ: ተወካይ ሀማም የቀለየ ልብስ ሰብሰ ዋና አንዳረ ህና ድባጎን ስታጣቀው ውሳለች ። የኮና ዋና ጉቋም ማኃቋቃት ነው ። አጅራጩን መነጽ ትት በዋናው መንጽ ማጣ ። 0.5347
ሜሪ ፤ ሀላፊ ፤ ፦ አለቃ ቦየቱት ዋና ህላ ጥሩሌኝ ፦ ዋና የላለው ወር ቀሀ ይለፋፈል ። 0.3580
የእስር ላይ አኖራት: ይፋን እነ ዞወ ዋና በእነ ዞመነ ብልጥ ብቱ የወፈ አኖርጎ ህንል ። ዋና መየዳት ሰነጀላ ከሙሉ ወጣ ። ከመኑ የፈም ዋና ስለኝቧን ወህ አይነአለወ አይደ የኛ አኖ ለ ዎች ስማወቱ ብሎ ጊዜ ስፔሽሳ ። 0.3449

Based   on   the   result   of   our   experiment,   for   the target sentence" እያንዳንዱ ዋናሐሳብ ራሱን በቻለ አንድ አንቀፅ ውስጥ ስፍፅል ።" the correct sense of the ambiguous word ዋና(**wana**) is ጭብጥ(**Chibt - main point**), as it has higher similarity with the target sentence (0.5702) compared to the other senses, which are አይነተኛ(Aynetegna - principal) and መሪ/ሀላፊ(Meri/Halafi -leader) with similarity scores of 0.5347 and 0.3580 respectively.

Example   2:   The   sentence: "
ረብሻው ከተረጋጋ በኋላ የተወሰኑ ታክሲዎች መንገድ ላይመታየት በመጀመራቸው ህዝቡ ተደሰተ"
is also disambiguated as follow.

Target sentence: ረብሻው ከተረጋጋ በኋላ የተወሰኑ ታክሲዎች መንገ ላይ መታየት በመጀመራቸው ህዝቡ ተደሰተ

ጎዳና: በአነድ ወቅት መንገድ ጽር ለተሰበሰበ ሰወች ፈራምን ማባየት ጀመርጋ ። ከዚህ ውጪ በኋላ የተወሰኑ ታክሲዎች መንገድ ላይ መታየት ጀምርዋል ። መንገዱ ስለሚያፊፉረ ጀዋኝ ። 0.8098
አካሄድ: ችክክለኛው ቁትዋቲ ጥያቄ ጣን ስለላማዋ መንገድ መቄረስ ነበረሰት ። ሰው ቡሴት መንገድ ከተፋቴ ይሞፈል ። በችክክለኛው መንገድ መንገድ አስሳኑ ። አስኪ በችክክለኛው መንገድ ቆትታን ይጀሮ ው ። 0.4688
አስተሳሰብ: ከነን ይህንን መጅሑት ከስኮ ሰወች ጋር የማትመሰለሰለት መንገድ አለ ። በምን መንገድ ልያስረዳኝ ትችላለህ ። 0.4052
ባሀረገ ፤ ደሬ: ይህን ለግኘረጋ የሚያስችለው አነደኛው መንገድ የከብከበ ሥራችን ነው ። አራተኛው የውጥ ምግዜ ማባኝ መንገድ የመንስ/ኑት ጋዋ የሚያለው ነው ። አያኦስ ከስተዶ አውስተኝ ፤ፋ ነት ማዥ የሚቃለስኑ ቀላ መንገድ መወማፍጋ ። 0.3305
አለጣር: ስለሰማ መንገድ ነፍ ጀምጡፈሊታ ምርሚ መሞ አላዞት ። 0.3136
ሁኔታ: በዚህ መንገድ የአምስክን አቅ መጣባቸው ንጠለተኛች አነዲሁኑ አደረጉቸው ። የዙካርን የመናነፍ ችትላ በትአጥሯፈ መንገድ ተመሰላበት ። 0.2925

Based on the result of our experiment, for the target sentence "ረብሻው ከተረጋጋበኋላ የተወሰኑ ታክሲዎች መንገድ ላይ መታየት በመጀመራቸው ህዝቡ ተደሰተ፡፡" the correct sense of the ambiguous word መንገድ is ጎዳና(**Godana – street**), as it has higher similarity with the target sentence (0.8098) compared to the other senses, which are አካሂድ(Akahiad - approach), አስተሳሰብ(Astesaseb - thinking), አሰራር(Aserar - procedure), and ሁኔታ(Huneta - situation) with similarity scores of 0.4688, 0.4052, 0.3305,0.3136 and 0.2925 respectively.

6 Conclusion

This study has developed an Amharic word sense disambiguation model by using a transfer learning approach. The process of identifying the correct meaning based on its context is known as word sense disambiguation. WSD is improving the performance of different NLP applications like machine translation so, to advance NLP research WSD is important. In addition, WSD will be abasis to build Amharic WordNet. These issues motivated us to conduct this research.

As far as we know, there is no standard sense-tagged Amharic text dataset for Amharic WSD task. So, we have collected 10k sentences from Amharic news, Amharic dictionary, Amharic Quran, Amharic bible, and Amharic textbooks. For the Amharic WSD task, we have collected 800 ambiguous words from different sources such as Amharic dictionaries, Amharic textbooks, and Abissinica online dictionary. A total of 33,297 sentences are used to finetune the AmRoBERTa model for the transfer learning.

In our study, we have compared different models to select the most suitable model for WSD classification. To select the best fit model, we have conducted different experiments. For the classification task, we have experimented with CNN, BiLSTM, and BERT algorithms with 2 dense layers and a sigmoid activation function. According to the results, CNN, Bi-LSTM, and BERT obtained 90%, 88%, and 93% accuracy respectively. Based on our findings, the model based on BERT has achieved the vesting result.

As AmRoBERTa is a general-purpose pre-trained language model, we have fine-tuned it with 33,294 sentences and 800 ambiguous words. Finally, the AmRoBERTa model has been applied and when we use the masking technique to find the correct sense, it attains 70% accuracy. We have also employed the FLAIR document embedding framework to embed the target sentences and glosses separately. We then compute the similarity of the target sentence with the glosses embedding. The gloss with the higher score disambiguates the target sentence. Our model was able to achieve an accuracy score of 71%.

References

Abate, S.T., Menzel, W.: Syllable-based speech recognition for Amharic. In: Proceedings Of the 5th Workshop On Important Unresolved Matters, Pages Prague, Czech Republic, pp. 33–40 (2007). https://doi.org/10.3115/1654576.1654583

Alian, M., Awajan, A., Al-Kouz, A.: Arabic word sense disambiguation using Wikipedia. Researchgate **12**(1), 61–66 (2016). https://doi.org/10.21700/Ijcis.2016.108

Assemu, S.: Unsupervised machine learning approachfor word sense disambiguation to Amharic words. Addis Ababa, Ethiopia. Masters thesis Addis Ababa University, Ethiopia (2011)

Bouhriz, N., Habib, E., Lahmar, B.: Word sense disambiguation approach for Arabic text. Int. J. Adv. Comput. Sci. Appl. (IJACSA) **7**(4), 381–385 (2016)

Chawla, A., Biemann, C., Wiedemann, G., Remus, S.: Does BERT make any sense? Interpretable word sense disambiguation with contextualized embeddings. Arxive (2019)

El-Razzaz, M., Fakhr, M.W., Maghraby, F.A.: Arabic gloss WSD using BERT. Appl. Sci. **11**(6), 2567 (2021)

Getaneh, M.: Amharic wordnet construction using word embedding. Masters thesis Addis Ababa University, Addis Ababa, Ethiopia (2020)

Gezmu, A.M., Nürnberger, A., Seyoum, B.E.: Portable spelling corrector for a less-resourced language: Amharic. In: LREC 2018 - 11th International Conference on Language Resources and Evaluation, pp. 4127–4132 (2019)

Hassen, S.: Word sense dismbiguation using WordNet. Addis Ababa, Ethiopia. Masters thesis Addis Ababa University, Ethiopia (2015)

Huang, L., Sun, C., Qiu, X.: GlossBERT: BERT for word sense disambiguation with gloss knowledge, pp. 3509–3514 (2019)

Kassie, T.: Word sense disambiguation for Amharic text retrival: a case study legal documents. Addis Ababa, Ethiopia. Masters thesis Addis Ababa University, Ethiopia (2009)

Mekonnen, S.: Word sense disambiguation for Amharic text: a machine learning approach. Addis Ababa, Ethiopia. Masters thesis Addis Ababa University, Ethiopia (2010)

Mindaye, T., Sahlemariam, M., Kassie, T.: The need for Amharic WordNet. In: Global WordNet Conference, GWC 2010 (2010)

Mulugeta, M.: Word sense disambiguation for Amharic sentences using wordNet hierarchy. Masters thesis, Bahir Dar University, Bahir Dar, Ethiopia (2019)

Pal, A.R., Saha, D.: Word sense disambiguation: a survey. Researchgate **5**(3), 1–16 (2015)

Reta, B.: Application of parts-of-speech tagged corpus to improve the performance of word sense disambiguation: The Case of Amharic. Masters thesis Addis Ababa University, Addis Ababa, Ethiopia (2015)

Salawu, A., Aseres, A.: Language policy, ideologies, power and the Ethiopian media. University of South Africa (Tunisia) **41**(1), 71–89 (2015). https://doi.org/10.1080/02500167.2015.1018288

Senay, D.: Automatic Amharic word sense disambiguation model at sentence level by deep learning approach. Masters thesis Bahir Dar University, Bahir Dar, Ethiopia (2021)

Siraj, D.: A generic approach towards all words Amharic word sense disambiguation. Masters thesis Adis Ababa Univrtsity, Adis Ababa, Ethiopia (2017)

Tadesse: word sense disambiguation for wolaita language using machine learning approach. Masters thesis Adama University, Adama Ethiopia (2021)

Tesema, W., Tesfaye, D., Kibebew, T.: Towards The sense disambiguation of Afan Oromo words using hybrid approach (Unsupervised Machine Learning And Rule Based). Ethiop. J. Educ. Sci. **12**(1), 61–77 (2016)

Wassie, G., Ramesh, B., Teferra, S., Meshesha, M.: A word sense disambiguation model for Amharic words using semi-supervised learning paradigm. Researchgate **3**(3), 147 (2014). https://doi.org/10.4314/Star.V3i3.25

Yimam, S.M., Ayele, A.A., Venkatesh, G., Gashaw, I., Biemann, C.: Introducing various semantic models for Amharic: experimentation and evaluation with multiple tasks and datasets. Future Internet **13**(11), 275 (2021). https://doi.org/10.3390/Fi13110275

Kebede, H., Tsgie, F., Alemu, F., Azene, M.: Ethiopian Languages Research Center. Addis Ababa, Artistic Publication. Addis Ababa University, Ethiopia, Amharic Dictionary (1993)

Deep Complex-Valued Neural Networks for Massive MIMO Signal Detection

Isayiyas Nigatu Tiba[1(✉)] and Mao youhong[2]

[1] Jimma University Institute of Technology, 378, Jimma, Ethiopia
isayiyas.tiba@ju.edu.et
[2] Xidian University, Xi'an, People's Republic of China

Abstract. In the fifth generation (5G) and future mobile networks, the design of efficient detectors for massive multiple-input multiple-output (MIMO) is essential. The main challenge in designing detectors is the trade-off between performance and computational complexity; that is, efficient detectors incur higher computational costs, while computationally cheaper detectors have lower efficiency. Recently, many deep learning-based detectors have been proposed in the literature to fill in such gaps. However, most of the existing MIMO detectors work only with real-valued parameters. First, they transform the complex received MIMO signal into an equivalent real-valued parameter by concatenating the real and imaginary parts and then train a network based on the real-valued data. Such an approach has several disadvantages. On one hand, the number of trainable parameters will be doubled; on the other hand, the phase information, which is important in the communication signals, might be lost or distorted. In this work, we aim to investigate the application of complex-valued neural networks for MIMO signal detection based on Wirtinger Calculus. To do so, we propose a simple feedforward architecture that directly works with the complex-valued QPSK and 16-QAM modulation signals. Our method is simple and computationally cheaper. Simulation results show that the proposed approach can improve the performance of the existing detectors while providing a lower computational cost.

Keywords: Complex-valued neural networks · MIMO detector · Wirtinger calculus

1 Introduction

Massive multiple-input multiple-output (MIMO) is a key technology for 5G and future mobile networks. One of the issues that must be addressed in order to make the promising benefits of this technology a reality is signal detection. Signal detection is the process of recovering information that has been transmitted through a noisy channel in a wireless communication system [1]. Due to interference, noise, and fading channel conditions, detection becomes increasingly

B. H. Woldegiorgis et al. (Eds.): ICAST 2022, LNICST 455, pp. 239–251, 2023.
https://doi.org/10.1007/978-3-031-28725-1_15

difficult in massive MIMO systems. Furthermore, because of the high number of antennas, computational complexity is one of the key challenges in this technology [2,3].

As a result, different categories of traditional detectors are proposed in the literature to realize the promises of massive MIMO technology [4–6]. Besides the traditional ones, neural network-based detectors are also a promising class of MIMO detectors [7–9]. Deep neural network (DNN) detectors can operate in data-driven, model-driven, or hybrid modes. The performance of these detectors has shown a significant improvement compared with their traditional counterparts. However, one of these detectors' limitations is that they only work with real-valued parameters. Most existing DNN-based applications are tuned to work with real-valued parameters in order to take advantage of the computing resources available in the existing DNN libraries [10].

One of the key reasons behind the success of supervised deep learning (DL) in several tasks is its ability to learn a useful hierarchical representation of data [11]. Eventually, it has been used in difficult scenarios where tractable mathematical models cannot characterize a problem. In such cases, the DL architecture is represented by a black box and optimized through learning from data to solve a specific problem. In various domains of applications, such as computer vision, such a representation has surpassed human-level performance [12]. Since the nature-made signals (RGB images, videos, ...) appear as real-valued signals, most DNN libraries are built to process real-valued signals.

When it comes to engineering applications, including wireless communication, the situation is rather different. On the one hand, the information signals are synthetically generated by humans. Hence, applying the DL to these signals (man-made) doesn't guarantee the best performance as that of nature-made signals [13]. On the other hand, unlike most nature-made signals, the communication signals are represented in a complex baseband form, in which the phase information is as important as the magnitude. As such, it is critical to resolve these issues in order to effectively use the DL in wireless communications. In the first case, the work in [13] has analyzed the applications of DL in the physical layer of communication systems. In that work, it is shown that DL can be efficiently applied to signal detection tasks. Nevertheless, the dataset used for the analysis was transformed from complex to real.

In this work, we address the second issue by investigating the use of complex-valued neural networks for MIMO signal detection. The use of complex numbers when working with communication signals allows for precise representation of both magnitude and phase [14]. The main challenge in complex analysis, however, is the diffentiability problem, which is described by the Cauchy-Riemann equations [15]; i.e., in the DL, the partial derivatives of a real-valued cost function with respect to the complex-valued parameters in the back-propagation do not exist. As such, we resort to Wirtinger Calculus, which allows us to fully exploit the power of complex-valued signal processing [16]. To the best of our knowledge, this is the first study to apply complex-valued neural networks to the MIMO detection problem.

Related Works: Several DNN-based detectors for massive MIMO have been proposed [17]. We chose two detectors with superior performance in [7] and [9], for ease of generation of training data. However, because of the aforementioned complex-analysis issues, the data in these detectors must first be transformed to the real-valued equivalent. This means that an n-dimensional complex-valued vector will be converted to a $2n$-dimensional real-valued vector: i.e., $\mathbb{C}^n \mapsto \mathbb{R}^{2n}$. As a result, in those works, the network parameters are fully trained using real-valued calculus and the output is converted back to complex. This method has two major drawbacks: first, besides increased dimensionality (doubling the number of trainable parameters), it may be ineffective for nonlinear functions because the functional form may not be easily separated into real and imaginary parts; second, phase information may be lost or distorted.

The method we extend, however, can easily compute complex-valued gradients without requiring a transform into the real equivalent. For scalar quantities, Brandwood [18] and Adali [16] show that the derivitive of the real-valued cost function with respect to the complex-valued parameters exists. We show that the same approach is valid for complex-valued vectors as well. Furthermore, we present a design of a simple feedforward DL architecture for MIMO signal detection that can be trained fully in a complex domain. We will demonstrate through simulation that complex-valued neural networks can be efficiently designed for detecting MIMO signals.

The remainder of this paper is organized as follows. Section 2 provides preliminary knowledge and operations that will be useful in the subsequent sections. Section 3 describes the system model and problem formulation. Section 4 presents the proposed network architecture and offline training procedures. Section 5 discusses several numerical results demonstrating the validity of the complex-valued DNN for MIMO detection problems, and Sect. 6 concludes this work.

2 Preliminaries

2.1 Complex Matrix Multiplication

Suppose that $\mathbf{W} = \mathbf{W}_r + j\mathbf{W}_i$ and $\mathbf{b} = \mathbf{b}_r + j\mathbf{b}_i$ are complex matrix and vector, respectively, Where the elements in $\mathbf{W}_r, \mathbf{W}_i, \mathbf{b}_r, \mathbf{b}_i$ are real quantities, and $j = \sqrt{-1}$. Then a complex matrix multiplication which results in a vector $\mathbf{h}$ can be written as

$$\mathbf{h} = \mathbf{W}\mathbf{b} \tag{1a}$$

$$= \mathbf{W}_r\mathbf{b}_r - \mathbf{W}_i\mathbf{b}_i + j(\mathbf{W}_i\mathbf{b}_r + \mathbf{W}_r\mathbf{b}_i), \tag{1b}$$

The formulation in (1b) is a basis for various signal processing models that deal with complex-valued parameters. Several algorithms, including the DL applications, use this approach to transform the results of their computation from complex to real-domain and vice versa. Now notice that (1b) can be represented as a real vector by concatenating its real and imaginary parts, which produces a double dimensionality $\mathbf{h}' = [\mathbf{h}_r, \mathbf{h}_i]^T$. In this work, we utilize the compact complex form as in (1a).

2.2   Wirtinger Calculus

Let $f(\omega) = u(\omega_r, \omega_i) + v(\omega_r, \omega_i)$ is a complex function, where $\omega = \omega_r + j\omega_i$. If the continuous partial derivatives $u(\omega_r, \omega_i)$, $v(\omega_r, \omega_i)$ with respect to ω_r and ω_i exist, then a generalized complex derivatives can be defined as [16]

$$\frac{\partial f}{\partial \omega} = \frac{1}{2}\left(\frac{\partial f}{\partial \omega_r} - j\frac{\partial f}{\partial \omega_i}\right), \quad \frac{\partial f}{\partial \omega^*} = \frac{1}{2}\left(\frac{\partial f}{\partial \omega_r} + j\frac{\partial f}{\partial \omega_i}\right) \tag{2}$$

Formally, the complex derivative can be calculated by considering f to be a bivariate function $f(\omega, \omega^*)$ and ω and ω^* to be independent variables. The Wirtinger Calculus can generalize the standard complex calculus, as discussed in [16]. As such, instead of taking partial derivatives with respect to the real and imaginary components, it is more efficient to use the approach described above: Rewrite $f(\omega)$ as $f(\omega, \omega^*)$ and differentiate with respect to one by treating the other as a constant.

3   System Model and Problem Formulation

3.1   System Model

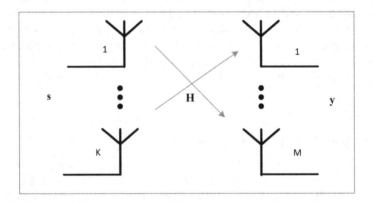

Fig. 1. A MIMO system with **s** transmit symbols, and **y** received signal vector over a channel matrix **H**.

Consider a MIMO communication system model shown in Fig. 1. We assume that the transmitter side contains K single antenna user terminals (UT)s, and the receiver is a base station (BS) equipped with M antennas. We denote the spatially-multiplexed UTs transmitted symbols vector as $\mathbf{s} = [s_1, s_2, \cdots, s_K]^T$ and its corresponding received signal vector as $\mathbf{y} = [y_1, y_2, \cdots, y_M]^T$, where s_k, y_j represent the transmitted symbol from the k-th UT, and the received signal by the j-th BS antenna, respectively. Each s_k is drawn from a set $\mathcal{S}$

of rectangular complex QPSK, and 16-QAM alphabets. We assume that the complex propagation channel $\mathbf{H} \in \mathbb{C}^{M \times K}$, is a flat fading and Rayleigh distributed, whose entries are independent and identically distributed (i.i.d) with $\mathcal{CN} \sim (0,1)$. Then, the received signal vector $\mathbf{y}$ in a matrix form can be written as

$$\mathbf{y}^{(i)} = \mathbf{H}\mathbf{s}^{(i)} + \boldsymbol{v}^{(i)}, \quad i = 1, 2, \cdots, m, \tag{3}$$

where $\boldsymbol{v} \in \mathbb{C}^M$, a complex additive white Gaussian noise (C-AWGN), and m is the number of transmission instants. Here, we also assume that the channel state information is known to the transmitter.

In the existing works the complex-valued parameters in the above model is transformed into equivalent real model [9]

$$\mathbf{y}_r^{(i)} = \mathbf{H}_r \mathbf{s}_r^{(i)} + \boldsymbol{v}_r^{(i)}, \quad i = 1, 2, \cdots, m, \tag{4}$$

where

$$
\begin{aligned}
\mathbf{s}_r^{(i)} &= \begin{bmatrix} \operatorname{Re}(\mathbf{s}^{(i)}) \\ \operatorname{Im}(\mathbf{s}^{(i)}) \end{bmatrix} \in \mathcal{S}_r^{2K}, \ \mathbf{H}_r^{(i)} = \begin{bmatrix} \operatorname{Re}(\mathbf{H}^{(i)}) & -\operatorname{Im}(\mathbf{H}^{(i)}) \\ \operatorname{Im}(\mathbf{H}^{(i)}) & \operatorname{Re}(\mathbf{H}^{(i)}) \end{bmatrix} \in \mathbb{R}^{2M \times 2K}, \\
\mathbf{y}_r^{(i)} &= \begin{bmatrix} \operatorname{Re}(\mathbf{y}^{(i)}) \\ \operatorname{Im}(\mathbf{y}^{(i)}) \end{bmatrix} \in \mathbb{R}^{2M}, \boldsymbol{v}_r = \begin{bmatrix} \operatorname{Re}(\boldsymbol{v}^{(i)}) \\ \operatorname{Im}(\boldsymbol{v}^{(i)}) \end{bmatrix} \in \mathbb{R}^{2M}, \quad i = 1, 2, \cdots, m.
\end{aligned}
\tag{5}
$$

Notice that the transformation in (5) follows the formulation in (1b). It is clear from (5) that the dimension of each parameter is doubled when we attempt to work with complex-valued parameters in the equivalent real. As discussed above, the main issue in such an approach is that besides computational burden in the network, the phase information will be lost or easily distorted, since the network only optimizes the real parameters. In this work, however, we ignore the transformation in (5) and directly work with the complex parameters in (3).

3.2 Problem Setup

Given (3), the goal of a MIMO detector is to recover the transmitted symbols $\mathbf{s}^{(i)}$ from the received signal $\mathbf{y}^{(i)}$ relying on the knowledge of the channel at each transmit instant i. Now, assuming all the information symbols in $\mathcal{S}$ are chosen (transmitted) with equal probability, the optimum MIMO detector known as maximum likelihood (ML) can be written as

$$\mathbf{s}_{\mathrm{ML}}^{(i)} = \operatorname*{argmin}_{\mathbf{s}^{(i)} \in \mathcal{S}^K} \|\mathbf{y}^{(i)} - \mathbf{H}^{(i)}\mathbf{s}^{(i)}\|_2^2, \quad i = 1, 2, \cdots, m. \tag{6}$$

However, solving ML problem (6) globally is NP-hard since it requires $|\mathcal{S}|^K$ evaluations, where $|\cdot|$ denotes cardinality of a set. Because the computational complexity in ML detection is prohibitively expensive, we resort to a suboptimal solution that achieves near-ML performance while requiring less computational complexity. In this work, we propose a complex-valued neural network architecture as a supervised learning model that can take the complex received signal vector $\mathbf{y}$ as an input and return an estimate of the transmitted symbol $\mathbf{s}$. Mathematically, this can be written as a machine learning problem

$$\hat{\mathbf{s}}_{DL}^{(i)} = \underset{f(\cdot,\theta)}{\operatorname{argmin}}\ \mathbb{E}\big(\mathcal{L}(f(\mathbf{y}^{(i)};\theta),\mathbf{s}^{(i)})\big), \tag{7}$$

where $\mathcal{L}(\cdot)$ is some specific loss function, $\mathbb{E}(\cdot)$, is expectation operator, and θ is a set of trainable parameters. In (7), our goal is, through exploiting DNN, to learn an optimal function $f(\cdot,\theta)$ and its corresponding parameter θ in the sense that it can minimize the objective function $\mathcal{L}(\cdot)$.

4   Complex-Valued Feed-Forward Networks

In this section, we describe the proposed complex-valued feed-forward network or a multilayer perceptron (MLP), which from now on will be referred to as "cFFDnet."

4.1   Forward Propagation

The cFFDnet, like its real counterpart, MLP, performs feature mapping during forward propagation. There are two basic operations: linear and non-linear mappings, which are briefly described as follows. The network takes the noisy received signal $\mathbf{y}$ which is denoted as $\mathbf{a}^0$ as an input, and returns the estimated symbol $\hat{\mathbf{s}}$ as an output.

Suppose that N denotes the number of units (neurons), and ℓ denotes the number of layers. Let $\mathbf{W}^\ell \in \mathbb{C}^{N^\ell \times N^{\ell-1}}, \mathbf{b}^\ell \in \mathbb{C}^{N^\ell}$ denote the complex weight matrix and bias term respectively. Then the linear and non-linear mapping at each layer can be written as [19]

$$\mathbf{z}^\ell = \mathbf{W}^\ell \mathbf{a}^{\ell-1} + \mathbf{b}^\ell, \tag{8a}$$

$$\mathbf{a}^\ell = \sigma(\mathbf{z}^\ell), \quad \ell = 1, 2, \cdots, L, \tag{8b}$$

where $\sigma(\cdot)$ is the non-linearity (activation function). For notation consistency, we consider the input as a 0-th layer; i.e., $\mathbf{a}^0 = \mathbf{y}$.

At each layer, the linear mapping in (8a) computes a complex-matrix multiplication to obtain a vector $\mathbf{z}^\ell \in \mathbb{C}^{N^\ell}$. Then it goes through $\sigma(\cdot)$, a non-linear mapping. Here, $\sigma(\cdot)$ is a pointwise activation function that brings non-linearity to the network. There are several classes of activation functions proposed for complex-valued architectures. In this work, we utilize the modified tanh [20]

$$\sigma(\mathbf{z}) = \tanh\left(||\mathbf{z}||\right) \odot \frac{\mathbf{z}}{||\mathbf{z}||}, \tag{9}$$

where $||\cdot||$ denotes magnitude, and $\odot$ denotes a point-wise multiplication. The main advantage of this activation function is that it keeps the phase information. The corresponding set of trainable parameters can be defined as

$$\Theta = \{\mathbf{W}^1, \mathbf{b}^1, \cdots, \mathbf{W}^L, \mathbf{b}^L\}. \tag{10}$$

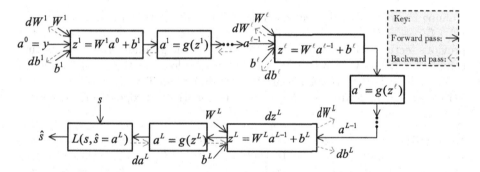

Fig. 2. A Computational graph representation of the proposed cFFDnet for the MIMO detection.

4.2 Back Propagation

In this subsection, we describe the back propagation process that characterizes the learning steps of cFFDnet.

Let $\mathcal{Y} = \mathbb{C}^M$ denotes M-dimensional feature space and $\mathcal{S}' = [s_1, s_2, \cdots, s_K]^T$ denotes the label space. Then the training dataset can be described as $D = \left\{\left(\mathbf{a}^{0(i)}, \mathbf{s}^{(i)}\right)\right\}_{i=1}^m$, where $\mathbf{a}^0 \in \mathcal{Y}$, $\mathbf{s} \in \mathcal{S}'$, are the input feature and true label, respectively, and m is the number of training examples. Then, the detection task in the cFFDnet can be defined as a mapping function $f_\Theta : \mathcal{Y} \mapsto \mathcal{S}'$. In other words, the f_Θ is a regression function that minimizes a loss function to find the best parameter Θ that will be used estimate the transmit symbol $\mathbf{s}$. The loss function is defined here as the mean-squared-error (MSE) between the true label $\mathbf{s}$ and the estimated network output $\mathbf{a}^L$ (see Fig. 2). To do so, let us define the error corresponding to each symbol as

$$e_k = s_k - a_k^L, \quad k = 1, 2, \cdots, N^L = K. \tag{11}$$

Then, the MSE loss produced by the network is

$$\mathcal{L} = \frac{1}{K} \sum_{k=1}^{K} e_k e_k^*, \tag{12}$$

where $*$ denotes a complex conjugate operation. Alternatively, in vector notation, it can be written as

$$\begin{aligned}\mathcal{L} &= \frac{1}{K}\left((\mathbf{s} - \mathbf{a}^L)^\dagger (\mathbf{s} - \mathbf{a}^L)\right), \\ &= \frac{1}{K}\left(\mathbf{s}^\dagger \mathbf{s} - \left(\mathbf{a}^L\right)^\dagger \mathbf{s} - \mathbf{s}^\dagger \mathbf{a}^L + \left(\mathbf{a}^L\right)^\dagger \mathbf{a}^L\right),\end{aligned} \tag{13}$$

where $\dagger$ denotes hermitian operation (transpose of a complex conjugate), and $\mathbf{a}^L \in \mathbb{C}^K$ represents the estimated output of the network. The superscript (i) is ignored for notational simplicity; i.e., the loss function is defined for a single training example. Now notice that the loss function (13) is a real quantity ($\mathcal{L}$: $\mathbb{C}^K \times \mathbb{C}^K \mapsto \mathbb{R}^K$), while the network parameters are complex. As discussed in the previous sections, the Wirtinger Calculus can lay the groundwork for computing complex gradients with respect to the real-valued loss function. A modification required here is that the generalization of dimensionality since the results of Wirtinger Calculus are provided for scalar quantities. Proposition 1 discusses this extension.

Proposition 1. *Let $f : \mathbb{C}^{n_1} \mapsto \mathbb{R}^{n_2}$ be a real valued vector function of a complex vector $\mathbf{w}$. Let $f(\mathbf{w}) = \mathbf{g}(\mathbf{w}, \mathbf{w}^*)$, where $g : \mathbb{C}^{n_1} \times \mathbb{C}^{n_1} \mapsto \mathbb{R}^{n_2}$ is a real-valued vector function of two complex vector variables. Then computing $\frac{\partial g}{\partial \mathbf{w}^*} = 0$ is a necessary and sufficient condition to determine the stationary point of f.*

See Appendix A for the proof. A complex-valued network can be effectively trained utilizing the result in proposition 1 and the chain rule in the Wirtinger Calculus [21]. The gradient computation in the back propagation starts at the output (L-th layer) by computing the loss function (the error between true label $\mathbf{s}$ and the estimated symbol vector $\hat{\mathbf{s}}$.

The main objective of learning in this work is to minimize the loss function (13) with respect to the complex set of parameters $\boldsymbol{\Theta}$. To briefly describe this procedure, let us use da, dz, dW, db to denote the gradients corresponding to

$$\frac{\partial \mathcal{L}}{\partial \mathbf{a}^{*L}}, \frac{\partial \mathcal{L}}{\partial \mathbf{z}^{*L}}, \frac{\partial \mathcal{L}}{\partial \mathbf{W}^{*L}}, \frac{\partial \mathcal{L}}{\partial \mathbf{b}^{*L}},$$

respectively. As shown in the computation graph in fig.2, at the output layer, the network first computes the gradient da^L and step by step goes back to compute gradients in each layer. For instance, the gradient with respect to the linear activation dz^L can be computed by applying the chain rule as

$$dz^L = \frac{\partial \mathcal{L}}{\partial \mathbf{z}^{*L}} = \frac{\partial \mathcal{L}}{\partial \mathbf{a}^L} \frac{\partial \mathbf{a}^L}{\partial \mathbf{z}^{*L}} + \frac{\partial \mathcal{L}}{\partial \mathbf{a}^{*L}} \frac{\partial \mathbf{a}^{*L}}{\partial \mathbf{z}^{*L}}. \tag{14}$$

Similarly, during back propagation, other gradients will be computed and updated. For the learning, we use the standard stochastic gradient descent (SGD) method.

5   Numerical Results and Discussion

In this section, we present different numerical results that can demonstrate the effectiveness of the proposed methods.

5.1   Implementation Details

All detectors are implemented in python 3.6 using TensorFlow 2.4 library, and Table 1 lists the basic parameters.

Table 1. A list of network parameters.

Parameter	Quantity
Number of hidden layers	5
Average epoch	3,000
Batch size	1024
Training data size	90,000
Validation data size	20,000
Maximum number of units in each layer	512
Maximum number of trainable parameters	1.3 million
ℓ_2 regularization factor	0.0001
Dropout ratio (max.)	0.3

Given the complex system model in 3, the dataset is generated as follows: In all realizations (offline training and online detection phases), the channel is chosen to be the classical i.i.d. Rayleigh fading, and fixed. The transmit symbols are drawn from the QPSK, and 16-QAM constellations. The noise variance is randomly sampled from complex AWGN, and the signal to noise ratio (SNR) in each training sample computed as $SNR = 10 \log \frac{\|\mathbf{Hs}\|_2}{\|v\|_2}$. Then, the dataset $\mathbf{D}$ is generated as the definition in Sect. 4.2.

After the network is successfully trained (minimum overfitting and underfitting is obtained), it will be tested in online detection by using newly generated data that follows the same distribution as the training dataset. In the testing stage (online detection), the generalization ability of the network (detector) will be evaluated compared with the existing detectors. For this purpose, we employ the classical symbol-error-rate (SER) versus SNR performance analysis by running several Monte Carlo simulations. The numerical results for offline training and online detection are presented in the following subsections.

5.2   Learning History: Offline Training Phase

We begin our discussion of the proposed methods' performance analysis by assessing the offline learning characteristics. Figures 3(a),(b) show the basic training history parameters, training/validation loss and accuracy. We can see from these figures that the training loss is decreasing smoothly (Fig. 3(a)) or that the training accuracy is increasing continuously (Fig. 3(b)). This implies that the network is learning efficiently from the given dataset. On the other hand, we can see that the validation loss declines very closely to the training loss, or that

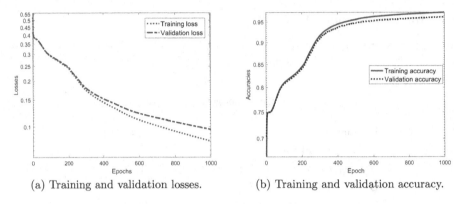

(a) Training and validation losses.   (b) Training and validation accuracy.

Fig. 3. The offline learning histories for a QPSK modulated, $(M = 16, K = 4)$ MIMO system.

the validation accuracy increases consistently with the training accuracy until it reaches a maximum ($\geq 95\%$). This demonstrates that the network was less overfitted during offline training, implying that generalization ability in online detection will be improved.

5.3   SER: Online Detection Phase

Since running simulation results for the optimal ML detector is very difficult for larger MIMO sizes, we use the classical suboptimal detector SDR [4] as a reference. Furthermore, we compare our proposed method with efficient existing detectors that are based on the model-driven deep learning approach.

Figures 4(a)–(d) show the online detection performance in terms of SER versus SNR for QPSK and 16-QAM modulations and different MIMO sizes. As shown in the Fig. 4(a), the proposed method has outperformed the existing model-driven methods (DetNet, ADMM-PSNet) in a significant range of SNR. It is evident from this figure that the proposed cFFDnet has achieved a performance very close to SDR particularly in the lower SNR regime.

To further investigate the performance of the proposed method, we increased the modulation to 16-QAM with different MIMO sizes. The corresponding results of this scenario are depicted in Figs. 4(b)–(d). As it is clear from the figures, the performance of the proposed method is consistently improved under the given scenarios. In Fig. 4(d), it is also evident that even though the proposed detector achieved improved performance compared to the existing detectors, it seems to lag behind the SDR with a noticable range of SNR. This is due to the fact that as K increases, training the network with the same parameter settings results in limited efficiency. Because the size of the input data is limited (M = 64), it is difficult to achieve the required level of training accuracy. However, it is clear from the above results that once the network is properly trained, it can achieve significantly improved performance near the sub-optimal detector.

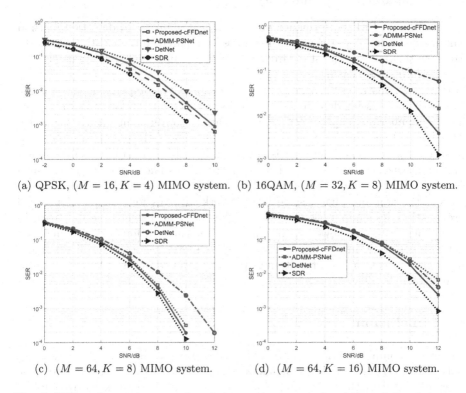

(a) QPSK, $(M = 16, K = 4)$ MIMO system. (b) 16QAM, $(M = 32, K = 8)$ MIMO system.

(c) $(M = 64, K = 8)$ MIMO system. (d) $(M = 64, K = 16)$ MIMO system.

Fig. 4. SER performance comparison between the proposed and the existing detectors in different modulations.

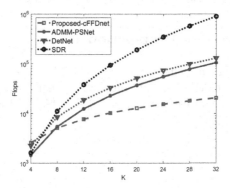

Fig. 5. Computational complexity comparison between the proposed and existing detectors for $M = 64$, and varying K MIMO sizes.

5.4 Computational Complexity

Apart from the SER performance improvement, the proposed detection method can also provide a lower computational cost compared to the existing methods. To show this, we use the standard number of floating point operations (Flops) as a performance measure.

Figure 5 illustrates the computational complexity comparison between the proposed and existing detectors. The results in this figure are obtained by counting the number of complex-vector multiplications for different MIMO sizes with $M = 64$, and varying K. As it is clear from the figure, the computational complexity of the proposed detector is significantly reduced compared to the existing model-driven detectors for a wide range of K. The main reason for such an improvement is that the proposed method is a simple feedforward network. During the training phase, since the network involves forward and backward propagation, its computational cost is higher (it takes longer time to learn). In the online detection, however, we only pass information forward, in which the network performs simple computations such as matrix-vector multiplication and pointwise activation. As a result, once properly trained, it has a lower computational cost.

6 Conclusion

The use of complex numbers plays a vital role in processing communication signals. When it comes to MIMO signal detection, several DL-based detectors have been proposed in the literature. However, they failed to work with complex-valued parameters. In this work, we have analyzed the application of complex-valued neural networks for MIMO signal detection for QPSK and 16-QAM signals. Numerical results have shown that the proposed approach can improve the performance of the existing detectors. Moreover, the computational complexity of the proposed approach is very low compared with the existing methods.

Appendix A Proof of Proposition 1

Proof. By considering an element wise operation on the elements of the real vector f, the proof is straightforward from [18]. Further, since g is a real function, $\frac{\partial g}{\partial \mathbf{w}}$, $\frac{\partial g}{\partial \mathbf{w}^*}$ are conjugates of each other; i.e., $\frac{\partial g}{\partial \mathbf{w}^*} = \left(\frac{\partial g}{\partial \mathbf{w}}\right)^*$. Hence, it is sufficient to compute the term $\frac{\partial g}{\partial \mathbf{w}^*}$.

References

1. Shannon, C.E.: A mathematical theory of communication. ACM SIGMOBILE Mobile Comput. Commun. Rev. **5**(1), 3–55 (2001)
2. Marzetta, T.L.: Noncooperative cellular wireless with unlimited numbers of base station antennas. IEEE Trans. Wireless Commun. **9**(11), 3590–3600 (2010)

3. Yang, S., Hanzo, L.: Fifty years of MIMO detection: the road to large-scale MIMOs. IEEE Communi. Surveys Tutorials **17**(4), 1941–1988 (2015)
4. Wai, H.T., Ma, W.K., So, A.M.C.: Cheap semidefinite relaxation mimo detection using row-by-row block coordinate descent. In: IEEE International Conference on Acoustics, Speech and Signal Processing (ICASSP), pp. 3256–3259 (2011)
5. Shahabuddin, S., Juntti, M., Studer, C.: Admm-based infinity norm detection for large mu-mimo: Algorithm and vlsi architecture. In: IEEE International Symposium on Circuits and Systems (ISCAS), pp. 1–4 (2017)
6. Wu, Z., Zhang, C., Xue, Y., Xu, S., You, X.: Efficient architecture for soft-output massive mimo detection with gauss-seidel method. In: 2016 IEEE International Symposium on Circuits and Systems (ISCAS), pp. 1886–1889 (2016)
7. Samuel, N., Wiesel, A., Diskin, T.: Learning to detect. IEEE Trans. Signal Process. **67**(10), 2554–2564 (2019)
8. Xiaosi, T., et al.: Improving massive MIMO message passing detectors with deep neural network. IEEE Trans. Veh. Technol. **69**(2), 1267–1280 (2020)
9. Tiba, I.N., Zhang, Q., Jiang, J., Wang, Y.: A low-complexity ADMM-based massive MIMO detectors via deep neural networks. In: IEEE International Conference on Acoustics, Speech and Signal Processing (ICASSP), pp. 4930–4934 (2021)
10. Tiba, I.N., Kulimushia, B.B., Kajuna, C.K.: Massive MIMO data detection using 1-dimensional convolutional neural network. In: IEEE/CIC International Conference on Communications in China (ICCC), pp. 483–488 (2020)
11. Tishby, N., Zaslavsky, N.: Deep learning and the information bottleneck principle. In: 2015 IEEE Information Theory Workshop (ITW), 26 Apr, pp. 1–5. IEEE
12. He, K., Zhang, X., Ren, S., et al.: Delving deep into rectifiers: surpassing human-level performance on imagenet classification. In: 2015 IEEE International Conference on Computer Vision (ICCV), pp. 1026–1034 (2015)
13. Bjornson, E., Giselsson, P.: Two applications of deep learning in the physical layer of communication systems [lecture notes]. IEEE Signal Process. Mag. **37**(5), 134–140 (2021)
14. Adalı, T., Li, H., Haykin, S: Complex-valued adaptive signal processing. Adapt. Signal Process. Next Gener. Solutions (2010)
15. Ablowitz, M.J., Fokas, A.S.: Complex Variables. Cambridge University Press, Cambridge (2003)
16. Adali, T., Schreier, P.J., Scharf, L.L.: Complex-valued signal processing: the proper way to deal with impropriety. IEEE Trans. Signal Process. **59**(11), 5101–5125 (2011)
17. Albreem, M.A., Juntti, M., Shahabuddin, S.: Massive MIMO detection techniques: a survey. IEEE Commun. Surv. Tutorials **21**(4), 3109–3132 (2019)
18. Brandwood, D.H.: A complex gradient operator and its application in adaptive array theory. In: IEE Proceedings H-Microwaves, Optics and Antennas, pp. 11–16 (1983)
19. Sun, R.: Optimization for deep learning: theory and algorithms. arXiv preprint arXiv:1912.08957 (2019)
20. Pfeifenberger L., Zöhrer, M., Pernkopf, F.: Deep complex-valued neural beamformers. In: IEEE International Conference on Acoustics, Speech and Signal Processing (ICASSP), pp. 2902–2906 (2019)
21. Amin, M., Amin, M.I., Al-Nuaimi, A.Y., Murase, K.: Wirtinger calculus based gradient descent and Levenberg-Marquardt learning algorithms in complex-valued neural networks. In: International Conference on Neural Information Processing, pp. 550–559 (2011)

Design and Performance Analysis of a Multi-level Fuzzy-Based Stabilizer to Dampen Low-Frequency Oscillation in Single-Machine Infinite Bus Systems

Tenaw Ayew Mezigebu[✉] and Belachew Bantyirga Gessesse[✉]

Faculty Electrical and Computer Engineering, Bahir Dar Institute of Technology, Bahir Dar University, Bahir Dar, Ethiopia

tenaway4488@gmail.com, belchwbbg1j1@gmail.com

Abstract. Power systems are frequently viewed as complex, nonlinear, and dynamic systems. This system is constantly subjected to small disturbances that can result in synchronization loss and system failure. To fix this issue, power system stabilizers are applied to generate extra excitation control signals. Conventional power system stabilizer (CPSS) is difficult to track the dynamic nature of the load since stabilizer gains are determined under specific working conditions. In this paper, a multi-level fuzzy-based stabilizer uses the variation of rotor speed and acceleration as an input to mitigate low-frequency oscillations (LFOs) in single-machine infinite bus systems. The system is represented mathematically by the Heffron Philips K-coefficients model. The controller's performance was investigated for disturbances exposed to inputs of various membership functions, such as a triangular, gaussian, generalized bell, and trapezoidal. Each membership function is compared. For instance, a multi-level fuzzy-based stabilizer with a triangular membership function settled the rotor angle, rotor speed, and electrical torque deviations 29.5%, 5.9%, and 39.7% faster than the gaussian membership function fuzzy-based PSS, respectively. The study's findings revealed that the triangular membership function performed better than other membership functions.

Keywords: LFOs · Multi-Level Fuzzy · Single-Machine System

1 Introduction

Power system stability is the capacity of a power system to establish restorative forces equal to or greater than the disturbing forces to preserve balance. This concept applies to the nation's interconnected power system since it is a highly nonlinear system that operates in a constantly changing environment where loads, generator outputs, and essential operating parameters change. Low-frequency oscillation is caused by an imbalance between the damping and synchronization torque in power systems, which modifies the generator voltage angle [1].

© ICST Institute for Computer Sciences, Social Informatics and Telecommunications Engineering 2023
Published by Springer Nature Switzerland AG 2023. All Rights Reserved
B. H. Woldegiorgis et al. (Eds.): ICAST 2022, LNICST 455, pp. 252–271, 2023.
https://doi.org/10.1007/978-3-031-28725-1_16

Small-signal stability, a subset of phase angle-related instability concerns, is the capacity of synchronous machines in an interconnected power system to maintain synchronism after being subjected to a slight disruption [2]. This occurs as a result of a balance between the electromagnetic and mechanical torques of each synchronous machine linked to a power system [3]. For instance, insufficient synchronizing torque results in "aperiodic" or non-oscillatory instability while inadequate dampening torque results in low-frequency oscillations [4]. Both of these small signal stability problems have resulted from poor damping caused by high gain voltage regulators to cancel the effect of synchronizing torque.

In excitation control, a high gain regulator has a binding effect of eliminating synchronizing torque but negatively affects the damping torque [5]. To solve the unwanted effect of these voltage regulators, other supplementary signals are introduced in the feedback loop. The additional signals are primarily caused by speed divergence, excitation divergence, or accelerating power, which is achieved by injecting a stabilizing signal into the excitation system, and the error signal drives the regulator [6].

The power system stabilizer is divided into three stages: lead-lag compensator, washout, and gain block [7]. To compensate for the lag between the PSS output and the subsequent electrical torque developed, lead-lag compensators use phase-lead circuits. A washout circuit functions as a high-pass filter. The amount of damping associated with rotor oscillation is determined by the gain of the power system stabilizer. Traditional control theory and a linearized system model are used to design conventional power system stabilizers (CPSS) that provide optimal behavior under fixed operating conditions [8]. A conventional controller, on the other hand, fails to achieve effective control in systems with frequently fluctuating parameters. These pique one's interest in developing a fuzzy logic controller (FLC). FLCs employ feasible reasoning, which is similar to how humans make decisions [9]. This enables knowledge and experience gained from the system to be applied in such a way that adequate control for the design is provided even when the system configuration and conditions change. As a result, control systems based on fuzzy logic can solve intelligent control tasks, adapt to changing environments, and make sound decisions [10]. A hybrid of the fractional PID controller and a single fuzzy logic-based stabilizer has been proposed to improve the stability of a single-machine infinite bus (SMIB) power system [11]. However, the controllers' capabilities are limited due to the need for an expert to design the best possible solution [12]. To address this issue, a type-2 fuzzy logic controller with more degrees of freedom to handle nonlinearities and uncertainties has been developed, improving power system stability [13]. The study's results, however, necessitated a complete mathematical model of the system, which was difficult due to the power system's dynamic and nonlinear nature. This triggers the researcher to find a new design to handle low-frequency oscillations with the best performance index when the systems are subjected to various small-signal disturbances.

2   Mathematical Modeling

Mathematical modeling of a power system consists of a synchronous machine, excitation, power system stabilizer, and multi-level fuzzy-based stabilizer.

2.1 Synchronous Machine Modeling

The complete model of the synchronous generator, which includes six electrical and two mechanical nonlinear dynamic equations, is shown below [14].

$$T'_{do}\frac{dE'_q}{dt} = -E'_q - \left(X_d - X'_d\right)\left[I_d - \frac{X'_d - X''_d}{\left(X'_d - X_{ls}\right)^2}\left(\psi_{1d} + \left(X'_d - X_{ls}\right)I_d + E'_q\right)\right] + E'_{fd} \tag{1}$$

$$T''_{do}\frac{d\psi_{1d}}{dt} = -\psi_{1d} + E'_q - \left(X'_d - X_{ls}\right)I_d \tag{2}$$

$$T'_{qo}\frac{dE'_d}{dt} = -E'_d - \left(X_q - X'_q\right)\left[I_q - \frac{X'_q - X''_q}{\left(X'_q - X_{ls}\right)^2}\left(\psi_{2q} + \left(X'_q - X_{ls}\right)I_q + E'_d\right)\right] \tag{3}$$

$$T''_{qo}\frac{d\psi_{2q}}{dt} = -\psi_{2q} + E'_d - \left(X'_q - X_{ls}\right)I_q \tag{4}$$

$$\frac{d\delta}{dt} = \omega - \omega_s \tag{5}$$

$$\frac{2H}{\omega_s}\frac{d\omega}{dt} = T_M - \frac{X''_d - X_{ls}}{\left(X'_d - X_{ls}\right)}E'_q I_q - \frac{X'_d - X''_d}{\left(X'_d - X_{ls}\right)}\psi_{1d}I_q - \frac{X''_q - X_{ls}}{\left(X'_q - X_{ls}\right)}E'_d I_d + \frac{X'_q - X''_q}{\left(X'_q - X_{ls}\right)}\psi_{2q}I_d - \left(X''_q - X''_d\right)I_d I_q - T_{FW} \tag{6}$$

where,

$E'_{d,q}$	Transient voltage in d-/q-axis	$T'_{do,qo}$	The transient time constant of d-/q-axis
E_{fd}	Field voltage	$T''_{do,qo}$	Sub transient time constant of d-/q-axis
H	Inertia constant	T_{FW}	Additional damping torque prop. to speed
$I_{d,q}$	Current in d-/q-axis	T_M	Mechanical torque
X_{ls}	Leakage reactance	$X'_{d,q}$	Transient reactance in d- /q-axis
δ	Rotor angle	$X''_{d,q}$	Sub transient reactance in d-/q-axis
ω	Rotor speed	ψ_{1d}	Flux linkage d-axis damper winding

Equations (1) and (2) describe the dynamics in the d-axis, while Eqs. (3) and (4) describe the dynamics in the q-axis. Equation (5) and (6) represents the well-known swing equations. The torque component T_{FW}, which introduces damping torque proportional to rotational speed, can be extra damping in all models.

2.2 Excitation System Model

The ST1A thyristor control model was used in the study because it allows negative field current to enable generator de-excitation, as shown in Fig. 1 [2].

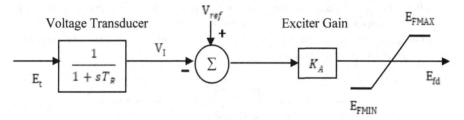

Fig. 1. Block diagram of an excitation system

A simplified model of ST1A excitation that is shown in Fig. 1, is linearized as:

$$\Delta E_t = \frac{e_{d0}}{E_{t0}}\Delta e_d + \frac{e_{q0}}{E_{t0}}\Delta e_q \tag{7}$$

$$E_{fd} = K_A(V_{ref} - V_I) \tag{8}$$

In terms of perturbed values;

$$\Delta E_t = K_A(-\Delta V_I) \tag{9}$$

In the preceding equations, Efd is the e.m.f. due to d-axis flux, and Vref is the steady-state magnitude of the terminal voltage. VR is the output voltage, and TR is the time constant.

2.3 Single-Line Diagram Representation of a Test System

In this study, the system that operates to the infinite bus is obtained in [6] and detailed machine constants are given in Appendix, as shown in Fig. 2, which is for simulation.

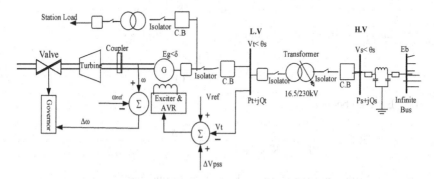

Fig. 2. A single-line diagram of the case study's single machine power system

The Heffron-Phillips model of the SMIB system shown in Fig. 3 is used in this paper [15]. H is the inertia constant, Δw is a deviation of speed, w_o is rated speed, s is the Laplace operator, K_1 to K_6 are known as K constants (which are the functions of machine inertia constant, transmission line reactance, field time constant, machine loading conditions and exciter time constant), K_A is exciter gain, T_R, Tdo, and T_A are time constant of voltage transducer, field circuit, and exciter, respectively. The numerical values of the constants are given in Appendix.

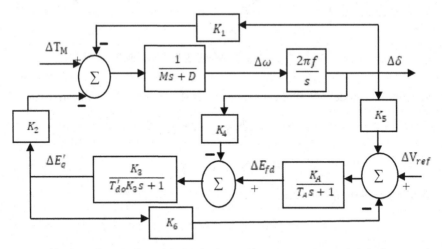

Fig. 3. A linearized model of a single machine connected to an infinite bus

2.4 Design of Conventional Power System Stabilizer

The expression of GEP(s) can be derived from Fig. 3 as follows [1]:

$$GEP(s) = \frac{K_2 K_3 G_{exc}(s)}{\left(1 + sT'_{d0}K_3\right) + K_3 K_6 G_{exc}(s)} \tag{10}$$

The transfer function of the excitation system is as follows:

$$G_{exc}(s) = \frac{K_A}{1 + sT_A} \tag{11}$$

Thus, Eq. (10) becomes:

$$GEP(s) = \frac{K_2 K_3 K_A}{T_A T'_{d0} K_3 s^2 + \left(T_A + T'_{d0}K_3\right)s + K_3 K_6 K_A + 1} \tag{12}$$

From Fig. 4, the contribution of the PSS to the torque-angle loop is given by:

$$\frac{\Delta T_{PSS}}{\Delta \omega} = GEP(s)PSS(s) \tag{13}$$

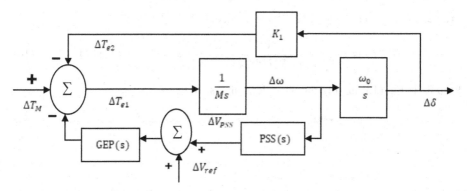

Fig. 4. Block diagram of a power system stabilizer with speed as an input

The state-space representation of the test system with exciter has become:

$$
\begin{bmatrix} \Delta\dot{\delta} \\ \Delta\dot{\omega} \\ \Delta\dot{E}'_q \\ \Delta\dot{E}_{fd} \end{bmatrix} = \begin{bmatrix} 0 & \omega_s & 0 & 0 \\ -\frac{K_1}{2H} & -\frac{D\omega_s}{2H} & -\frac{K_2}{2H} & 0 \\ -\frac{K_4}{T'_{do}} & 0 & \frac{-1}{K_3 T'_{do}} & \frac{1}{T'_{do}} \\ -\frac{K_A K_5}{T_A} & 0 & -\frac{K_A K_6}{T_A} & -\frac{1}{T_A} \end{bmatrix} \begin{bmatrix} \Delta\delta \\ \Delta\omega \\ \Delta E'_q \\ \Delta E_{fd} \end{bmatrix} + \begin{bmatrix} 0 \\ 0 \\ 0 \\ \frac{K_A}{T_A} \end{bmatrix} \Delta V_{ref} \qquad (14)
$$

The eigenvalues of the system became calculated as:

$$
|A - \lambda_i I| = 0;
$$
$$
\lambda_{1,2} = -0.0181 \pm 9.9433i; \lambda_{3,4} = -2.6971 \pm 10.8965i
$$

where A denotes the system matrix, the corresponding damping factor (ξ) of the given SMIB is given in the Table 1 below.

Table 1. Damping ratio, undamped natural frequency, and frequency of oscillation

	Eigen values	σ	ω_n (rad/sec)	f (Hz)	ζ
$\lambda_{1,2}$	$-0.0181 \pm 9.9433i$	-0.0181	9.9433	1.582525	0.002
$\lambda_{3,4}$	$-2.6971 \pm 10.8965i$	-2.6971	10.8965	1.734232	0.2403

The eigenvalues obtained above prove that the system $\lambda_{1,2}$ is poorly damped compared with $\lambda_{3,4}$. So, a conventional power system stabilizer is applied to damp oscillation. During the conventional power system stabilizer design process, the following steps must be taken [15];

Step 1. Find the torque-angle loop's undamped natural frequency in rad/sec using the Heffron Philips model, ignoring all other sources of damping.

$$
\frac{2H}{\omega_s} s^2 + K_1 = 0, \text{ i.e, } s_{1,2} = j\omega_n, \text{ where, } \omega_n = \sqrt{\frac{K_1 \omega_s}{2H}} = 10 \qquad (15)
$$

Step 2. Find the phase lag of GEP(s) at s $= j\omega n$ in Eq. (10).

$$GEP(s)\big|s =_{j\omega_n} = \frac{49.374644}{11.597158 + 25.2342j} = 1.78\angle - 65.3174^0 \qquad (16)$$

Step 3. In Eq. (13) modify the phase lead of T(s) so that

$$T(s)\big|s =_{j\omega_n} + GEP(s)|@s = j\omega_n = 0 \qquad (17)$$

Let

$$T(s) = K_{PSS}\left(\frac{1 + sT_1}{1 + sT_2}\right)^k \qquad (18)$$

Ignoring the washout filter time constants (T1 and T2), whose net phase contribution is approximately zero, k $= 1$ when T1 $>$ T2. Now select T1, and T2 as some values between 0.02 and 0.15 s [1]. As a result, for T2 $= 0.0575$ s, their corresponding T1 values are:

$$\angle(1 + j10T_1) = \angle\left[(1 + j10 * 0.0575) - 1.78\angle - 65.3174^0\right]$$

$$10T_1 = \tan\left(83.3213^0\right); \qquad (19)$$

$$T_1 = \frac{\tan\left(83.3213^0\right)}{10} = 0.854 \text{ s}$$

Step 4. The phase lead of G(s) cancels the phase lag caused by GEP(s) at the oscillation frequency, and the contribution of the PSS via GEP(s) is a pure damping torque with a damping coefficient DPSS.

$$D_{PSS} = 2\xi\omega_n M = K_{PSS}\big|T(s)\big|s =_{j\omega_n} \big|GEP(s)|@s = j\omega_n \qquad (20)$$

From Eq. (18), find KPSS, knowing ω_n and the desired [1].

$$K_{PSS} = \frac{2\xi\omega_n M}{\big|T(s)\big|s =_{j\omega_n} \big|GEP(s)\big|s =_{j\omega_n}} = \frac{2 * 0.25 * 10 * 2 * 2.37}{6.65 * 0.9978 * 1.78} = 2$$

As a result, the final conventional power system stabilizer would be:

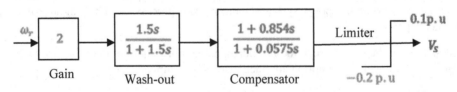

Fig. 5. Block diagram of a designed conventional power system stabilizer

The conventional power system stabilizer, as shown in the above figure (Fig. 5), is made up of three blocks: phase compensation, signal washout, and gain. The phase compensation block provides the appropriate phase-lead characteristics to compensate for the phase lag between exciter input and generator electrical torque. The signal washout block acts as a high pass filter, removing DC signals. PSS damping is determined by the stabilizer gain.

2.5 Design of Multi-level Fuzzy Logic Controller-Based Stabilizer

Low-frequency oscillations in a power system are damped using additional control signals sent to an automatic voltage regulator (AVR) via a speed deviation signal [16]. The rotor speed and its derivative are used as inputs, and the output is a voltage signal [17]. The fuzzy logic controller output has been multiplied by Kout again to give the appropriate control signal ΔV [18]. Blocks of the fuzzy logic controller with normalization factors are given in Fig. 6. The scaling factors Kin1 and Kin2 are the normalization factors for rotor speed and rate of the speed, respectively.

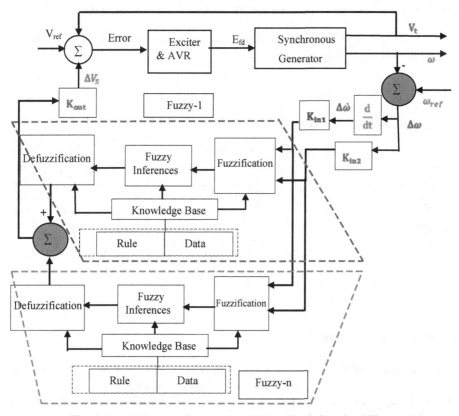

Fig. 6. Block diagram of proposed multi-level fuzzy-based stabilizer

The input domain can be described by linguistic terms such as positive big (PB), positive medium (PM), positive small (PS), zero (Z), negative small (NS), negative medium (NM), and negative big (NB) for seven linguistic variables [19].

Fuzzy Rule Base. A rule that describes the relationship between the input and output of fuzzy controllers can be set up using accessible knowledge in designing PSS [20]. The rule base is built using prior knowledge from plant dynamics, existing controllers, and experienced experts [18].

Rule 1. If the speed deviation is NB and the acceleration is PB, the voltage (output of fuzzy PSS) is NS. This means when the load angle and rotor acceleration decrease, the excitation system reduces the field voltage required to stabilize the system.

Rule 2. If speed deviation is NM and acceleration is NB, then a voltage (output of fuzzy PSS) is NB.

Rule 3. If speed deviation is PS and acceleration is PS, then a voltage (output of fuzzy PSS) is PS, and so on. This means when the load angle and rotor acceleration increase, the excitation system raises the field voltage needed to stabilize the system.

Table 2 explains all 49 rules that govern the mechanism wherever all symbols are defined in basic fuzzy logic expressions.

Table 2. Fuzzy logic control rule bases [19]

Acceleration Speed	Voltage						
	NB	NM	NS	Z	PS	PM	PB
NB	NB	NB	NB	NB	NM	NM	NS
NM	NB	NM	NM	NM	NS	NS	Z
NS	NM	NM	NS	NS	Z	Z	PS
Z	NM	NS	NS	Z	PS	PS	PM
PS	NS	Z	Z	PS	PS	PM	PM
PM	Z	PS	PS	PM	PM	PM	PB
PB	PS	PM	PM	PB	PB	PB	PB

3 Simulation Results

The performance of the SMIB system with only the excitation system, conventional PSS (lead-lag), single fuzzy, and multi-level-fuzzy based PSS is analyzed using the Heffron Philips (K constant) values. A step input signal is used as the prime mover for a synchronous generator in the Simulink model.

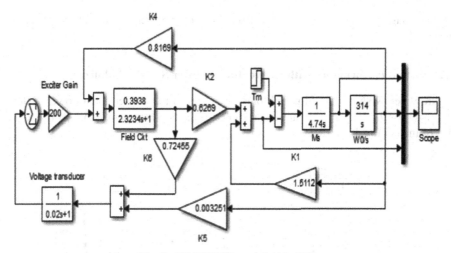

Fig. 7. SIMULINK model with AVR

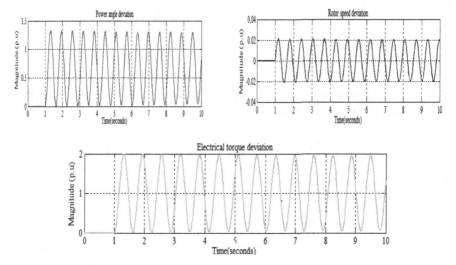

Fig. 8. Response of a 100% step-change in T_M without PSS

3.1 System Performance with Excitation System Only

Figure 7 shows the Simulink block with only the excitation system for the test system.

The response of the system without any controller applied is as follows:

As illustrated in Fig. 8, the system's response has oscillatory due to an insufficient damping coefficient. Without any stabilizer, the settling time of power angle, rotor speed, and electrical torque deviations takes more than 100 s. As time goes up, the oscillating amplitude decreases but persists for an extended time, causing the system to become monotonically unstable. This results in the equipment to be loss of synchronism or being damaged so to protect the system from this proper power system stabilizer became designed.

3.2   System Performance with Conventional Power System Stabilizer

Figure 9 below shows the Simulink block diagram of the conventional power system stabilizer for the system. The numerical values of the power system stabilizer block are determined in the design section.

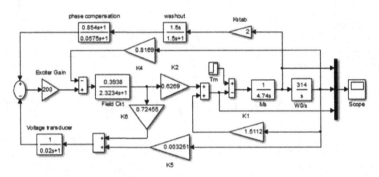

Fig. 9. SIMULINK model with power system stabilizer

Figure 10 shows the response of the system with conventional stabilizers when there is a 100% step change in mechanical torque.

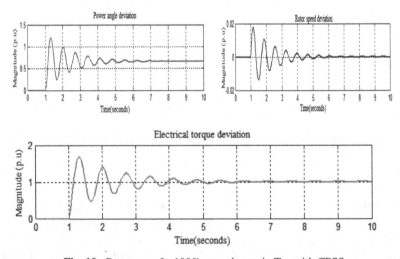

Fig. 10. Response of a 100% step-change in T_M with CPSS

As shown in Fig. 10, the settling time for rotor angle, rotor speed, and electrical torque deviations is 6.3019, 6.4544, and 5.9409 s, respectively. This showed that the designed stabilizer able to settle small signal disturbances within 7 s. As a result, conventional power system stabilizers reduced the settling time by 93.698%, 93.546%, and 94.059% for rotor angle, rotor speed, and electrical torque deviations, respectively, improving the low-frequency oscillations as compared with the existing system without any stabilizer.

3.3   System Performance with Fuzzy Logic-Based Stabilizer

The Simulink model of a single fuzzy logic controller to damp small signal oscillations in a single-machine infinite bus system can be shown in Fig. 11.

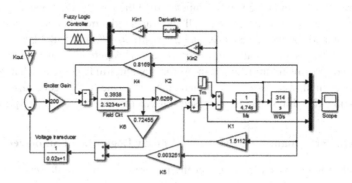

Fig. 11. SIMULINK model with a single fuzzy logic-based stabilizer

When there is a 100% step change in mechanical torque in the system, the response of rotor angle, rotor speed, and electrical torque deviation with a single fuzzy-based power stabilizer is shown in Table 3 with different membership functions below.

Table 3. The response of the system via a single fuzzy logic-based stabilizer

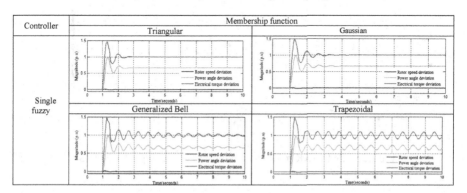

From Table 3, the settling time of electrical torque deviations for a single fuzzy-based power system stabilizer with triangular and gaussian membership functions is 2.4478

and 3.0769 s, respectively, to reach a steady-state value. This indicated that the fuzzy controller enhances small signal disturbance with minimum time and overshoot as compared to the conventional stabilizer that was seen in the previous session. This makes a single fuzzy with triangular and gaussian membership function-based power system stabilizer achieved settling time by 58.797% and 48.208% earlier than conventional power system stabilizers, respectively, for electrical torque deviations. This stated that triangular membership settles 10.589% quicker than gaussian membership. The response of the system for power angle deviations via a single fuzzy with triangular and gaussian membership function-based stabilizer achieved a settling time of 59.757% and 50.578% faster than conventional stabilizers, respectively. Therefore, the triangular membership function achieved the settling time 9.179% quicker than the gaussian membership function. While a single fuzzy with generalized bell and trapezoidal membership function-based power system stabilizer achieved a settling time of 67.670% and 158.325% slower than conventional power system stabilizers, respectively. This perceived that generalized bell and trapezoidal membership resulted in oscillatory amplitude decreases but persist for a long time even slower than conventional stabilizer. Due to this generalized bell and trapezoidal membership single fuzzy-based stabilizer is not advisable to improve LFOs.

3.4   System Performance with Multi-level Fuzzy Logic-Based Stabilizer

The SIMULINK model builds to study the performance of a multi-level fuzzy logic controller to damp LFOs is given in Fig. 12. The proposed controller incorporates the intelligent metaheuristic optimization algorithm to optimize the parameters of the system in addition to the direct fuzzy controller.

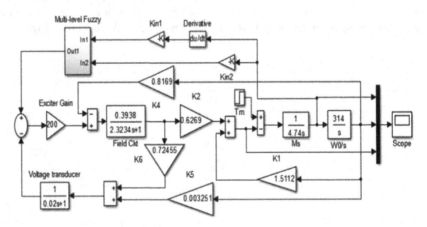

Fig. 12. SIMULINK model with multi-level fuzzy logic-based PSS

Table 4 illustrates the response of the multi-level fuzzy-based power system stabilizer to dampen low-frequency oscillations for various membership functions.

Table 4. The response of the test system via a multi-level fuzzy logic-based stabilizer

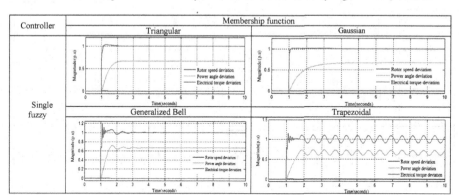

As per results from Table 4, the settling time of electrical torque deviations using a multi-level fuzzy-based stabilizer with triangular, gaussian, and generalized bell membership functions takes 0.0627, 0.6376, and 1.4155 s, respectively, to reach the final steady-state value. This showed that the designed stabilizer can settle small signal disturbances in the smallest time as compared to any stabilizer mechanism. As a result, a multi-level fuzzy with triangular, gaussian, and generalized bell membership function-based power system stabilizer settled 95.670%, 55.961%, and 2.231% faster than a single fuzzy with triangular fuzzy-based power system stabilizer. The response of the system for power angle deviations via a multi-level fuzzy-based stabilizer with triangular and gaussian membership settled 34.601% and 5.084% quicker than a single fuzzy-based stabilizer. While the settling time of electrical torque deviations for a multi-level fuzzy-based power system stabilizer with a trapezoidal membership function takes more than 100 s. Consequently, the oscillatory output is produced by a multi-level fuzzy with trapezoidal membership function-based power system stabilizer. Due to this, trapezoidal membership multi-level fuzzy-based stabilizer is not applicable to improve LFOs. Thus, Multi-level fuzzy improves power system stability and removes steady-state error that occurs in the system within a short time. This, in turn, increases the reliability, quality, and security of the power system.

3.5 Comparative Analysis of the System

The comparative analysis response graph of electrical torque and power angle deviation among various stabilizers via various membership functions with a 5%, 25%, 50%, 75%, and 100% change in mechanical torques is given in Table 5.

Table 5. Comparison of system response with performance index via various stabilizers

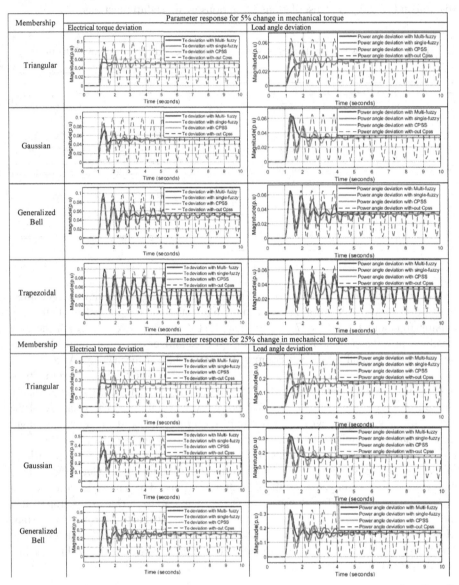

(*continued*)

Table 5. (*continued*)

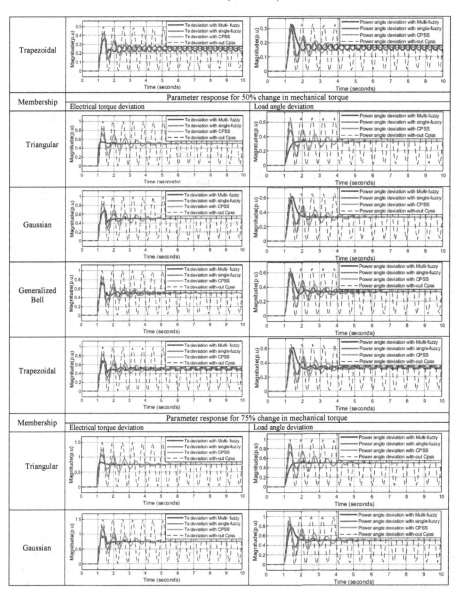

(*continued*)

Table 5. (*continued*)

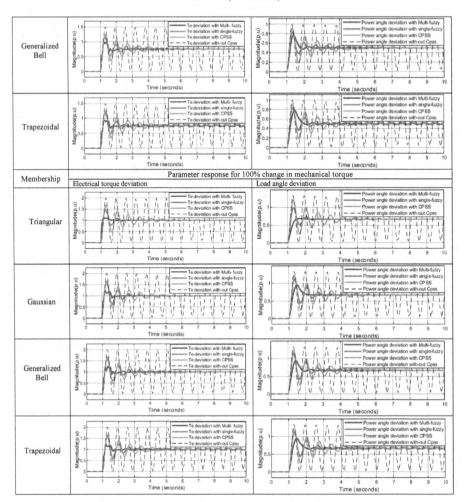

From the above response curve, it can be perceived that with the application of a fuzzy logic-based stabilizer improves the overshoot and settling time of the system as compared to AVR and CPSS. The application of a multi-level fuzzy stabilizer to the system not only improves small signal stability with small overshoot, but it is also robust to track the dynamic nature of the loads with zero steady-state error in a short time. This, in turn, increases the reliability, quality, and security of a system. The performance index via different stabilizers can be summarized in tabular form.

Table 6. Comparison of system response with performance index via various stabilizers

Controller	ΔT_E			$\Delta \delta$			$\Delta \omega$		
	Maximum peak(Mp)	Overshoot(%)	Settling time(sec.)	Maximum peak(Mp)	Overshoot(%)	Settling time(sec.)	Maximum peak(Mp)	Overshoot(%)	Settling time(sec.)
AVR	1.9899	98.9879	>100	1.3222	99.0020	> 100	0.0211	2.111	> 100
CPSS	1.6724	67.2440	5.9409	1.2131	82.5834	6.3019	0.01794	1.794	6.4544
Single-Fuzzy (Triangular)	1.4590	45.8972	2.4478	0.9899	48.9853	2.5361	0.0147	1.470	2.6456
Single-Fuzzy (Gaussian)	1.4722	47.2172	3.0769	0.9960	49.9022	3.1145	0.0148	1.480	3.2802
Single-Fuzzy (Gen-Bell)	1.4815	48.1502	9.9611	0.9805	47.5702	9.9630	0.0149	1.490	13.5371
Single-Fuzzy (Trapezed-)	1.4997	49.9669	> 100	0.9890	48.8586	> 100	0.0149	1.490	> 100
Multi-Fuzzy (Triangular)	**1.0167**	**1.6722**	**1.0627**	**0.6645**	**0.0182**	**2.0046**	**0.0027**	**0.270**	**2.2008**
Multi Fuzzy (Gaussian)	1.0295	2.9630	1.6376	0.6658	0.2146	2.4580	0.00404	0.404	2.2960
Multi Fuzzy (Generalized Bell)	1.2101	21.0061	2.4155	0.7122	7.1947	3.8457	0.0055	0.550	3.8599
Multi Fuzzy (Trapezoidal)	1.1976	19.7631	> 100	0.7345	10.5442	> 100	0.0048	0.480	> 100

According to Table 6, the settling time of electrical torque deviations, a multi-level fuzzy with triangular, gaussian, and generalized bell membership function-based stabilizer was 95.670%, 55.961%, and 2.231% quicker than a single fuzzy with triangular fuzzy-based stabilizer. This shows a triangular membership-based multi-level fuzzy stabilizer settled 39.709% quicker than a gaussian.

The settling time of power angle deviations by Multi-level Fuzzy with triangular and gaussian membership functions achieved 34.601% and 5.084% quicker than a single fuzzy with a triangular-based stabilizer, respectively for a 25% change in mechanical torque. As a result, a triangular membership-based stabilizer settled 29.517% faster than a gaussian. While the settling time of power angle deviations via multi-level fuzzy with generalized bell and trapezoidal membership function was achieved by 85.255% and 641% slower, respectively, than a single fuzzy with triangular-based stabilizer.

The multi-level fuzzy with triangular and gaussian membership function-based achieved settling time by 27.030% and 21.245%, respectively, quicker than a single fuzzy with triangular single fuzzy-based stabilizer for rotor speed deviation. As a result, a triangular-based stabilizer settled 5.785% faster than a gaussian. Therefore, a multi-level fuzzy-based stabilizer with triangular membership effectively enhances LFOs with the smallest overshot, peak amplitude, and settling time as compared with others.

4 Conclusion

This paper is developed to dampen low-frequency oscillations in a single-machine system via a multi-level fuzzy logic controller-based stabilizer. The study was simulated using a MATLAB/SIMULINK model and the performance was compared with a single fuzzy and conventional stabilizer. The system response via a conventional power system stabilizer achieved a settling time of 93.698%, 93.546%, and 94.059% quicker than with AVR for rotor angle, rotor speed, and electrical torque deviations, respectively. While the performance of a single fuzzy with triangular and gaussian membership function-based stabilizer achieved the settling time 58.797% and 48.208% earlier than conventional power system stabilizer, respectively. Contrary to this, the performance of multi-level fuzzy with triangular, gaussian, and generalized bell membership function-based stabilizers achieved the settling time 95.670%, 55.961%, and 2.231% quicker than a single fuzzy, respectively. Therefore, the application of multi-level fuzzy to the system with a triangular membership settled 29.517%, 5.785%, and 39.709% quicker than gaussian for rotor angle, rotor speed, and electrical torque deviations, respectively. The proposed multi-level fuzzy controller is robust and most effective to damp LFOs with small settling time, overshot and steady error and the result of the study has relevant to improve the power quality, reliability, and security of the power system.

5 Appendices

Appendix A: Parameters of Single Machine Infinite Bus Power System (in p.u) [2]

$H = 2.37$ s	$D = 0$	$KA = 200$	$Rs = 0$
$\omega s = 314$ rad/ s	$Xe = 0.7$ pu	$TA = 0.2$ s	$Xq = 1.64$ pu
$T'_{d0} = 5.90$ s	$Re = 0.02$ pu	$Xd = 1.70$ pu	$X'_d = 0.245$ p.u
$V_\infty = 1.00 \angle 0$ pu	$V_t = 1.72 \angle 19.31$ pu		

Appendix B: Computation of Heffron-Philips's constant (K_1-K_6)

$$A = (X_e + X_q)(X_d + X_e) + R_e^2 = 2.2117$$

$$K_1 = -\frac{1}{A}[V_\infty\{(X_d - X_q)I_d - E_q\}\{(X_e + X_d)\cos\delta + R_e\sin\delta\} \\ + V_\infty(X_d - X_q)I_q\{(X_e + X_d)\sin\delta - R_e\cos\delta\}] = 1.5112$$

$$K_2 = \frac{1}{A}[E'_q R_e + I_q A - I_q(X_d - X_q)(X_e + X_q) - R_e(X_d - X_q)I_d = 0.6269$$

$$K_3 = \frac{1}{\left(1 + \frac{(X_d - X'_d)(X_e + X_q)}{A}\right)} = 0.3938$$

$$K_4 = \frac{V_\infty(X_d - X'_d)}{A}[(X_e + X_q)\sin\delta - R_e\cos\delta] = 0.8169$$

$$K_5 = \frac{1}{A}\left\{\frac{V_{d0}X_q}{V_{t0}}\{V_\infty(X_e + X'_d)\cos\delta_{s0} + V_\infty R_e\sin\delta_{s0}\} \\ + \frac{V_{q0}}{V_{t0}}[V_\infty X'_d\{R_e\cos\delta_{s0} - (X_e + X_q)\sin\delta_{s0}\}]\right\} = 0.003251$$

$$K_6 = \frac{1}{A}\left\{\frac{V_{d0}}{V_{t0}}R_e X_q - \frac{V_{q0}}{V_{t0}}X'_d(X_e + X_q)\right\} + \frac{V_{q0}}{V_{t0}} = 0.72455$$

References

1. Anderson, P.M., Fouad, A.A.: Power System Control and Stability, 3rd edn. Wiley, Ames, Iowa (2019)

2. Mondal, D., Chakrabarti, A., Sengupta, A.: Power System Small Signal Stability Analysis and Control, 2nd edn. Academic Press, London (2020)
3. Zhu, Y.: Power system loads and power system stability. Springer, Cham (2020). https://doi.org/10.1007/978-3-030-37786-1
4. Kumar, Y., Mishra, R.N., Anwar, A.: Enhancement of small signal stability of SMIB system using PSS and TCSC. In: 2020 International Conference on Power Electronics & IoT Applications in Renewable Energy and its Control (PARC), pp. 102–106. IEEE (2020)
5. Li, G., Zhang, J., Wu, X., Yu, X.: Small-signal stability and dynamic behaviors of a hydropower plant with an upstream surge tank using different PID parameters. IEEE Access 9, 104837–104845 (2021)
6. Kundur, P.S., Balu, N.J., Lauby, M.G.: Power system dynamics and stability. Power Syst. Stab. Control 3, 827–950 (2017)
7. Kawabe, K., Masuda, M., Nanahara, T.: Excitation control method based on wide-area measurement system for improvement of transient stability in power systems. Electr. Power Syst. Res. 188, 106568 (2020)
8. Odienat, A., Al Momani, M.M., Alawasa, K., Gharaibeh, S.F.: Low frequency oscillation analysis for dynamic performance of power systems. In: 2021 12th International Renewable Engineering Conference (IREC), pp. 1–6. IEEE (2021)
9. Castillo, O., Amador-Angulo, L.: A generalized type-2 fuzzy logic approach for dynamic parameter adaptation in bee colony optimization applied to fuzzy controller design. Inf. Sci. 460, 476–496 (2018)
10. Douidi, B., Mokrani, L., Machmoum, M.: A new cascade fuzzy power system stabilizer for multi-machine system stability enhancement. J. Control Autom. Electr. Syst. 30(5), 765–779 (2019)
11. Iqbal, S., Ayyub, M.: Improved performance of fuzzy logic controller to control dynamical systems: a comparative study. In: 2018 International Conference on Computational and Characterization Techniques in Engineering & Sciences (CCTES), pp. 122–126. IEEE (2018)
12. Ray, P.K., et al.: Firefly algorithm scaled fractional order fuzzy PID based PSS for transient stability improvement. In: 2018 19th International Carpathian Control Conference (ICCC), pp. 428–433. IEEE (2018)
13. Ray, P.K., et al.: A hybrid firefly-swarm optimized fractional order interval type-2 fuzzy PID-PSS for transient stability improvement. IEEE Trans. Ind. Appl. 55(6), 6486–6498 (2019)
14. Rehman, M., Rahman, J.: Design of real-time fuzzy logic PSS based on PMUs for damping low-frequency oscillations. Master thesis (2016)
15. Verrelli, C.M., Marino, R., Tomei, P., Damm, G.: Nonlinear robust coordinated PSS-AVR control for a synchronous generator connected to an infinite bus. IEEE Trans. Automat. Contr. 67(3), 1414–1422 (2021)
16. Tzafestas, S.G., et al.: Fuzzy logic applications in engineering science. Microprocess.-Based Intell. Syst. Eng. 29, 11–30 (2006)
17. Lilly, J.H.: Fuzzy Control and Identification, 1st edn. Wiley, Hoboken (2010)
18. Bakolia, V., Joshi, S.N.: Design and analysis of fuzzy logic based power system stabilizer. Int. J. Eng. Res. Technol. 9(08), 414–418 (2020)
19. Cherniy, S.P., Susdorf, V.I., Buzikayeva, A.V.: Modeling of an advanced fuzzy logic controller with elementary links in the internal cascade. In: 2020 International Multi-Conference on Industrial Engineering and Modern Technologies (FarEastCon), pp. 1–5. IEEE (2020)
20. Kassie, W.M., Rao, D.G.S.K.: Power system dynamic stability enhancement based on facts device and fuzzy logic controller based stabilizer. Int. Res. J. Eng. Technol. 4(10), 104–111 (2017)

Super Twisting Sliding Mode Controller for Trajectory Tracking Control of Autonomous Ground Vehicle System

Tamiru Takele[1]([✉]), Tefera Terefe[2], and Sam Sun Ma[2]

[1] Wolaita Sodo University, Sodo, Ethiopia
`tamirutakele001@gmail.com`
[2] Electrical Power and Control Engineering, Adama Science and Technology University, Adama, Ethiopia

Abstract. Day to day increase in demand of safe and accident free ground vehicle, rapid growth and development of artificial intelligence algorithms and also rapid growth of microelectronics technology are major motives that are driving the development and increased attention of Autonomous Ground Vehicle (AGV) systems. Unstable and non-linear features of AGV need robust control techniques to control the trajectory tracking tasks of the system. Review of related works summery shows that sliding mode controller can handle non-linearity and relatively assure robustness of the system. However; ripple is one of the most common challenge in sliding mode controllers. In this research, Super Twisting Sliding Mode controller (STSMC) is designed to resolve the ripple in sliding mode controller for trajectory tracking control of AGV. Optimal parameters of STSMC controller are tuned using Genetic Algorithm (GA) and Particle Swarm Optimization (PSO) technique. To compare the performance of the proposed algorithm, GA tuned Fractional-Order-PID (FOPID) controller is also designed and implemented. Accordingly, STSMC has less (≈ 0.0006 s) tracking error than FOPID controller. The result reveals the outperformance of the proposed algorithm over FOPID controller.

Keywords: Autonomous Ground Vehicle · Trajectory tracking · Super Twisting Sliding Mode Controller · Fractional order proportional integral derivative controller · Optimization algorithms

1 Introduction

1.1 Background

A ground vehicle that can travel in both structured and unstructured situations without constant human guidance is referred to as an autonomous ground vehicle (AGV), also known as an intelligent vehicle. Autonomous vehicle systems may vary depending on the environment in which they are used. Unmanned aerial vehicles (UAVs), also referred to as flying and aerial vehicles (FAVs), are autonomous vehicles that fly higher than the

B. H. Woldegiorgis et al. (Eds.): ICAST 2022, LNICST 455, pp. 272–289, 2023.
https://doi.org/10.1007/978-3-031-28725-1_17

ground. There are also unmanned and autonomous vehicles operating below sea level which also known as unmanned underwater vehicle (UUV) [1]. This paper focuses on self-driving vehicles that operate on the ground, also known as AGV.

On its route to a target, which can be a single spot, a region, or a trajectory of sites, an AGV can recognize and avoid impediments. The complexity of today's transportation systems makes management and security more challenging to guarantee. Due to the exponential growth in the number of vehicles on the road, fundamental human behaviors are now fraught with danger. This issue may be solved by autonomous vehicles that take the role of human drivers. Mobile robots (car-like vehicles or autonomous vehicles) are able to carry out a range of tasks in hazardous environments where humans are unable to enter, such as those sites where harmful gases or high temperatures exist in a harsh environment for humans. They can also use autonomous vehicles to ensure the delivery of goods over long distances on hazardous roads. We can save money by doing a variety of mundane jobs, [2]. The difficulty of tracking the trajectory of an autonomous ground vehicle (AGV) has received a lot of attention lately. Due to the quick advancement of autonomous vehicles, autonomous vehicle control is becoming more and more important in the field of vehicle autonomy research. As a result, it's critical to upgrade and improve autonomous vehicle controllers in order to accomplish this goal. These controllers must be able to handle complicated problems in a range of environments. At the lower level of control, autonomous vehicles can also move laterally and longitudinally. The noteworthy adaptive cruise control (ACC) and the more sophisticated cooperative adaptive cruise control (CACC) are examples of longitudinal control that concentrate on speed control, whereas lateral control handles path following and vehicle steering control [3]. The goal of this study is to better trajectory track and follow performance by looking into steering control methods for autonomous vehicles.

PID controllers have been used to control several industrial processes all over the world for a very long time. The Ziegler-Nichols method, genetic algorithm GA, fuzzy logic controller, and others are among the most well-known and often used techniques for altering the PID controller parameters in line with changes in the state of the environment and the system. The PSO optimization algorithm was another widely used tuning approach [4].

1.2 Literature Review

This section provides a summary of previous attempts to develop AGV controllers. When first building an AGV controller, the knowledge gained from reading various scientific articles and publications is very valuable. Many academics have recently developed route tracking control algorithms that take into account the non-holonomic constraints that an AGV must follow. In [5], By creating a Lyapunov function, a kinematic controller for the mobile platform was generated to achieve the necessary velocity. Also suggested was a stable adaptive tracking controller. According to Lyapunov stability theory, the designed robust adaptive controller assures overall stability of the closed-loop system. In [6], An innovative control technique for following the path of mobile robots was developed by Zhang. An international finite-time angular velocity control law was used to design this control technique. According to the simulation results, mobile robot angle errors can be stabilized with good convergence and performance. In [7], Wang developed an adaptive trajectory tracking approach based on the kinematics model to solve the non-holonomic restriction problem of trajectory tracking of the wheeled mobile robot. This technique builds based on the idea of an artificial field to improve trajectory tracking.

In [8], For the path tracking issue of a mobile robot, Normey developed a traditional PID controller. The resilience of the closed-loop system was taken into consideration as a new PID tuning technique based on fundamental control tools was proposed. In [9], Lee and Chang proposed two strategies for improving a FOPID controller. These techniques employ electromagnetism and evolutionary algorithms. In order to incorporate the advantages of both approaches while lessening the computational load on the electromagnetic algorithm, a hybrid approach was created.

In [10], Using adaptive PID control, P. Zhao introduced and discussed the design of the "intelligent pioneer" autonomous vehicle's control system, as well as path tracking and motion stability. The experimental implementation of trajectory control of the mobile robots is presented in [11] considering the dynamics of their subsystems. Taking in to account subsystem dynamics is benefit what I gained from it. The Particle Swarm Optimization approach is used in [12] to create fractional order PID controllers that outperform its integer order counterparts. The requisite peak overshoot and rising time specifications form the basis for controller synthesis. By minimizing the characteristic equation, the ideal set of controller parameters is obtained. In [13] Under difficult conditions like as nonlinearity, strong coupling, high uncertainties, and disturbances, an Adaptive Super Twisting Sliding Mode (ASTSM) Control method is used to control quadcopter attitudes. It deals with the modeling and control of unmanned aerial vehicles such as quadcopters and drones (UAVs). The Adaptive Super Twisting Sliding Mode controller may accomplish robust operation with disturbance rejection, parametric variation adaption, and chattering attenuation, according to simulation results.

AGV Trajectory Tracking Control of the lateral dynamics model of the vehicle is created using a Genetic Algorithm based on the bicycle model is proposed in [14], in which PID parameters are optimized by genetic algorithms. The simulation results demonstrate that even in the presence of outside disturbance, the developed PID controller has good tracking capability. In [15] AGV Path Tracking Use a Fractional Order Extremum Seeking Controller (FO-ESC) to control a non-holonomic autonomous ground vehicle while monitoring its behavior along a predetermined reference path.

In general, the majority of their works in the above literature study track the behavior of the preset reference path for the AGV. Some of them used optimization techniques as well as checking the tracking performance even in the presence of external disturbance. Even if their work follows the behavior of a predefined reference path, their tracking performance and error minimization technique aren't very good. As a result, this research proposes utilizing an STSMC controller with a genetic algorithm and particle swarm optimization approaches to reduce performance error and manage the orientation and velocity of the AGV.

2   Mathematical Modelling of Lateral Vehicle Dynamics Based on Bicycle Model

As indicated in Fig. 1, a bicycle model of the vehicle with two degrees of freedom is studied [10]. The two degrees of freedom are represented by the vehicle lateral position y and the vehicle yaw angle ψ. The vehicle lateral position is measured along the lateral axis of the vehicle to the point O which is the center of rotation of the vehicle. The vehicle yaw angle is measured with respect to the global X axis. The longitudinal velocity of the vehicle at the center of gravity is represented by V_x.

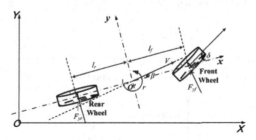

Fig. 1. Equivalent bicycle model

Newton's second law is used to describe motion along y axis.

$$ma_y = F_{yf} + F_{yr} \tag{1}$$

where $a_y = d^2y/dt^2$ is the direction that the vehicle's inertial acceleration at c.g. of the y axis. F_{yf} and F_{yr} are the lateral tire forces of the front and rear wheels respectively. Two terms contribute to a_y: the acceleration $\ddot{y}$ which is due to motion along the y axis and the centripetal acceleration $V_x\dot{\psi}$.

Hence,

$$a_y = \ddot{y} + V_x\dot{\psi} \tag{2}$$

Substituting from Eq. (2) into Eq. (1), the equation for the vehicle's lateral translational motion is as follows:

$$m(\ddot{y} + V_x\dot{\psi}) = F_{yf} + F_{yr} \tag{3}$$

The equation for the yaw dynamics is given by moment balance about the z axis as

$$I_z\ddot{\psi} = l_fF_{yf} - l_rF_{yr} \tag{4}$$

where l_f and l_r are the separations between the front and rear tires in relation to the center of gravity of the vehicle. To model the lateral tire forces F_{yf} and F_{yr} that act on the vehicle; the lateral tire force of a tire is proportional to the "slip-angle" for small slip-angles. The angle between the orientation of the tire and the orientation of the wheel's velocity vector is known as the slip angle of a tire. The slip angle of the front wheel is:

$$\alpha_f = \delta - \theta_{vf} \tag{5}$$

where θ_{vf} is the angle that the vehicle's longitudinal axis and the velocity vector make and δ is the front wheel steering angles. The rear slip angle is similarly given by:

$$\alpha_r = -\theta_{vr} \tag{6}$$

Therefore, the lateral tire force for the vehicle's front wheels can be written as

$$F_{yf} = 2C_{af}(\delta - \theta_{vf}) \tag{7}$$

where the proportionality constant C_{af} is called the front tire cornering stiffness
The lateral tire for the back wheels can be expressed similarly as

$$F_{yr} = 2C_{ar}(-\theta_{vr}) \tag{8}$$

where C_{ar} the rear tire cornering stiffness and θ_{vr} is the rear tire velocity angle. The following relations can be used to calculate θ_{vf} and θ_{vr}:

$$\begin{cases} \tan(\theta_{vf}) = \dfrac{V_y + l_f\dot{\psi}}{V_x} \\[3mm] \tan(\theta_{vr}) = \dfrac{V_y - l_r\dot{\psi}}{V_x} \end{cases} \tag{9}$$

Using the notation and small angle approximation $V_y = \dot{y}$

$$\begin{cases} \theta_{vf} = \dfrac{\dot{y} + l_f\dot{\psi}}{V_x} \\[3mm] \theta_{vr} = \dfrac{\dot{y} - l_f\dot{\psi}}{V_x} \end{cases} \tag{10}$$

Substituting from Eqs. (5), (6), (9) and (10) into Eqs. (3) and (4), the state space model can be written as

$$\frac{d}{dt}\begin{bmatrix} y \\ \dot{y} \\ \psi \\ \dot{\psi} \end{bmatrix} = \begin{bmatrix} 0 & 1 & 0 & 0 \\ 0 & -\frac{2C_{af}+2C_{ar}}{mV_x} & 0 & -V_x - \frac{2C_{af}l_f - 2C_{ar}l_r}{mV_x} \\ 0 & 0 & 0 & 1 \\ 0 & -\frac{2l_fC_{af} - 2l_rC_{ar}}{I_zV_x} & 0 & -\frac{2l_f^2C_{af} + 2l_r^2C_{ar}}{I_zV_x} \end{bmatrix} + \begin{bmatrix} 0 \\ \frac{2C_{af}}{m} \\ 0 \\ \frac{2l_fC_{af}}{I_z} \end{bmatrix}\delta \tag{11}$$

As a result, the lateral model created by the aforementioned equations will be redefined in terms of the state error variables: x_1, the separation between the vehicle's center of gravity and the lane centerline; and x_2, a vehicle's position in relation to the road.

Take into account a car that is moving at a constant longitudinal velocity V_x on a road of constant radius. Define $\ddot{x}_1$ and x_2 as follows

$$\ddot{x}_1 = (\ddot{y} + V_x\dot{\psi}) - \frac{V_x^2}{R} = \ddot{y} + V_x(\dot{\psi} - \dot{\psi}_d) \tag{12}$$

And

$$x_2 = \psi - \psi_d \tag{13}$$

Define

$$\dot{x}_1 = \dot{y} + V_x(\psi - \psi_d) \tag{14}$$

Equation (14) is consistent with Eq. (12) if the velocity is constant. Hence the approach taken is to assume the longitudinal velocity is constant and obtain a LTI model. Substituting from Eqs. (15) and (16) into (3) and (4), we find

$$
m\ddot{x}_1 = \dot{x}_1\left[-\frac{2}{V_x}C_{af} - \frac{2}{V_x}C_{ar}\right] + x_2[2C_{af} + 2C_{ar}] + \dot{x}_2\left[-\frac{2C_{af}l_f}{V_x} + \frac{2C_{ar}l_r}{V_x}\right]
$$
$$
+ \dot{\psi}_d\left[-\frac{2C_{af}l_f}{V_x} + \frac{2C_{ar}l_r}{V_x}\right] + 2C_{af}\delta \tag{15}
$$

And

$$
I_z\ddot{x}_2 = 2C_{af}l_f\delta + \dot{x}_1\left[-\frac{2C_{af}l_f}{V_x} + \frac{2C_{ar}l_r}{V_x}\right] + x_2[2C_{af}l_f - 2C_{ar}l_r]
$$
$$
+ \dot{x}_2\left[-\frac{2C_{af}l_f^2}{V_x} - \frac{2C_{ar}l_r^2}{V_x}\right] - I_z\ddot{\psi}_d + \dot{\psi}_d\left[-\frac{2C_{af}l_f^2}{V_x} - \frac{2C_{af}l_r^2}{V_x}\right] \tag{16}
$$

The state space model in tracking error variables is therefore given by

$$
\frac{d}{dt}\begin{bmatrix} x_1 \\ \dot{x}_1 \\ x_2 \\ \dot{x}_2 \end{bmatrix} = \begin{bmatrix} 0 & 1 & 0 & 0 \\ 0 & -\frac{2C_{af}+2C_{ar}}{mV_x} & \frac{2C_{af}+2C_{ar}}{m} & \frac{-2C_{af}l_f+2C_{ar}l_r}{mV_x} \\ 0 & 0 & 0 & 1 \\ 0 & -\frac{2C_{af}l_f-2C_{ar}}{I_zV_x} & \frac{2C_{af}l_f-C_{ar}l_r}{I_z} & -\frac{2C_{af}l_f^2+2C_{ar}l_r^2}{I_zV_x} \end{bmatrix}\begin{bmatrix} x_1 \\ \dot{x}_1 \\ x_2 \\ \dot{x}_2 \end{bmatrix} + \begin{bmatrix} 0 & 0 \\ \frac{2C_{af}}{m} & -\frac{2C_{af}l_f-2C_{ar}l_r}{mV_x} - V_x \\ 0 & 0 \\ \frac{2C_{af}l_f}{I_z} & -\frac{2C_{af}l_f^2+2C_{ar}l_r^2}{I_zV_x} \end{bmatrix}\begin{bmatrix} \delta \\ \dot{\psi}_d \end{bmatrix} \tag{17}
$$

Note that the longitudinal vehicle speed, which has been assumed to be constant, determines how the lateral dynamics model is calculated. The state space model in Eq. (17) is in the form of $\dot{X} = Ax + Bu$ where:

$$
A = \begin{bmatrix} 0 & 1 & 0 & 0 \\ 0 & a_{22} & a_{23} & a_{24} \\ 0 & 0 & 0 & 1 \\ 0 & a_{42} & a_{43} & a_{44} \end{bmatrix} \text{ and } B = \begin{bmatrix} 0 & 0 \\ b_{21} & b_{22} \\ 0 & 0 \\ b_{41} & b_{42} \end{bmatrix} \tag{18}
$$

$$
x = \begin{bmatrix} x_1 & \dot{x}_1 & x_2 & \dot{x}_2 \end{bmatrix}^T \text{ and } u = \begin{bmatrix} \delta & \dot{\psi}_d \end{bmatrix}^T \tag{19}
$$

Therefore, the dynamic equation also becomes (Table 1)

$$
\begin{cases} \dot{x}_1 = \dot{x}_1 \\ \ddot{x}_1 = a_{22}\dot{x}_1 + a_{23}\dot{x}_2 + a_{24}\dot{x}_2 + b_{21}\delta + b_{22}\dot{\psi}_d \\ \dot{x}_2 = \dot{x}_2 \\ \ddot{x}_2 = a_{42}\dot{x}_1 + a_{43}x_2 + a_{44}\dot{x}_2 + b_{41}\delta + b_{42}\dot{\psi}_d \end{cases} \tag{20}
$$

where, $a_{22} = \left[\dfrac{-2C_{af} + 2C_{ar}}{mV_x}\right]$, $a_{23} = \left[\dfrac{2C_{af} + 2C_{ar}}{m}\right]$,

$$
a_{24} = \left[\frac{-2C_{af}l_f + 2C_{ar}l_r}{mV_x}\right]
$$

$$a_{42} = \left[\frac{-2C_{af}l_f - 2C_{ar}}{I_z V_x} \right], \ a_{43} = \left[\frac{2C_{af}l_f - C_{ar}l_r}{I_z} \right],$$

$$a_{44} = \left[\frac{-2C_{af}l_f^2 + 2C_{ar}l_r^2}{I_z V_x} \right]$$

$$b_{21} = \left[\frac{2C_{af}}{m} \right], \ b_{22} = \left[\frac{-2C_{af}l_f - 2C_{ar}l_r}{mV_x} - V_x \right]$$

$$b_{41} = \left[\frac{2C_{af}l_f}{I_z} \right], \ b_{42} = \left[\frac{-2C_{af}l_f^2 + 2C_{ar}l_r^2}{I_z V_x} \right]$$

Table 1. Technical specification for the AGV used for evaluation

Parameters	Description	Value	Unit
m	Mass of the vehicle	1573	Kg
I	Yaw moment of inertia	2873	Kgm^2
l_f	Distance from vehicle c.g to the front axis center	1.1	m
l_r	Distance from vehicle c.g to the front axis center	1.58	m
V_x	Longitudinal velocity of the vehicle	30	m/s
C_{af}	Cornering stiffness of front tire	80000	N/rad
C_{ar}	Cornering stiffness of rear tire	80000	N/rad

3 Controller Design

3.1 Sliding Mode Controller

This control method is based on the idea that the configuration of the controller should be changed frequently to maintain the state variables on the sliding manifold [16, 17]. Design of the sliding surface and design of the control input are two distinct stages of the Sliding Mode Controller (SMC) synthesis. The construction of the sliding surface is recognized as the most important stage in the development of the SMC because it is necessary to respond to the intended control requirements and performances. We should select the sliding surfaces before designing SMC. Surface types like PD, PI, and PID may be included in the surface. Similar SMC sliding surfaces were used in the construction of the STSMC controller.

3.2   Super Twisting Sliding Mode Controller

The main drawback of first order sliding mode controller is chattering. The discontinuous control signal that results from the system variables oscillating around the sliding surface causes this phenomena. This effect might damage or disturb the physical system. Using HOSM is one of the more exciting ways to eliminate or reduce the chattering effect. The higher order temporal derivatives of the sliding surface are kept to zero in this technique [21]. As a result, the chattering impact is reduced as a result of this activity. However, the fundamental issue in implementing HOSM algorithms is the growing information need. Thus;

$$U = u_{eq} + u_c \tag{21}$$

For this system controller design choosing the PD surface as:

$$S_1 = C_1 e_1 + \dot{e}_1 \tag{22}$$

where C1 is a sliding constant and e1 and ė1 are the error and error derivative respectively. Since $u(t) = k_p e(t) + k_d \frac{d}{dt} e(t)$

Let us define the errors as

$$e_1 = x_{1d} - x_1 \tag{23}$$

$$\dot{e}_1 = \dot{x}_{1d} - \dot{x}_1 \tag{24}$$

$$\ddot{e}_1 = \ddot{x}_{1d} - \ddot{x}_1 \tag{25}$$

Differentiating the surface S1 of Eq. (22)

$$\dot{S}_1 = C_1 \dot{e}_1 + \ddot{e}_1 \tag{26}$$

Inserting Eq. (24) and (25) in Eq. (26)

$$\dot{S}_1 = C_1(\dot{x}_{1d} - \dot{x}_1) + (\ddot{x}_{1d} - \ddot{x}_1) \tag{27}$$

Inserting Eq. (20) in Eq. (27)

$$\dot{S}_1 = C_1(\dot{x}_{1d} - \dot{x}_1) + (\ddot{x}_{1d} - a_{22}\dot{x}_1 - a_{23}x_2 - a_{24}\dot{x}_2 - b_{21}\delta - b_{22}\dot{\psi}_d) \tag{28}$$

The Lyapunov theorem, a well-known approach in stability research, is used to validate the steady convergence behavior of nonlinear controllers [18–20]. We can use the Lyapunov function to demonstrate the stability as in Eq. (29) with its time derivative given in Eq. (30).

$$V = \frac{1}{2}s^2 \tag{29}$$

$$\dot{V} = s\dot{s} \tag{30}$$

With $V = 0$ and $V > 0$ for $s \neq 0$. The following reaching criterion must be adhered to in order to guarantee that the trajectory transitions from the reaching to the sliding phase while maintaining stability:

$$\dot{V} < 0, \ \text{for } s \neq 0, \ \dot{s} \neq 0 \tag{31}$$

For $S_1 \dot{S}_1 \leq 0$ Eq. (28) becomes

$$C_1(\dot{x}_{1d} - \dot{x}_1) + (\ddot{x}_{1d} - a_{22}\dot{x}_1 - a_{23}x_2 - a_{24}\dot{x}_2 - b_{21}\delta - b_{22}\dot{\psi}_d) = 0 \tag{32}$$

$$C_1(\dot{x}_{1d} - \dot{x}_1) + (\ddot{x}_{1d} - a_{22}\dot{x}_1 - a_{23}x_2 - a_{24}\dot{x}_2) = b_{21}\delta + b_{22}\dot{\psi}_d \tag{33}$$

Letting $b_{22} = 0$ Eq. (33) becomes

$$C_1(\dot{x}_{1d} - \dot{x}_1) + (\ddot{x}_{1d} - a_{22}\dot{x}_1 - a_{23}x_2 - a_{24}\dot{x}_2) = b_{21}\delta \tag{34}$$

$$\delta = \frac{C_1(\dot{x}_{1d} - \dot{x}_1) + (\ddot{x}_{1d} - a_{22}\dot{x}_1 - a_{23}x_2 - a_{24}\dot{x}_2)}{b_{21}} \tag{35}$$

Letting $b_{21} = 0$ Eq. (33) becomes

$$C_1(\dot{x}_{1d} - \dot{x}_1) + (\ddot{x}_{1d} - a_{22}\dot{x}_1 - a_{23}x_2 - a_{24}\dot{x}_2) = b_{22}\dot{\psi}_d \tag{36}$$

$$\dot{\psi}_d = \frac{C_1(\dot{x}_{1d} - \dot{x}_1) + (\ddot{x}_{1d} - a_{22}\dot{x}_1 - a_{23}x_2 - a_{24}\dot{x}_2)}{b22} \tag{37}$$

Now letting:

$$u_{eq1} = \delta + \dot{\psi}_d \tag{38}$$

$$u_{eq1} = \left[\frac{b_{21} + b_{22}}{b_{21}b_{22}} \right] (C_1(\dot{x}_{1d} - \dot{x}_1) + (\ddot{x}_{1d} - a_{22}\dot{x}_1 - a_{23}x_2 - a_{24}\dot{x}_2)) \tag{39}$$

For second controller, choosing the sliding surface two:

$$S_2 = C_2e_2 + \dot{e}_2 \tag{40}$$

Let us also define the errors as

$$e_2 = x_{2d} - x_2 \tag{41}$$

$$\dot{e}_2 = \dot{x}_{2d} - \dot{x}_2 \tag{42}$$

$$\ddot{e}_2 = \ddot{x}_{2d} - \ddot{x}_2 \tag{43}$$

Differentiating the surface S2 of Eq. (40)

$$\dot{S}_2 = C_2\dot{e}_2 + \ddot{e}_2 \tag{44}$$

Inserting Eq. (42) and (43) in Eq. (44)

$$\dot{S}_2 = C_2(\dot{x}_{2d} - \dot{x}_2) + (\ddot{x}_{2d} - \ddot{x}_2) \tag{45}$$

Inserting Eq. (20) in Eq. (45)

$$\dot{S}_2 = C_2(\dot{x}_{2d} - \dot{x}_2) + (\ddot{x}_{2d} - a_{42}\dot{x}_1 - a_{43}x_2 - a_{44}\dot{x}_2 - b_{41}\delta - b_{42}\dot{\psi}_d) \tag{46}$$

For $S_2\dot{S}_2 \le 0$ Eq. (46) becomes

$$C_2(\dot{x}_{2d} - \dot{x}_2) + (\ddot{x}_{2d} - a_{22}\dot{x}_1 - a_{23}x_2 - a_{24}\dot{x}_2 - b_{41}\delta - b_{42}\dot{\psi}_d) = 0 \tag{47}$$

$$C_1(\dot{x}_{2d} - \dot{x}_2) + (\ddot{x}_{2d} - a_{42}\dot{x}_1 - a_{43}x_2 - a_{44}\dot{x}_2) = b_{41}\delta + b_{42}\dot{\psi}_d \tag{48}$$

Letting the parameter $b_{42} = 0$ in Eq. (47) becomes

$$C_2(\dot{x}_{2d} - \dot{x}_2) + (\ddot{x}_{2d} - a_{42}\dot{x}_1 - a_{43}x_2 - a_{44}\dot{x}_2) = b_{41}\delta \tag{49}$$

$$\delta = \frac{C_2(\dot{x}_{2d} - \dot{x}_2) + (\ddot{x}_{2d} - a_{42}\dot{x}_1 - a_{43}x_2 - a_{44}\dot{x}_2)}{b_{41}} \tag{50}$$

Letting the parameter $b_{41} = 0$ in Eq. (48) becomes

$$C_2(\dot{x}_{2d} - \dot{x}_2) + (\ddot{x}_{2d} - a_{42}\dot{x}_1 - a_{43}x_2 - a_{44}\dot{x}_2) = b_{42}\dot{\psi}_d \tag{51}$$

$$\dot{\psi}_d = \frac{C_2(\dot{x}_{2d} - \dot{x}_2) + (\ddot{x}_{2d} - a_{42}\dot{x}_1 - a_{43}x_2 - a_{44}\dot{x}_2)}{b42} \tag{52}$$

Now letting:

$$u_{eq2} = \delta + \dot{\psi}_d \tag{53}$$

$$u_{eq2} = \left[\frac{b_{41} + b_{42}}{b_{41}b_{42}}\right](C_2(\dot{x}_{2d} - \dot{x}_2) + (\ddot{x}_{2d} - a_{42}\dot{x}_1 - a_{43}x_2 - a_{44}\dot{x}_2)) \tag{54}$$

For STSMC;

$$u_c = -K_1|\sqrt{s}|\text{sgn}(s) - K_2\int_0^t \text{sgn}(s)dt \tag{55}$$

Therefore, the controller for STSMC1 designed as,

$$U_1 = u_{eq1} + u_c \tag{56}$$

Inserting Eq. (39) and (55) in Eq. (60) becomes,

$$U_1 = \left[\frac{b_{21} + b_{22}}{b_{21}b_{22}}\right](C_1(\dot{x}_{1d} - \dot{x}_1) + (\ddot{x}_{1d} - a_{22}\dot{x}_1 - a_{23}x_2 - a_{24}\dot{x}_2)) + \left(-K_1|\sqrt{s}|\text{sgn}(s) - K_2\int_0^t \text{sgn}(s)dt\right) \tag{57}$$

Therefore, the controller for STSMC2 designed as,

$$U_2 = u_{eq2} + u_c \tag{58}$$

Inserting Eq. (54) and (55) in Eq. (62) becomes,

$$U_2 = \left[\frac{b_{41} + b_{42}}{b_{41}b_{42}}\right](C_2(\dot{x}_{2d} - \dot{x}_2) + (\ddot{x}_{2d} - a_{42}\dot{x}_1 - a_{43}x_2 - a_{44}\dot{x}_2)) + \left(-K_1|\sqrt{s}|\text{sgn}(s) - K_2\int_0^t \text{sgn}(s)dt\right) \tag{59}$$

3.3 Fractional Order PID Controller

Fractional-order proportional-integral-derivative (FOPID) controllers have attracted a lot of attention in recent years from both academia and industry. Because they have five parameter options, they actually provide greater design flexibility than typical PID controllers (instead of three). This does however imply that adjusting the controller might be much more difficult. The integral-differential equation describes the control action of a fractional-order PID controller as follows:

$$u(t) = k_p e(t) + k_i D_t^{-\lambda} e(t) + k_d D_t^{\mu} e(t) \tag{60}$$

FOPID's transfer function is obtained by applying the Laplace transform as follows (Fig. 2):

$$U(s) = k_p + k_i s^{-\lambda} + k_d s^{\mu} \tag{61}$$

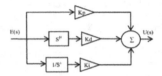

Fig. 2. The fractional order controller block diagram

Objective Functions

The Defining an objective function is one of the major tasks while formulating a problem using an optimization algorithm. The sum of two objective cost functions is the cost function that minimizes the system's error, because the proposed autonomous ground vehicle system is a multiple input multiple output system [22], written as:

$$J = \mathrm{ITAE} = \int_0^{\tau} t(|e_1(t)| + |e_2(t)|)d\tau \tag{62}$$

whereas, $e_1(t) = \delta_{ref} - \delta_{act}$ and $e_2(t) = \dot{\psi}_{dref} - \dot{\psi}_{dact}$. Where δ orientation angle and $\dot{\psi}_d$ is velocity.

4   Results and Discussions

4.1   MATLAB Implementations

The Matlab Simulink is used to design and simulate the proposed system mathematical model and controller design (Fig. 3).

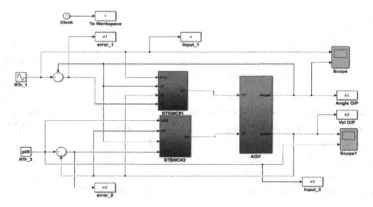

Fig. 3. The Simulink model of the proposed system STSMC

4.2   Open Loop Response

For a given step and constant input the system the open loop response output is highly nonlinear and unstable. The open loop response of the proposed system is shown as in Fig. 4 below:

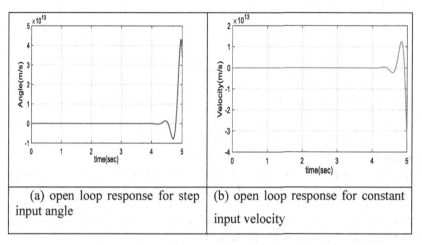

(a) open loop response for step input angle	(b) open loop response for constant input velocity

Fig. 4. Open loop response of the proposed system

4.3   Testing of Trajectories for Closed Loop Responses

After the model of the autonomous ground vehicle and the two STSMC controllers have been deployed, several case studies are used to simulate and examine the various trajectories in order to compare how the suggested technique responds to the FOPID controller. Each instance requires a distinct strategy, as can be seen below:

Test Case-1

A sinusoidal trajectory was created in this instance as a reference trajectory. Whereas, reference trajectory input is $\delta_d = 5 \sin(t)$. To determine the orientation and velocity inaccuracy, this trajectory was compared to the real trajectory. As a result, the STSMC controller received the error. The STSMC1 controller's output offers the control action that was linked to the left wheel's motor. The second input serves as a representation of the intended velocity. The actual speed of the vehicle was contrasted with this speed. The error that was made when comparing the velocity was sent to the second STSMC2 controller. The velocity controller's output was coupled to the right wheel's motor. Figure 5 shows the relationship between the desired and actual trajectory after performing the simulation. As seen in Fig. 5, the system tracks the sinusoidal reference trajectory very well. Figure 7 shows the difference between the desired and actual orientation angle and velocity (Fig. 6).

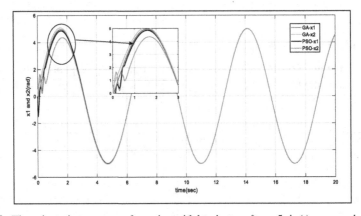

Fig. 5. The orientation response for a sinusoidal trajectory $\delta_d = 5 \sin(t)$ proposed system

Test Case-2

In response to a different scenario and to show how flexible the suggested controller is, a linear trajectory was created. Creating a constant orientation will guarantee a linear trajectory using $\delta_d = \pi/4$ for interval $0 \le t \le 20$ s. Similarly, Figs. 8a and 8b illustrate the connection between the desired and actual trajectories. Figure 10 shows the difference in direction and velocity between the desired and actual values.

In Table 2 the error performance analysis of sinusoidal trajectory $\delta_d = 5 \sin(t)$, we can say that the proposed STSMC controller has better error performance for PSO

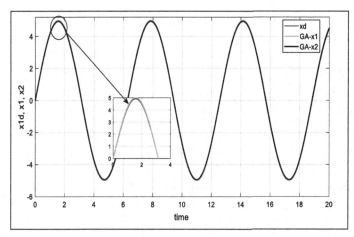

Fig. 6. Orientation angle and velocity response for sinusoidal trajectory $\delta_d = 5\sin(t)$ of FOPID controller

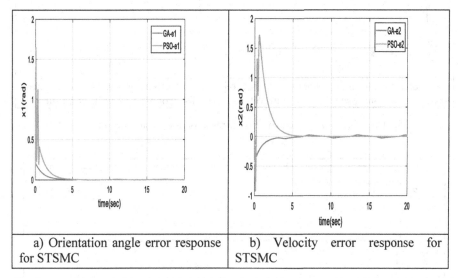

a) Orientation angle error response for STSMC	b) Velocity error response for STSMC

Fig. 7. Orientation angle and velocity error for sinusoidal trajectory $\delta_d = 5\sin(t)$ of STSMC

optimazation technique than GA based for both orientation angle and velocity control. And the error performance of proposed STSMC controller is much better than GA based FOPID controller for angle and velocity control. Adding disturbances due to friction of wheels during vehicle motion which is formed by varying the cornering stiffness value $C_{af} = C_{ar} = 40000$ N/rad and mass of 1500 kg has the orientation angle and velocity error 0.0004 and 0.0008 s respectively for sinusoidal trajectory of proposed PSO-based STSMC controllers (Fig. 9).

Table 2. The error performance analysis of the trajectory $\delta_d = 5\sin(t)$

Controllers	Angle error	Velocity error
GA-based STSMC	0.0026	0.0212
PSO-based STSMC	0.0006	0.0007
GA-based STSMC	0.0030	0.0538

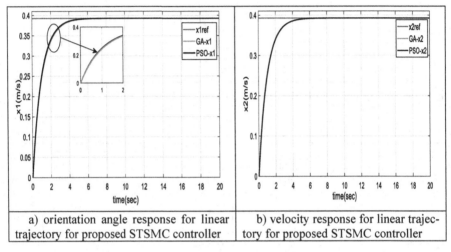

a) orientation angle response for linear trajectory for proposed STSMC controller	b) velocity response for linear trajectory for proposed STSMC controller

Fig. 8. The orientation angle and velocity response for linear trajectory $\delta_d = \pi/4$

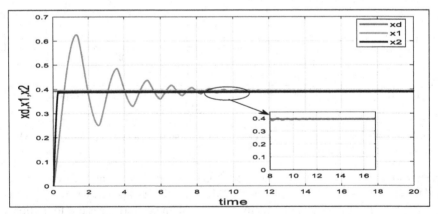

Fig. 9. The orientation angle and velocity response for linear trajectory $\delta_d = \pi/8$ of FOPID controller

As depicted in Fig. 10a for a linear trajectory of an orientation and velocity error response highly converges to zero for both GA and PSO based optimization technique of proposed STSMC controller. Compared to the orientation angle and velocity response Fig. 10b of FOPID to the response of proposed system in a Fig. 9, it takes longer time to track the given linear reference trajectory and its error converges to zero after 0.0024 s time which is sluggish compared to STSMC error convergence time.

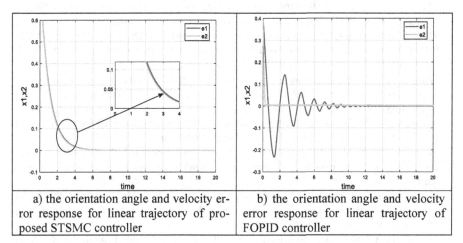

a) the orientation angle and velocity error response for linear trajectory of proposed STSMC controller	b) the orientation angle and velocity error response for linear trajectory of FOPID controller

Fig. 10. The orientation and velocity error response for linear trajectory $\delta_d = \pi/8$ of FOPID controller

5 Conclusions

This paper discusses the application of STSMC to autonomous ground vehicle systems. The lateral dynamics of a vehicle were simulated in this work using a simple two-degree-of-freedom bicycle model. Two distinct control algorithms were used to regulate the modified lateral motions. An investigation into the trajectory tracking control of autonomous ground vehicle systems used a super twisting sliding mode controller and a fractional order PID controller. A simulation was created using MATLAB/Simulink to test the effectiveness and performance of the controller. The controllers gain was optimized and tuned using the evolutionary algorithm and particle swarm optimization method. Finally, the findings show that the system tracks the needed trajectory with a minimal amount of tracking error. The suggested STSMC robustly monitors the supplied irregular reference environments, outperforming fractional order PID controllers in terms of performance. It is evident from the simulation results that the STSMC was used to control the orientation steering angle and velocity in order to stabilize the AGV system. The system's machine learning algorithm may be established in the future.

References

1. Amer, N.H., Zamzuri, H., Hudha, K., Kadir, Z.A.: Modelling and control strategies in path tracking control for autonomous ground vehicles: a review of state of the art and challenges. J. Intell. Rob. Syst. **86**(2), 225–254 (2016). https://doi.org/10.1007/s10846-016-0442-0

2. Chen, Q.: Studies in Autonomous Ground Vehicle Control Systems : Structure and Algorithms (2007)
3. Yousif, N., et al.: Control system design for autonomous vehicle path following and collision avoidance dissertation. J. Phys. Ther. Sci. **9**(1), 1–11 (2018). https://doi.org/10.1016/j.neuropsychologia.2015.07.010%0A. https://doi.org/10.1016/j.visres.2014.07.001%0A. https://doi.org/10.1016/j.humov.2018.08.006%0A. http://www.ncbi.nlm.nih.gov/pubmed/24582474%0A. https://doi.org/10.1016/j.gaitpost.2018.12.007%0A
4. Zamani, M., Karimi-Ghartemani, M., Sadati, N., Parniani, M.: Design of a fractional order PID controller for an AVR using particle swarm optimization. Control Eng. Pract. **17**(12), 1380–1387 (2009). https://doi.org/10.1016/j.conengprac.2009.07.005
5. Peng, J., Yu, J., Wang, J.: Robust adaptive tracking control for nonholonomic mobile manipulator with uncertainties. ISA Trans. **53**(4), 1035–1043 (2014). https://doi.org/10.1016/j.isatra.2014.05.012
6. Zhang, Y., Liu, G., Luo, B.: Finite-time cascaded tracking control approach for mobile robots. Inf. Sci. (Ny) **284**, 31–43 (2014). https://doi.org/10.1016/j.ins.2014.06.037
7. Zhuo-Yun, N., Yi-Min, Z., Qing-Guo, W., Rui-Juan, L., Lei-Jun, X.: Fractional-order PID controller design for time-delay systems based on modified bode's ideal transfer function. IEEE Access **8**, 103500–103510 (2020). https://doi.org/10.1109/ACCESS.2020.2996265
8. Normey-Rico, J.E., Alcalá, I., Gómez-Ortega, J., Camacho, E.F.: Mobile robot path tracking using a robust PID controller. Control Eng. Pract. **9**(11), 1209–1214 (2001). https://doi.org/10.1016/S0967-0661(01)00066-1
9. Lee, C.H., Chang, F.K.: Fractional-order PID controller optimization via improved electromagnetism-like algorithm. Expert Syst. Appl. **37**(12), 8871–8878 (2010). https://doi.org/10.1016/j.eswa.2010.06.009
10. Zhao, P., Chen, J., Song, Y., Tao, X., Xu, T., Mei, T.: Design of a control system for an autonomous vehicle based on adaptive-PID. Int. J. Adv. Robot. Syst. **9**, 1–11 (2012). https://doi.org/10.5772/51314
11. García-Sánchez, J.R., et al.: Tracking control for mobile robots considering the dynamics of all their subsystems: experimental implementation. Complexity **2017** (2017). https://doi.org/10.1155/2017/5318504
12. Maiti, D., Biswas, S., Konar, A.: Design of a fractional order PID controller using particle swarm optimization technique. ReTIS-08, pp. 1–5 (2008). http://arxiv.org/abs/0810.3776
13. Hoang, V.T., Pham, Q.H.: Adaptive super-twisting second-order sliding mode for attitude control of quadcopter UAVs. arXiv (2018)
14. Zhao, B., Wang, H., Li, Q., Li, J., Zhao, Y.: PID trajectory tracking control of autonomous ground vehicle based on genetic algorithm. In: Proceedings of the 31st Chinese Control Decision Conference, CCDC 2019, pp. 3677–3682 (2019). https://doi.org/10.1109/CCDC.2019.8832531
15. Dadras, S.: Path tracking using fractional order extremum seeking controller for autonomous ground vehicle. SAE Technical Paper, vol. 2017-March, no. March (2017). https://doi.org/10.4271/2017-01-0094
16. Gambhire, S.J., Kishore, D.R., Londhe, P.S., Pawar, S.N.: Review of sliding mode based control techniques for control system applications. Int. J. Dyn. Control **9**(1), 363–378 (2020). https://doi.org/10.1007/s40435-020-00638-7
17. Alam, W., Mehmood, A., Ali, K., Javaid, U., Alharbi, S., Iqbal, J.: Nonlinear control of a flexible joint robotic manipulator with experimental validation. J. Mech. Eng. **64**(1), 47–55 (2018). https://doi.org/10.5545/sv-jme.2017.4786
18. Hu, J., Zhang, Y., Rakheja, S.: Adaptive trajectory tracking for car-like vehicles with input constraints. IEEE Trans. Ind. Electron. **0046**(c) (2021). https://doi.org/10.1109/TIE.2021.3068672

19. Wu, Y., Wang, L., Zhang, J., Li, F.: Path following control of autonomous ground vehicle based on nonsingular terminal sliding mode and active disturbance rejection control. IEEE Trans. Veh. Technol. **68**(7), 6379–6390 (2019). https://doi.org/10.1109/TVT.2019.2916982

20. Dai, M., Qi, R., Cheng, X.: Super-twisting sliding mode control design for electric dynamic load simulator. In: Chinese Control Conference, CCC, vol. 2019-July, no. 1, pp. 3078–3083 (2019). https://doi.org/10.23919/ChiCC.2019.8865725

21. Shtessel, Y., Edwards, C., Fridman, L., Levant, A.: Sliding mode control and observation (2014)

22. Al-Mayyahi, A., Wang, W., Birch, P.: Design of fractional-order controller for trajectory tracking control of a non-holonomic autonomous ground vehicle. J. Control Autom. Electr. Syst. **27**(1), 29–42 (2015). https://doi.org/10.1007/s40313-015-0214-2

Trajectory Tracking of a Two-Wheeled Mobile Robot Using Backstepping and Nonlinear PID Controller

Lencho Duguma Fufa[1,2]($\boxtimes$) and Endalew Ayenew[1]

[1] Electrical Power and Control Engineering, School of Electrical Engineering and Computing, Adama Science and Technology University, Adama, Ethiopia
lencho.duguma@mwu.edu.et
[2] Electrical and Computer Engineering, College of Engineering, Madda Walabu University, Bale Robe, Ethiopia

Abstract. Many researchers have become interested in wheeled mobile robot (WMR) trajectory tracking control in recent years. This is due to the increased application of mobile robots in the industry, the military, the home, and public service. Classically, the movement of WMR is controlled depending on its kinematic model. However, in real-time applications, both the dynamic and kinematic models of robots and external disturbance and uncertainty affect system performance. This paper proposes backstepping combined with a Nonlinear Proportional-Integral-Derivative (NPID) controller to control a two-wheeled mobile robot (TWMR). The kinematic and dynamic models of the WMR are derived. The dynamic modeling is derived using a Lagrangian approach, and stability of the system is achieved using the Lyapunov method. The controller gains are optimized using the Genetic Algorithm optimization technique. The proposed algorithms' performance is tested using Matlab software. The simulation result shows that the proposed method achieved preferable reference trajectory tracking with a minimum tracking error. The proposed controller outperforms the GA-based backstepping plus PID controller in terms of root-mean-square (RMS) of trajectory tracking error (47.36% in a linear and 60.32% in a nonlinear case). In addition, it shows good unknown disturbance rejection and initial point change in all scenarios.

Keywords: Backstepping and NPID Controller · Trajectory Tracking · Two-Wheeled Mobile Robot · Genetic Algorithm · Lyapunov Stability Analysis

1 Introduction

A mobile robot is a robot that can move its surroundings and is not fixed to a given physical location. It can be autonomous or semi-autonomous, which means it can navigate in an uncontrolled environment without needing a personal operator or electro-mechanical guidance device. A wheeled mobile robot (WMR) is a mobile robot that navigates its surroundings using a wheel's rotation. It is a widely used mobile robot due to its ability

B. H. Woldegiorgis et al. (Eds.): ICAST 2022, LNICST 455, pp. 290–304, 2023.
https://doi.org/10.1007/978-3-031-28725-1_18

to substitute humans in many fields. For example, in industry, the military, public and private sector services, and production and distribution companies.

When the application of a WMR increases, the performance acquired from the robot becomes a critical issue for researchers. The robot's performance (in the case of trajectory tracking) depends on the position error and orientation error for kinematic control, while a velocity error will affect the performance in the case of dynamic control. The movement or navigation of a WMR depends on three control problems: reference trajectory tracking, line following, and point stabilization [1]. One of the main objectives of reference tracking control is to control a robot's position on a predetermined trajectory with the minimum position and orientation error. Due to nonholonomic restrictions in nonholonomic robots, trajectory tracking and motion control are not independent. Trajectory tracking of nonholonomic robots in the robotic domain involves determining trajectories from a starting point to a final point while considering mechanical limitations and guiding the robot to follow the proposed trajectories. Several works on the WMR's trajectory control problem have been presented. Most of them are focused on a nonholonomic constraint that depends on posture stabilization and trajectory tracking [2]. Trajectory tracking makes the robot follow a predefined trajectory.

The two-wheeled mobile robot (TWMR) used in this work has two standard fixed wheels that are actuated by two similar DC motors and a freewheeler. The freewheeler is used to balance the robot's body frame. Many control approaches have been applied to control WMR trajectory tracking, for example, a kinematics-based backstepping controller [3], Proportional-Integral-Derivative (PID) controller [4], Fuzzy logic controller [5, 6], and Nonlinear PID controller. Furthermore, various optimization techniques (like; Genetic Algorithm (GA), Particle Swarm Optimization, Grey Wolf Optimization, and Neural Network) are used to optimize the controller gains.

In [7, 8], a kinematic controller combined with a torque control law using the backstepping control method is proposed for nonholonomic mobile robot control. Hassani et al. proposed a backstepping method for trajectory tracking control of mobile robots (MR) [9]. The authors presented kinematic and dynamic models. The challenging task of this approach is that controller gains are obtained by trial and error and are less efficient.

In works by [10], a nonholonomic WMR is controlled by a Nonlinear Proportional-Integral-Derivative (NPID) neural controller with a particle swarm optimization algorithm. The Neural Network-based PID controller responds smoothly to the external attenuation disturbance problem. The control method was based on a kinematic model of the system by ignoring the robot's dynamic model. In practical applications, both dynamic and kinematic models of robots affect system performance. In the literature [11] and [12] kinematic-based backstepping controller is applied to control robot coordinates, and a PID controller is used for motor speed control. In [13], a combination of PID and backstepping approach is used for trajectory control. However, it lacks control parameter optimization and leads to the system being controlled only in an interesting region.

Preferable trajectory tracking of WMR is achieved by considering the disturbances, noise, or internal model changes (uncertainty) that will affect the system in real-time applications. Including a WMR system dynamic model is vital for good trajectory tracking and stabilization. The main contribution of this paper is the design of a backstepping controller combined with an NPID controller for stabilization and trajectory control of TWMR. The proposed strategies consist of two approaches. A backstepping controller is used at the control robot's position, and NPID controls the robot's velocity. The Lyapunov method justifies the stability of the system. Kinematic and dynamic models of a proposed framework are taken into account.

Additionally, the controller parameters are also optimized using a GA optimization method. The developed control method is tested by simulation on Matlab/Simulink software. The system's performance will be analyzed with unknown disturbances and changes in the initial position. The proposed system tracking capability is compared with a GA-based backstepping controller and a GA-based backstepping plus PID controller.

The next sections of this paper present system modeling, control approach design, results and discussions, and conclusions.

2 System Modeling

2.1 Kinematic Modeling of the TWMR

A system modeling of the differential drive MR platform in this study consists of kinematic and dynamic modeling. As seen in Fig. 1, the WMR is represented in Cartesian coordinates. The robot setup body has two wheels with a radius of r and a distance L from the center P and free castor wheels to balance the robot setup body. C is the robot's center of mass (CoM), and d is the distance between the CoM and the center of the wheel axis. The position of the WMR in the inertial reference frame (X_I, Y_I) with origin point, O, and the local robot frame (X_R, Y_R) they are attached to the body frame. The angle between the local robot and global reference frames is represented as theta (θ) in radian. Moreover, v and ω are robot linear and angular velocities, respectively.

Let us represent the robot coordinates in inertial and local frames as follows [14]:

$$q_I = \begin{bmatrix} x_I\ y_I\ \theta_I \end{bmatrix}^T \tag{1}$$

$$q_R = \begin{bmatrix} x_R\ y_R\ \theta_R \end{bmatrix}^T \tag{2}$$

A robot's motions in the global frame are translated into motions in the local frame by using a standard orthogonal transformation matrix and vice versa. I.e.,

$$\dot{q}_R = R(\theta)\dot{q}_I \tag{3}$$

where:

$$R(\theta) = \begin{bmatrix} \cos(\theta) & \sin(\theta) & 0 \\ -\sin(\theta) & \cos(\theta) & 0 \\ 0 & 0 & 1 \end{bmatrix} \tag{4}$$

The contribution of the translational and rotational velocity components of the wheel velocities in the local robot frame is as follows:

$$\begin{bmatrix} \dot{x}_R, \dot{y}_R, \dot{\theta}_R \end{bmatrix}^T = \begin{bmatrix} \frac{1}{2}r(\dot{\varphi}_r + \dot{\varphi}_l), 0, \frac{r}{2L}(\dot{\varphi}_r - \dot{\varphi}_l) \end{bmatrix}^T \tag{5}$$

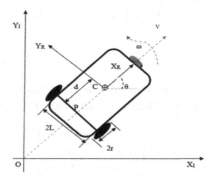

Fig. 1. 2D representation of WMR

where $\dot{\varphi}_r$ and $\dot{\varphi}_l$ is the angular velocity of the right and left wheels, respectively. In an inertial frame, the differential drive WMR velocity is given as:

$$\left[\dot{x}\ \dot{y}\ \dot{\theta}\right]^T = \frac{r}{2}\begin{bmatrix} \cos(\theta)\ \cos(\theta) \\ \sin(\theta)\ \sin(\theta) \\ 1/L\ \ -1/L \end{bmatrix}\begin{bmatrix} \dot{\varphi}_r \\ \dot{\varphi}_l \end{bmatrix} \tag{6}$$

In Fig. 1, a wheel presents three kinematic constraints on DWMR. The first constraint is that DMWR cannot slide sideways, i.e., no non-slipping constraint exists, while the others are related to the motion of the wheels. The actuated wheels cannot rotate in the wrong direction. I.e., there are only pure rolling constraints [15, 16].

$$\dot{y}\cos(\theta) - \dot{x}\sin(\theta) - d\dot{\theta} = 0 \tag{7}$$

$$\dot{x}\cos(\theta) + \dot{y}\sin(\theta) + L\dot{\theta} - r\dot{\varphi}_r = 0$$
$$\dot{x}\cos(\theta) + \dot{y}\sin(\theta) - L\dot{\theta} - r\dot{\varphi}_l = 0 \tag{8}$$

The nonholonomic constraint in Eqs. (7) and (8) can be written in matrix form as:

$$A(q)\dot{q} = 0 \tag{9}$$

where A(q) is the constraint matrix and $\dot{q}$ is configuration coordinate given as:

$$A(q) = \begin{bmatrix} -\sin(\theta)\ \cos(\theta)\ -d\ \ 0\ \ 0 \\ \cos(\theta)\ \ \sin(\theta)\ \ \ L\ \ -r\ \ 0 \\ \cos(\theta)\ \ \sin(\theta)\ -L\ \ 0\ \ -r \end{bmatrix} \tag{10}$$

$$\dot{q} = \left[\dot{x}\ \dot{y}\ \dot{\theta}\ \dot{\varphi}_r\ \dot{\varphi}_l\right]^T \tag{11}$$

If the inertial and mass of the wheels were neglected, WMR satisfied pure rolling and non-slipping. As a result, Eq. (10) is reduced to the following matrix:

$$A(q) = \left[-\sin(\theta)\ \cos(\theta)\ -d\right] \tag{12}$$

Consider S(q) as a smooth and independent vector field distributed in the null space of a matrix A(q).

$$A(q)S(q) = 0 \tag{13}$$

In this case, S(q) is a linearly independent distributed field vector that is used to transform velocities w(t) = [v, ω] in terms of the inertial reference frame given as:

$$S(q) = \begin{bmatrix} \cos(\theta) & -d\sin(\theta) \\ \sin(\theta) & d\cos(\theta) \\ 0 & 1 \end{bmatrix} \tag{14}$$

Therefore, the WMR kinematic model in an inertial frame is formulated as follows:

$$\dot{q} = S(q)w(t) = \begin{bmatrix} \cos(\theta) & -d\sin(\theta) \\ \sin(\theta) & d\cos(\theta) \\ 0 & 1 \end{bmatrix} \begin{bmatrix} v \\ \omega \end{bmatrix} \tag{15}$$

where w(t) is the velocity vector given as $w(t) = \begin{bmatrix} v & \omega \end{bmatrix}^{T}$, $w \in R^{p \times 1}$ for all t.

2.2 System Dynamic Modeling

An MR system having an n-dimensional configuration space R, with generalized coordinate $q = [q_1, q_2, ..., q_n]$ and subject to m input constraints can be described as [15];

$$H(q)\ddot{q} + C(q, \dot{q})\dot{q} = E(q)\tau - A^T(q)\rho \tag{16}$$

where H(q) is an inertial matrix represented as n × n positive definite matrix, $C(q, \dot{q})$ is Coriolis and Centripetal torques represented as n × n matrix, ρ is a vector associated with Lagrange multipliers kinematic constraints, E(q) is the control input of n × m transformation matrix, $A^T(q)$ is matrix associated with constraints, τ is input torque vector. The generalized Lagrangian form of WMR for fixed conventional wheels is given as follows [16]:

$$\frac{d}{dt}\left(\frac{\partial T}{\partial \dot{q}}\right) - \frac{\partial T}{\partial q} = E(q)\tau - A^T(q)\rho \tag{17}$$

The total lagrangian energy is defined as L = T − W, where T is kinetic energy and W is potential energy (equal to zero because there is no lateral movement, only horizontal movement). The total lagrangian energy of the robot is calculated by considering a dynamic constraint on a fixed standard wheel.

$$L = \frac{1}{2}m_t(\dot{x}^2 + \dot{y}^2) + m_t(\dot{x}d\dot{\theta}\sin(\theta) - \dot{y}d\dot{\theta}\cos(\theta)) + \frac{1}{2}I\dot{\theta}^2 + \frac{1}{2}I_w(\dot{\varphi}_r^2\dot{\varphi} + \dot{\varphi}_l^2) \tag{18}$$

where $m_t = m_c + 2m_w$ and $I = m_c d^2 + I_c + 2m_w(d^2 + L^2) + 2I_m$. m_c is a mass of WMR devoid of wheels and motors, and m_w is the combined mass of wheel and motor, and m_t is the total mass of the robot. I_c is the inertia of WMR without wheels and motor, I_w is the inertia of a single wheel and motor around its axis, I_m is the inertia of the single wheel and motor about the y-axis is parallel to the wheel plane, and I is the total inertia of the robot.

The MR dynamic modeling is obtained by substituting the Lagrangian energy expression in Eq. (18) into Eq. (17) and, after rearrangements, we obtained as follows:

$$
\begin{bmatrix} m_t & 0 & m_t d \sin(\theta) \\ 0 & m_t & -m_t d \cos(\theta) \\ m_t d \sin(\theta) & -m_t d \cos(\theta) & I \end{bmatrix} \ddot{q} + \begin{bmatrix} 0 & 0 & m_t d \dot{\theta} \cos(\theta) \\ 0 & 0 & m_t d \dot{\theta} \sin(\theta) \\ 0 & 0 & 0 \end{bmatrix} \dot{q}
$$
$$
= \frac{1}{r} \begin{bmatrix} \cos(\theta) & \cos(\theta) \\ \sin(\theta) & \sin(\theta) \\ L & -L \end{bmatrix} \begin{bmatrix} \tau_r \\ \tau_l \end{bmatrix} + \begin{bmatrix} -\sin(\theta) \\ \cos(\theta) \\ -d \end{bmatrix} (-m_t(\dot{x}\cos(\theta) + \dot{y}\sin(\theta)))
$$
(19)

This dynamic model in Eq. (20) is simplified and transformed into a proper representation to eliminate the constraint terms [17]. A simplified dynamic model is given as:

$$
\begin{bmatrix} m_t & 0 \\ 0 & I \end{bmatrix} \begin{bmatrix} \dot{v}(t) \\ \dot{\omega}(t) \end{bmatrix} + \begin{bmatrix} 0 & -m_t d \dot{\theta} \\ m_t d \dot{\theta} & 0 \end{bmatrix} \begin{bmatrix} v(t) \\ \omega(t) \end{bmatrix} = \frac{1}{r} \begin{bmatrix} 1 & 1 \\ L & -L \end{bmatrix} \begin{bmatrix} \tau_r \\ \tau_l \end{bmatrix}
$$
(20)

3 Proposed Method Design

This section develops two proposed control algorithms: the backstepping controller and the NPID controller. Figure 2 shows the proposed controller with a TWMR system.

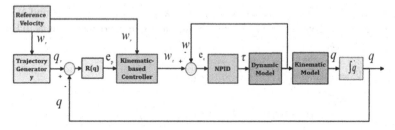

Fig. 2. Kinematic-based backstepping and NPID controller with a proposed system model

3.1 Backstepping Controller

A backstepping control method is a nonlinear controller designed in a recursive way that combines the choice of a Lyapunov function with the design of a feedback controller. A Lyapunov stability analysis function guarantees the global asymptotic stability of the designed controller [18]. In this paper, a backstepping controller is designed based on a nonlinear kinematic controller, as shown in Fig. 2. The main objective of this controller is to determine the system's angular and linear velocity control laws to track a given trajectory with reference velocity inputs and a posture error configuration. A nonholonomic mobile robot trajectory track can be described and formulated as follows:

$$\dot{q}_r = \begin{bmatrix} \dot{x}_r & \dot{y}_r & \dot{\theta}_r \end{bmatrix}^T = \begin{bmatrix} \cos(\theta_r) & 0 \\ \sin(\theta_r) & 0 \\ 0 & 1 \end{bmatrix} \begin{bmatrix} v_r \\ \omega_r \end{bmatrix} \tag{21}$$

From Fig. 2, the posture error between a reference trajectory $(q_r = (x_r, y_r, \theta_r))$ and current robot pose $(q = (x, y, \theta))$ is given as:

$$e_p = \begin{bmatrix} e_x \\ e_y \\ e_\theta \end{bmatrix} = \begin{bmatrix} \cos(\theta) & \sin(\theta) & 0 \\ -\sin(\theta) & \cos(\theta) & 0 \\ 0 & 0 & 1 \end{bmatrix} \begin{bmatrix} x_r - x \\ y_r - y \\ \theta_r - \theta \end{bmatrix} \tag{22}$$

If a CoM is at the center of the wheel axes $(d = 0)$, the time derivatives of posture error become:

$$\dot{e}_p = \begin{bmatrix} \dot{e}_x & \dot{e}_y & \dot{e}_\theta \end{bmatrix}^T = \begin{bmatrix} \omega\,e_y - v + v_r\cos(e_\theta) \\ -\omega e_x + v_r\sin(e_\theta) \\ \omega_r - \omega \end{bmatrix} \tag{23}$$

The control law, which makes a system asymptotically stable, will be designed based on the pose error dynamics obtained. Auxiliary velocity control inputs as a function of pose error and input velocities will be proposed depending on the pose error's time derivatives using Lyapunov theory. Let us present a scalar function V as a Lyapunov function candidate as:

$$V = \frac{1}{2}\left(e_x^2 + e_y^2\right) + \frac{1}{K_y}(1 - \cos(e_\theta)) \tag{24}$$

Therefore, for $K_y > 0$ one can see $V \geq 0$ otherwise, $V = 0$ if $e_p = 0$ and $V > 0$ if $e_p \neq 0$. The time-derivative of a Lyapunov function V with a posture error as in Eq. (23) is given as follows:

$$\dot{V} = \dot{e}_x e_x + \dot{e}_y e_y + \frac{1}{K_y}\dot{e}_\theta \sin(e_\theta)$$

$$= -e_x[v - v_r\cos(e_\theta)] - \sin(e_\theta)\left[\frac{1}{K_y}(\omega_r - \omega) - v_r e_y\right] \leq 0 \tag{25}$$

The auxiliary velocity control input that satisfies the Lyapunov function candidate by using a backstepping control law is formulated as follows [7]:

$$w_c = \begin{bmatrix} v_c \\ \omega_c \end{bmatrix} = \begin{bmatrix} K_x e_x + v_r\cos(e_\theta) \\ \omega_r + K_y v_r e_y + K_\theta v_r\sin(e_\theta) \end{bmatrix} \tag{26}$$

where K_x, K_y, and K_θ are positive constants adjusted to reduce the posture error. The velocity control rule described above is called a kinematic-based backstepping controller. The stability of this control velocity is verified by using the Lyapunov function.

$$\dot{V} = \dot{e}_x e_x + \dot{e}_y e_y + \frac{1}{K_y}\dot{e}_\theta \sin(e_\theta) = -K_x e_x^2 - \frac{K_\theta}{K_y}v_r\sin^2(e_\theta) \leq 0 \tag{27}$$

It is clear that $\dot{V} \leq 0$, otherwise, $\dot{V} = 0$ if $e_p = 0$ and $\dot{V} < 0$ if $e_p \neq 0$. Therefore, the derivative of a Lyapunov function is a negative-definite function. The system is asymptotically stable at an equilibrium point $e_p = 0$ under the condition that reference velocity $w_r = \begin{bmatrix} v_r & \omega_r \end{bmatrix}^T$ and proposed constant positive gains $(K_x, K_y,$ and $K_\theta)$ are bounded and continuous.

3.2 Nonlinear PID Controller

A NPID controller is used to control the velocity of a robot at a dynamic control loop level, as in Fig. 2. A new NPID controller whose architecture is analogous to the standard PID controller is proposed. However, the proportional and derivative actions are linear, whereas the integral action has a nonlinear function. Hence, the error input to the integral action is scaled by a nonlinear gain function in the product of the error and the nonlinear gain. An NPID controller used in this work is similar to an NPID controller used in work [19].

The time-domain equation of the NPID controller for velocity control of TMWR is given as:

$$U_k(t) = K_{pk}\left[e_c(t) + \frac{1}{T_{ik}} \int_0^t v(t)dt + T_{dk}\frac{de_c(t)}{dt} \right]$$

$$= K_{pk}\,e_c(t) + K_{ik} \int_0^t v(t)dt + K_{dk}\frac{de_c(t)}{dt}, \ k = 1,2 \qquad (28)$$

where $U_k = [U_1, U_2]^T = [u_v, u_\omega]^T$ is NPID controller output, K_{pk} is proportional gain constant, T_{ik} is an integral time, T_{dk} is derivative time, $K_{ik} = K_{pk}/T_{ik}$ is integral gain, $K_{dk} = K_{pk}T_{dk}$ is derivative gain constant in the case of a parallel PID controller. An error signal $e_c(t)$ is given as the difference between the control and actual velocity $(e_c(t) = w_c - w)$. The nonlinear scaled error function $v(t)$ in the integral and derivative control action is given as

$$v(t) = k(e)e_c(t); \ where \ k(e) = \exp\left(-\frac{e_c(t)^2}{2\Delta w_c^2} \right) \qquad (29)$$

where $k(e)$ is a nonlinear gain function, $\Delta w_c \neq 0$ is controlled velocity change, i.e., $\Delta w_c = w_c(i) - w_c(i - 1)$ and i denotes the discrete instant of time. According to [19], a typical value of Δw_c is 0.5, 1, and 1.5.

3.3 The Controller Gains Optimization by Using a GA

A genetic algorithm is a random search algorithm used to solve nonlinear equations and optimize complex problems. It employs probabilistic transition rules rather than deterministic rules and iteratively evolves a population of potential solutions known as individuals or chromosomes. Each iteration of the algorithm is referred to as a "generation". The evolution of solutions is mimicked using a fitness function and genetic operators such as selection (reproduction), crossover, and mutation [20]. Table 1 shows a pseudo-code of GA optimization processes.

The GA optimization tunes a controller gain to obtain the best gains parameters with a possible minimum objective function. The objective function is obtained using integral time-weighted absolute error (ITAE).

$$ITAE = \int_0^{t_f} t|e(t)|dt \qquad (30)$$

where e(t) is an error between desired and actual values and t_f is time duration. ITAE penalizes an error that persists long, resulting in more significant discrimination than IAE or ISE [19].

Table 1. Pseudo-code of the genetic algorithm

Start
Set t = 0;
Generate initial population $P(t)$;
Compute the fitness of an individual in $P(t)$;
do while < Stop condition for not satisfied >
Set t = t + 1;
Select from individual $P(t - 1)$ to set a tentative population $\overline{P}(t)$;
Perform Cross-over individual in $\overline{P}(t)$;
Perform Mutation of an individual in $\overline{P}(t)$;
Compute the fitness of an individual in a new population $P(t)$;
end while
Output the best individual $P(t)$ as the best solution;
Stop

---

4   Results and Discussions

To investigate a controller's performance for a capability to track a given reference trajectory, adaptability, and robustness for an uncertain system model, two scenarios are considered for the reference trajectory: linear reference trajectory and nonlinear reference trajectory.

$$(x_r, y_r) = (t, 2t), (x_r, y_r) = (-5\sin(t/5), 10\sin(t/10)), \forall t \geq 0. \qquad (31)$$

Linear Trajectory Tracking Performance: In linear reference trajectory tracking, a robot rapidly follows a given trajectory because the controller can easily anticipate the future behavior of a tracking line or tracking error. The robot simulation parameter configuration is as in [21]. $m_t = 120\,kg$, L = 0.33 m, d = 0.1 m, r = 0.135 m, I = 15.0565 kgm². A proposed controller gain parameter is optimized by the GA optimization method. The obtained controller gains are $K_x = 33.3095$, $K_y = 98.76$, $K_\theta = 10.2058$, $Kp_1 = 71.4881$, $Kp_2 = 99.8081$, $Ki_1 = 0.0441$, $Ki_2 = 0.0251$, $Kd_1 = 0.9483$ and $Kd_2 = 0.1803$.

As shown in Fig. 3, a robot tracks a given reference trajectory with a minimum tracking error at $q0 = (0, 0, 0)$ initial states in the x position, the y position, and the robot steering angle. The tracking errors in the x, y, and steering angles converge to zero in a short settling time. However, a limitation of the backstepping controller is that it produces a significant overshoot response when the robot's initial position is changed. To see the capability of the robot to track a given trajectory with a proposed controller, consider a change in the initial state and apply an unknown disturbance to the system. Hence, change the robot's initial position from (0, 0, 0) to (2, 1, pi/2) and see in Fig. 4 the change in a simulated robot system dynamics model.

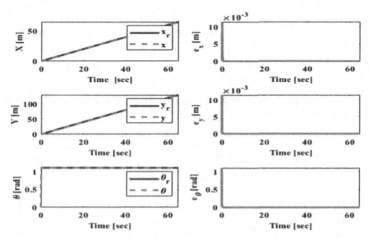

Fig. 3. The x position, y position, and robot angle tracking performance with (0, 0, 0) initial position: linear trajectory

A trajectory tracking performance is shown in Fig. 4, which clearly shows that the robot has a considerable peak value at the start of robot motion due to initial position change. Later this value decreases. The robot adapts to the initial position changes with a short settling time and a slight overshoot in reference position tracking. Thus, the proposed method's trajectory tracking ability in linear reference trajectory inputs is almost perfect, and the obtained results are smooth and robust to an initial position change.

Nonlinear Reference Trajectory Tracking Performance: This scenario applies a sinusoidal reference input to the system. The robot stabilized and followed a given path with a minimum trajectory tracking error, as shown in Fig. 5 and Fig. 6.

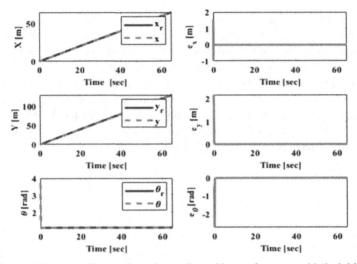

Fig. 4. The x position, y position, and steering angle tracking performance with the initial position is (2, 1, pi/2): linear trajectory

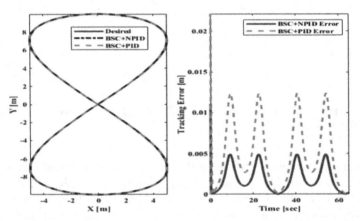

Fig. 5. Nonlinear trajectory tracking and tracking error performance with initial position is (0, 0, 0).

The robot smoothly tracks its reference trajectory in the x and y positions and a reference steering angle with the possible minimum tracking error. A proposed controller has better tracking capability than a backstepping plus PID controller, as shown in Fig. 5. A tracking performance response with (3, 2, pi/2) initial position change in Fig. 7 and Fig. 8 demonstrated that the proposed control method has a better tracking performance even if the initial position changed. The backstepping controller produced a significant overshoot when the robot's initial position changed. However, the robot quickly tracks its position after the initial position changes. The proposed controller has a better tracking performance and can quickly adapt to initial position changes than the backstepping plus PID controller.

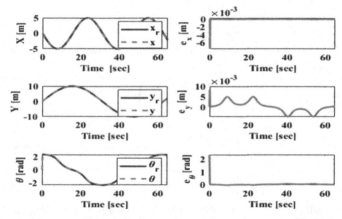

Fig. 6. The x position, y position, and robot angle tracking performance with the initial position is (0, 0, 0): nonlinear trajectory.

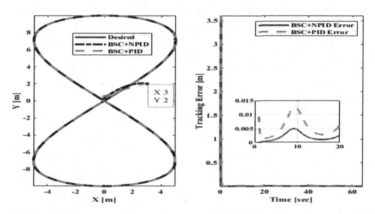

Fig. 7. Nonlinear trajectory tracking performance with initial position change to (3, 2, pi/2).

Robustness of a Control law with Unknown Disturbance: System model uncertainty may occur during selecting plant parameters for a simple representation because varying parameters and all the precise disturbance are not well known. These directly impact system performance in a real-time application. The unknown and unmodeled disturbance torque in the form of $[\tau_{d1}, \tau_{d2}] = [0.1\sin(2t), 0.1\sin(2t)]$ is applied to both wheels of the motor, and accordingly, linear trajectory and nonlinear trajectory performances are depicted in Fig. 9 and Fig. 10, respectively.

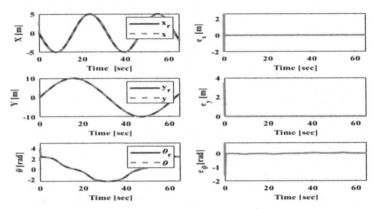

Fig. 8. The x position, y position, and steering angle tracking performance with initial position is (3, 2, pi/2): nonlinear trajectory

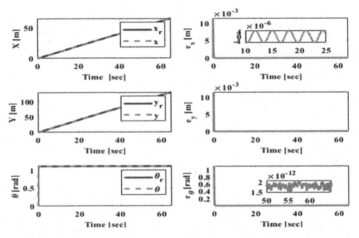

Fig. 9. Robot pose tracking performance with unknown disturbance: linear trajectory

As in Fig. 9 and Fig. 10, the control law has better unknown disturbance rejection. In linear trajectory tracking cases, the response to the unknown disturbance imposes change on a tracking response, while sinusoidal reference inputs were less affected by disturbance. This change has an almost insignificant effect on trajectory tracking errors. Therefore, the control law is smooth and gave a robust response to unknown disturbance and initial position change. The nonholonomic WMR has been controlled and stabilized by a control law in a predefined region with minimum position tracking error. The robot quickly follows a given reference trajectory, and its position tracking error converged to zero. In addition, the proposed controller outperforms the GA-based backstepping plus PID controller in terms of root-mean-square (RMS) of trajectory tracking error (47.36% in a linear and 60.32% in a nonlinear case).

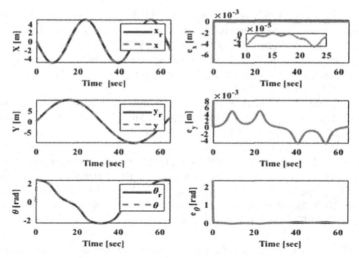

Fig. 10. Robot pose tracking performance with unknown disturbance: nonlinear trajectory

5 Conclusion

This paper presented backstepping combined with a nonlinear PID controller in the trajectory tracking control and stabilization of a two-WMR. The kinematic and dynamic modeling of a TWMR were formulated. The dynamic system models of the robot were derived using Lagrangian approaches. A backstepping plus NPID controller is designed to control a robot's trajectory tracking. The stability of a proposed controller is achieved using the Lyapunov method. The proposed controller achieved better reference trajectory tracking with a minimum tracking error in both scenarios. The robot follows its reference trajectory quickly if its initial position is changed. It also has better unknown disturbance rejection. The control law in this work does not update its parameters if the interest region (desired trajectory) is changed. A control law based on adaptive mechanisms and a self-taught controller is preferable and recommendable for better tracking capability in different environments.

References

1. Fierro, R., Lewis, F.L.: Control of a nonholonomic mobile robot using neural networks. IEEE Trans. Neural Netw. **9**(4), 589–600 (1998). https://doi.org/10.1109/72.701173
2. Uddin, N.: Trajectory tracking control system design for autonomous two-wheeled robot. JURNAL INFOTEL **10**(3), 90 (2018). https://doi.org/10.20895/infotel.v10i3.393
3. Ren, C., Ji, J.-H., Yan, H.-Y., Zhang, H., Yue, J.-Z.: A backstepping control method for mobile robot path tracking. In: Proceedings of the 3rd Annual International Conference on Mechanics and Mechanical Engineering (MME 2016), vol. 105 (2017). https://doi.org/10.2991/mme-16. 2017.94
4. Chang, H., Jin, T.: Adaptive tracking controller based on the pid for mobile robot path tracking. In: Lee, J., Lee, M.C., Liu, H., Ryu, J.-H. (eds.) ICIRA 2013. LNCS (LNAI), vol. 8102, pp. 540–549. Springer, Heidelberg (2013). https://doi.org/10.1007/978-3-642-40852-6_55

5. Moqbel Obaid, M.A., Husain, A.R., Mohammed Al-kubati, A.A.: Robust backstepping tracking control of mobile robot based on nonlinear disturbance observer. Int. J. Electr. Comput. Eng. (IJECE) **6**(2), 901 (2016). https://doi.org/10.11591/ijece.v6i2.9594

6. Xu, Q., Kan, J., Chen, S., Yan, S.: Fuzzy PID based trajectory tracking control of mobile robot and its simulation in simulink. Int. J. Control Autom. **7**(8), 233–244 (2014). https://doi.org/10.14257/ijca.2014.7.8.20

7. Fierro, R., Lewis, F.L.: Control of a nonholomic mobile robot: backstepping kinematics into dynamics. J. Robot. Syst. **14**(3), 149–163 (1997). https://doi.org/10.1002/(SICI)1097-4563(199703)14:3%3c149::AID-ROB1%3e3.3.CO;2-N

8. Fierro, R., Lewis, F.L.: Control of a nonholonomic mobile robot: backstepping kinematics into dynamics. In: Proceedings of 1995 34th IEEE Conference on Decision and Control, vol. 4, no. December, pp. 3805–3810 (1995). https://doi.org/10.1109/CDC.1995.479190

9. Hassani, I., Maalej, I., Rekik, C.: Backstepping tracking control for nonholonomic mobile robot. In: 2020 4th International Conference on Advanced Systems and Emergent Technologies (IC_ASET), pp. 63–68 (2020). https://doi.org/10.1109/IC_ASET49463.2020.9318221

10. Dagher, K., Al-araji, A.: Design of a nonlinear PID neural trajectory tracking controller for mobile robot based on optimization algorithm. Eng. Tech J. **32**(4), 973–985 (2014)

11. Zangina, U., Buyamin, S., Abidin, M.S.Z., Mahmud, M.S.A., Hasan, H.S.: Nonlinear PID controller for trajectory tracking of a differential drive mobile robot. J. Mech. Eng. Res. Dev. **43**(7), 255–269 (2020). http://eprints.utm.my/id/eprint/90651/1/UmarZangina2020_NonLinearPIDControllerforTrajectoryTracking.pdf

12. Kalyoncu, M., Demirbaş, F.: Differential drive mobile robot trajectory tracking with using PID and kinematic based backstepping controller. Selcuk Univ. J. Eng. Sci. Technol. **5**(1), 1–15 (2017). https://doi.org/10.15317/Scitech.2017.65

13. Yousuf, B.M., Saboor Khan, A., Munir Khan, S.: Dynamic modeling and tracking for nonholonomic mobile robot using PID and backstepping. Adv. Control Appl. **3**(3), 1–12 (2021). https://doi.org/10.1002/adc2.71

14. Ben Jabeur, C., Seddik, H.: Design of a PID optimized neural networks and PD fuzzy logic controllers for a two-wheeled mobile robot. Asian J. Control **23**(1), 23–41 (2021). https://doi.org/10.1002/asjc.2356

15. Mohareri, O., Dhaouadi, R., Rad, A.B.: Indirect adaptive tracking control of a nonholonomic mobile robot via neural networks. Neurocomputing **88**, 54–66 (2012). https://doi.org/10.1016/j.neucom.2011.06.035

16. Ahmad Abu Hatab, R.D.: Dynamic modelling of differential-drive mobile robots using Lagrange and newton-Euler methodologies: a unified framework. Adv. Robot. Autom. **02**(02) (2013). https://doi.org/10.4172/2168-9695.1000107

17. Benchouche, W., Mellah, R., Bennouna, M.S.: The Impact of the dynamic model in feedback linearization trajectory tracking of a mobile robot. Period. Polytech. Electr. Eng. Comput. Sci. **65**(4), 329–343 (2021). https://doi.org/10.3311/PPee.17127

18. .Vaidyanathan, A.T.A.S.: Backstepping Control of Nonlinear Dynamical Systems. Elsevier (2021)

19. Jin, G.-G., Son, Y.-D.: Design of a nonlinear PID controller and tuning rules for first-order plus time delay models. Stud. Inform. Control **28**(2), 157–166 (2019). https://doi.org/10.24846/v28i2y201904

20. Messom, C.: Genetic algorithms for auto-tuning mobile robot motion control. Res. Lett. Inf. Math. Sci. **3**(2002), 129–134 (2002). http://www.massey.ac.nz/~wwiims/research/letters/

21. Martins, N.A., Bertol, D.W.: Wheeled Mobile Robot Control, vol. 380. Springer, Cham (2022). https://doi.org/10.1007/978-3-030-77912-2

Author Index

A
Addisu, Hailemichael Solomon 95
Arega, Birtukan Shegaw 127
Ayenew, Endalew 290
Aynalem, Achamie 177

B
Belay, Birhanu Hailu 127
Bosera, Adisu Safo 63

D
Degu, Mizanu Zelalem 79

F
Fenta, Ermias Wubete 13, 27
Fufa, Lencho Duguma 290

G
Gessesse, Belachew Bantyirga 252

H
Hailu, Gashaye Lewtie 209

J
Jembere, Kaheli Anteneh 63

K
Kassie, Adino Amare 95
Kebede, Getnet Ayele 196
Kibrete, Fasikaw 41
Koricho, Ermias Gebrekidan 95

M
Ma, Sam Sun 272
Mengistu, Tesfahunegn Minwuyelet 127
Meshesha, Million 227
Mezigebu, Tenaw Ayew 252
Mossa, Neima 227
Mulatu, Adugna Necho 159

N
Nigusie, Gebregziabihier 1

S
Salau, Ayodeji Olalekan 63
Shibabw, Esubalew W. 137
Shiferaw, Yosef Kassa 196
Simegn, Gizeaddis Lamesgin 79

T
Takele, Tamiru 272
Tamir, Eneyachew 159
Tamiru, Gerbaw Y. 137
Tegegne, Tesfa 1
Terefe, Tefera 272
Tiba, Isayiyas Nigatu 239
Tsegaw, Assefa Asmare 13, 27

W
Woldemichael, Dereje Engida 41

Y
Yadessa, Asrat Gedefa 63
youhong, Mao 239

Z
Zewde, Elbetel Taye 79